SOIL QUALITY
and
SOIL EROSION

Edited by
Rattan Lal

SOIL
AND WATER
CONSERVATION
SOCIETY

Soil and Water Conservation Society
Ankeny, Iowa

CRC Press
Taylor & Francis Group
Boca Raton London New York

CRC Press is an imprint of the
Taylor & Francis Group, an **informa** business

CRC Press
Taylor & Francis Group
6000 Broken Sound Parkway NW, Suite 300
Boca Raton, FL 33487-2742

© 1999 by Soil and Water Conservation Society
CRC Press is an imprint of Taylor & Francis Group, an Informa business

First issued in paperback 2019

No claim to original U.S. Government works

ISBN-13: 978-0-367-44771-7 (pbk)
ISBN-13: 978-1-57444-100-0 (hbk)

Visit the Taylor & Francis Web site at
http://www.taylorandfrancis.com

and the CRC Press Web site at
http://www.crcpress.com

Library of Congress Cataloging-in-Publication Data

Catalog information may be obtained from the Library of Congress

Table of Contents

SECTION V: CONCLUSIONS

This volume is based on the proceedings of the conference on Soil Erosion and Soil Quality held at Keystone, Colorado, in July 1996. The conference was sponsored by the Soil and Water Conservation Society and organized by the NC-174 and NC-59 regional committees. The conference organizing committee consisted of R. Lal (chair), D. Mokma, K. Olson, G. Steinhardt, J. Doran, D. Stott, and C. Rice. The program advisors were Dr. G. Ham and Dr. B. Schmidt.

Foreword

The midwestern United States includes one of the largest areas of intensively cultivated agricultural cropland in the nation, with large amounts of soil tilled annually to continuous production of corn, soybeans, wheat, and other crops. Soil erosion by water and wind continues to be a major problem on these midwestern soils, especially when under intensive cropping, which is typical in this region. The National Soil Erosion–Soil Productivity Planning Committee (1981) concluded that, for many soils, soil erosion is occurring at a higher rate than formation, resulting in potentially irreversible losses in soil productivity. According to the *National Resources Inventory of 1982*, all cropland in the North Central Region of the United States had an overall soil erosion rate in excess of 5 tons/acre/year soil loss tolerance level (T-value), which was established as the maximum level of soil erosion that will permit a high level of crop production to be sustained economically and indefinitely.

The effect of soil erosion on productivity has been recognized as a major agricultural and natural resources issue since 1986 by the directors of the state agricultural experiment stations. A regional research project, North Central Regional Research Project NC-174, Soil Productivity and Erosion, was initiated in 1983 by 11 participating states in the North Central Region, with the objectives of strengthening the scientific foundation for describing the potential crop production of selected soils in the region, predicting changes in their production potential as soil erosion occurs, and evaluating the potential for extrapolating the information to other soils not studied.

From 1983 to 1992, the NC-174 project research identified and documented the effects of erosion on soil properties and corn and small grain yields under rain-fed conditions. Field experiments were established by all participating states using uniform procedures on soils with slight, moderate, and severe degrees of prior erosion, with data collected on soil properties, crop yields, and climatic parameters. Crop yields obtained on moderately or severely eroded soils were consistently lower than on comparable slightly eroded soils. Following the enlargement of the database on the selected soils, emphasis was placed on selection of management and restoration alternatives. Simulation models were used to identify and quantify the factors limiting crop productivity on these soils and, along with field testing, to evaluate those models for their usefulness in extrapolating these result to related soils.

This Midwest regional team of scientists from the 11 participating states has been instrumental in developing a multidisciplinary and scientifically sound understanding of the relationship between soil erosion and productivity. This research has evaluated practices including cropping systems, soil amendments, tillage, and water management, which can rehabilitate soils degraded by erosion. Factors most frequently limiting the productivity of eroded soils were (1) precipitation, (2) available water-holding capacity, and (3) topsoil depth. Conservation tillage, residue management, and the application of manures and municipal wastes have improved surface characteristics of eroded soils and minimized future erosion. These findings emphasize the critical importance of protecting soil against erosion, since reductions in topsoil depth and water-holding capacity are difficult to restore. Preliminary research by the NC-174 Committee scientists on restoration of eroded soils has shown that their surface properties can be improved; however, long periods of time are required when using commonly available management practices.

The relationship of soil erosion to crop productivity has been shown and clarified by this research. Soil erosion also has significant economic costs to society. In 1984, estimated monetary loss in the United States from soil erosion was $40 million per year due to loss of cropland productivity; other studies have estimated economic losses due to extra fertilizer costs to restore lost soil from erosion. In addition to productivity and economic losses, erosion can have major effects on the environment and soil quality — issues of increasing national and global concern in recent years. Sediment from soil erosion is one of the largest water pollutants in terms of volume and transports with it, in surface runoff, nutrients such as nitrogen and phosphorus and agricultural chemicals, pesticides, and animal wastes.

The National Research Council (1989) and the State Agricultural Experiment Station Committee on Organization and Policy (1990) both recognized the importance of research on soil erosion for the protection of the environment as well as productivity. Therefore, the objective of the NC-174 Regional Research Project — to determine the threshold soil property values, including available water-holding capacity, aggregation, bulk density, organic carbon, and infiltration, for restoration of soil productivity and quality of eroded soils — directly relates to the impacts of soil erosion on soil quality, water quality, and air quality. Because of this direct environmental as well as productivity and economic impact of soil erosion, the soil erosion scientists work closely with the researchers in soil quality and other environmental areas. Therefore, the NC-174 Regional Research Committee met and planned jointly with the NCR-59 Regional Research Committee on Soil Organic Matter and Soil Quality and jointly held a symposium on "Soil Quality and Soil Erosion Interaction," co-sponsored by the Soil and Water Conservation Society. Also, scientists from both groups contributed to a manual entitled *Methods for Assessing Soil Quality,* published by the Soil Science Society of America in 1996.

The research results and new knowledge being developed by these programs have been published in over 40 refereed publications from 1993 to 1997, as well as the symposium proceedings currently in press. These important investigations will continue to evaluate tillage, soil amendment, and water management effects on soil quality, water quality, air quality, and soil productivity and provide information required for predictive models and determination of threshold soil property values. These data will provide users, practitioners, and policymakers with scientifically based information on the effects of soil erosion on soil productivity and environmental quality.

Berlie Schmidt   *George Ham*
USDA-CSREES, NRE   Kansas State University
Washington, D.C.   Manhattan, Kansas

Preface

Accelerated soil erosion has plagued the earth since the dawn of settled agriculture. However, it became a major issue in the United States during the 1930s because of the increasing awareness about its adverse impact on sustainable use of natural resources, especially with regard to agricultural productivity. During the 1970s and 1980s, soil erosion was an issue with regard to its impact on water quality with particular reference to nonpoint source pollution. With increasing demand on the limited prime soil resources and shrinking per capita arable land area in densely populated regions of the world, erosion became a global issue during the 1990s with regard to its on-site impact on productivity and agricultural sustainability. However, the on-site impact of erosion remains a debatable issue because of the confounding impact of weather and the compensatory effects of high yields with technological improvements (e.g., new varieties, fertilizer use, soil water conservation) and on depositional sites. It was also during the 1980s and 1990s that soil scientists revisited the concept of soil capability and defined it in terms of soil quality. The importance of developing the concept of soil quality was enhanced because of the need to apply soil science to address the problems of nonagricultural uses of soil (e.g., mineland restoration, urban uses and disposal of urban wastes, soil contamination and pollution by industrial activities, athletic and recreational uses of soil, and environmental regulatory functions with particular reference to water quality and the greenhouse effect). A strong need, therefore, arose to develop appropriate indicators of soil quality in relation to specific soil function (e.g., agricultural, urban, industrial, recreational, athletic, environmental, and waste disposal).

The soil quality concept also has strong application in establishing the cause–effect relationship between soil erosion and productivity. Understanding the complex relationship can be simplified by evaluating and quantifying erosion-induced changes in soil quality (e.g., soil physical, chemical, and biological quality).

Two separate regional committees have been conducting research on these two important but interrelated themes. The NC-174 Committee has been working on "Soil Erosion and Productivity" issues since the mid-1980s and NC-59 on "Soil Quality" since the early 1990s. Because of their mutual interest in issues pertaining to sustainability and environmental quality, it was natural that these two committees work together. Therefore, a symposium was organized to review the progress made on soil quality and erosion-induced changes in soil quality and productivity. The

Organizing Committee of the symposium included members of both committees (R. Lal, D. Mokma, K. Olson, G. Steinhardt, J. Doran, D. Stott, and C. Rice). The symposium was organized in conjunction with the 51st Annual Meeting of the Soil and Water Conservation Society, held at Keystone, Colorado, in July 1996.

This volume is based on the invited and contributory papers presented at the Keystone symposium. The symposium was successful because of the excellent direction provided by Dr. G. Ham and Dr. B. Schmidt, advisors of NC-174 and NC-59 committees. It was their support and visionary direction that provided the incentive for both committees to work together. Most authors produced high-quality manuscripts that included state-of-the-art knowledge on the subject concerned and made the needed revisions to improve scientific quality. However, problems were encountered in submission and revision of some manuscripts, which delayed publication by about a year. Nonetheless, the support and cooperation received from members of the Organizing Committee and from all authors are gratefully acknowledged and much appreciated.

The cooperation and support received from the staff of the Soil and Water Conservation Society was indispensable in preparing this volume. In this regard, special thanks are due to Nancy Hercelius, Tim Kautza, and Sue Ballantine. The logistic support provided by Doug Klein is greatly appreciated.

The tedious task of typing and providing logistic support was done by Ms. Brenda Swank of the School of Natural Resources, The Ohio State University. Her tireless efforts are greatly appreciated. As usual, the staff of CRC Press was cooperative and helpful in producing a high-quality volume.

Rattan Lal, Editor
Columbus, Ohio

About the Editor

Dr. R. Lal is a professor of soil science in the School of Natural Resources at The Ohio State University. Prior to joining Ohio State in 1987, he served as a soil scientist for 18 years at the International Institute of Tropical Agriculture, Ibadan, Nigeria.

While based in Africa, Professor Lal conducted long-term experiments on soil erosion processes as influenced by rainfall characteristics, soil properties, methods of deforestation, soil tillage and crop residue management, cropping systems including cover crops and agroforestry, and mixed/relay cropping methods. He established critical limits of soil properties in relation to the severity of soil degradation and assessed effectiveness of different restorative measures. Data from these long-term experiments facilitated identification of indicators of soil quality and development of the concepts of soil resilience. He also assessed the impact of soil erosion on crop yield and related erosion-induced changes in soil properties to crop growth and yield.

Since joining The Ohio State University in 1987, he has continued research on erosion-induced changes in soil quality and developed a new project on soils and global warming. He has demonstrated that accelerated soil erosion is a major factor affecting emission of carbon from soil to the atmosphere. Soil erosion control and adoption of conservation-effective measures can lead to carbon sequestration and mitigation of the greenhouse effect. Impacts of severity of soil erosion on crop yield are evaluated at the landscape scale to quantify the compensatory effects of depositional sites. Erosion-induced changes in soil quality are related to crop growth and yield. The research has helped establish indicators of soil quality and resilience in relation to land use and management practices.

Professor Lal is a fellow of the Soil Science Society of America, American Society of Agronomy, Third World Academy of Sciences, American Association for the Advancement of Sciences, Soil and Water Conservation Society, and Indian Academy of Agricultural Sciences. He is the recipient of the International Soil Science Award, the Soil Science Applied Research Award of the Soil Science Society of America, the International Agronomy Award of the American Society of Agronomy, and the Hugh Hammond Bennett Award of the Soil and Water Conservation Society. He is past president of the World Association of the Soil and Water Conservation and the International Soil Tillage Research Organization. He is a member of the U.S. National Committee on Soil Science of the National Academy of Sciences. He has served on the Panel on Sustainable Agriculture and the Environment in the Humid Tropics of the National Academy of Sciences. He is a member of the Committee on Sustainable Agriculture in the Developing Countries of the World Resources Institute.

Contributing Authors

M.A. Arshad
Research Scientist
Agriculture and Agri-Food Canada
Alberta, Canada

D.L. Beck
Professor
Dakota Lakes Research Station
South Dakota State University
Pierre, South Dakota

J.M. Bradford
Research Leader
USDA-ARS
Weslaco, Texas

Bobby G. Brock
Conservation Agronomist
USDA-NRCS
Raleigh, North Carolina

J.R. Brown
Professor
Soil and Atmospheric Sciences
    Department
University of Missouri
Columbia, Missouri

C.A. Cambardella
Soil Microbiologist
USDA-Agricultural Research
    Service
National Soil Tilth Laboratory
Ames, Iowa

Larry Cihacek
Associate Professor
Department of Soil Science
North Dakota State University
Fargo, North Dakota

J.A. Delgado
Soil Scientist
USDA-ARS
Fort Collins, Colorado

John W. Doran
Soil Scientist
USDA-ARS
University of Nebraska
Lincoln, Nebraska

J.H. Edwards
Soil Scientist
USDA-ARS
National Soil Dynamics Laboratory
Auburn University
Auburn, Alabama

William J. Elliot
Project Leader
Intermountain Research Station
USDA-Forest Service
Moscow, Idaho

T.E. Fenton
Professor
Department of Agronomy
Iowa State University
Ames, Iowa

R.F. Follett
Research Leader
USDA-ARS
Fort Collins, Colorado

J.E. Gilley
Agricultural Engineer
USDA-ARS
University of Nebraska
Lincoln, Nebraska

G.L. Hart
Postdoctoral Fellow
Department of Soil Science
University of Wisconsin-Madison
Madison, Wisconsin

V.L. Hauser
Agricultural Engineer
Mitre Corporation
San Antonio, Texas

J.E. Herrick
USDA-ARS
Jornada Experimental Range
Las Cruces, New Mexico

C. Huang
Soil Scientist
USDA-ARS
National Soil Erosion Research
   Laboratory
Purdue University
West Lafayette, Indiana

Alice J. Jones
Professor of Agronomy
Agronomy Department
University of Nebraska
Lincoln, Nebraska

O.R. Jones
Soil Scientist
USDA-Agricultural Research Service
Bushland, Texas

A.C. Kennedy
Research Leader
USDA-Agricultural Research Service
Land Management and Water
   Conservation Unit
Washington State University
Pullman, Washington

K-J.S. Kung
Associate Professor
Department of Soil Science
University of Wisconsin-Madison
Madison, Wisconsin

R. Lal
Professor
School of Natural Resources
The Ohio State University
Columbus, Ohio

M.J. Lindstrom
Soil Scientist
USDA-Agricultural Research Service
North Central Soil Conservation
   Research Laboratory
Morris, Minnesota

B. Lowery
Professor
Department of Soil Science
University of Wisconsin-Madison
Madison, Wisconsin

M.J. Mausbach
Deputy Chief
Soil Survey and Resource Assessment
USDA-NRCS
Washington, D.C.

Robert McCallister
Department of Geography and
   Geology
University of Wisconsin-Madison
Madison, Wisconsin

D.L. Mokma
Professor
Department of Crop and Soil Sciences
Michigan State University
East Lansing, Michigan

Darrell Norton
Director, Research Leader
USDA-ARS
National Soil Erosion Research
  Laboratory
Purdue University
West Lafayette, Indiana

Peter Nowak
Professor
Department of Rural Sociology
University of Wisconsin-Madison
Madison, Wisconsin

K.R. Olson
Professor
Department of Natural Resources
  and Environmental Studies
University of Illinois
Urbana, Illinois

Deborah Page-Dumroese
Research Soil Scientist
Intermountain Research Station
USDA-Forest Service
Moscow, Idaho

J.D. Reeder
Soil Microbiologist
USDA-ARS
Crops Research Laboratory
Fort Collins, Colorado

D.C. Reicosky
Soil Scientist
USDA-Agricultural Research Service
North Central Soil Conservation
  Research Lab
Morris, Minnesota

R.R. Riggenbach
Conservation Agronomist
USDA-NRCS-SLVWQDP
Monte Vista, Colorado

Peter R. Robichaud
Research Engineer
Intermountain Research Station
USDA-Forest Service
Moscow, Idaho

T.E. Schumacher
Professor
Department of Plant Science
South Dakota State University
Brookings, South Dakota

G.E. Schuman
Soil Scientist and Research Leader
USDA-ARS
High Plains Research Station
Cheyenne, Wyoming

Issac Shainberg
Research Scientist
The Volcani Center
ARO
Bet Dagan, Israel

J.L. Sharkoff
Conservation Agronomist
USDA-NRCS-SLVWQDP
Monte Vista, Colorado

J.R. Simanton
Hydrologist
USDA-ARS
Southwest Watershed Research
  Center
Tucson, Arizona

S.J. Smith
Soil Scientist (retired)
USDA-ARS
Durant, Oklahoma

L.M. Southwick
Research Chemist
USDA-ARS
Baton Rouge, Louisiana

R.T. Sparks
Area Resource Conservationist
USDA-NRCS
Alamosa, Colorado

D.E. Stott
Soil Microbiologist
USDA-Agricultural Research Service
National Soil Erosion Research
   Laboratory
West Lafayette, Indiana

M.A. Weltz
Hydrologist and Research Leader
USDA-ARS
Southwest Watershed Research Center
Tucson, Arizona

Section I

Basic Concepts

1

Soil Quality and Food Security: The Global Perspective

R. Lal

INTRODUCTION

World per capita grain production increased from 247 kg in 1950 to 342 kg in 1984 and has progressively decreased to 320 kg in 1996 (Figure 1.1a) (Brown et al., 1997). The increase in per capita grain production is due to a steady increase in world average grain yield from about 1 ton/ha in 1950 to 2.5 ton/ha in 1995. The increase in world grain yield is due to substantial increase in rice and wheat yield in Asia and grain yield in Europe and North America since 1950. Coincidentally, trends in per capita grain production are similar to those in per capita fertilizer consumption, which increased from 5.5 kg in 1950 to 28.1 kg in 1989 and have progressively decreased to 22.2 kg in 1996 (Figure 1.1b). There is a good correlation between per capita grain production and fertilizer consumption (Figure 1.2). Per capita grain production has increased despite the progressive decline in per capita grain harvested area from 0.23 ha in 1950 to 0.12 ha in 1990 and will be 0.08 ha in 2030 (Brown, 1997). There has also been a steady decline in world irrigated area per person, which peaked at 0.047 ha in 1980, decreased to 0.043 ha in 1995, and will be only 0.031 ha in 2030.

Despite these impressive gains in productivity, hunger and malnutrition persist in several parts of the world. The total number of underfed in the world has progressively decreased from 976 million in 1975 to 786 million in 1990 (Uvin, 1994; Young, 1997) (Table 1.1). Increase in grain (agricultural) production since 1975 has decreased the incidence of hunger and malnutrition despite an additional 1.1 billion people.

Analyses of the data presented in Figure 1.1 and Table 1.1 indicate two principal observations that need further discussion:

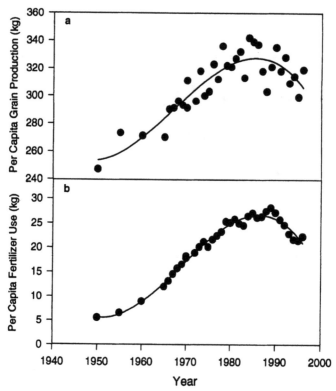

FIGURE 1.1 Global per capita fertilizer use (a) and grain production (b) from 1950 to 1966. (Redrawn from Brown et al., 1997.)

1. Increase in use of chemical fertilizers has at least partly been responsible for increase in world average yield of grain and other crops
2. Low crop yields have led to widespread hunger and malnutrition, in sub-Saharan Africa and regions of rain-fed agriculture in South Asia.

The underlying causes of overall increase in global crop yields due partly to increase in fertilizer consumption and persistence of low crop yields and hunger in sub-Saharan Africa and parts of South Asia may be related to soil quality, soil degradation, and soil resilience. Improvements in crop yields due to increase in fertilizer input may be, in part at least, due to improvement in soil chemical quality. Concomitantly, persistence of low yields in sub-Saharan Africa and South Asia may be due to widespread problems of soil degradation due to accelerated erosion and other degradative processes.

SOIL DEGRADATION AND SOIL EROSION

Soil degradation is the loss of actual or potential productivity or utility as a result of natural or anthropogenic factors (Lal, 1993, 1997). Soil degradation is a global

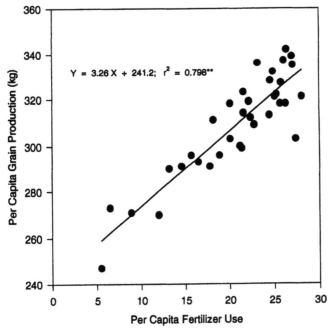

FIGURE 1.2   Correlation between per capita grain production and fertilizer use. (Recalculated from Brown et al., 1997.)

threat (Lal and Stewart, 1990), and it has strong impacts on food and energy resources (Pimentel et al., 1976; Lal, 1988) and environments (Lal, 1997), especially in relation to water quality (Lal and Stewart, 1994) and the greenhouse effect (Lal et al. 1995a,b, 1997a,b). There is a strong link among soil degradation, environment quality, food security, and energy use (Figure 1.3). Soil degradation is driven by poverty (Lal et al., 1989) and affects all aspects of human society through its adverse

TABLE 1.1
Regional Trends in Malnourished and Underfed Population of the World

	Absolute number (10^6)				Percent of the population			
Year	World	Sub-Saharan Africa	South Asia	China	World	Sub-Saharan Africa	South Asia	China
1970	942	94	255	406	36	35	34	46
1975	976	112	289	395	33	37	34	40
1980	846	128	285	290	26	36	30	22
1990	786	175	277	189	20	37	24	16

Modified from Uvin, 1994; Young, 1997.

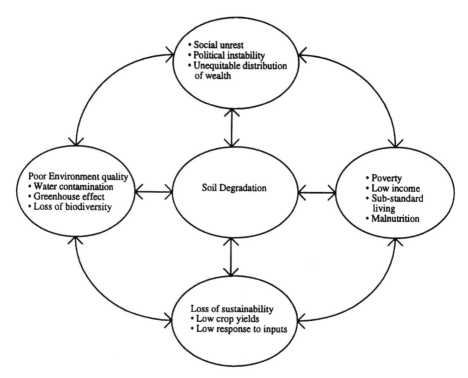

FIGURE 1.3 The vicious cycle of soil degradation with adverse impact on agricultural sustainability, environment quality, and social and political fabric of human society.

impacts on (1) agricultural productivity and returns per unit input of the essential resources; (2) environmental quality including loss in biodiversity; (3) income, caloric intake, human nutrition, health, and standard of living; and (4) social and political stability and equitable distribution of wealth.

Soil degradation is set in motion by the overall environmental degradation. Environmental degradation is any change or disturbance to the environment perceived to be deleterious or undesirable, and land degradation (soil, water, vegetation) is a subset of environmental degradation (Johnson et al., 1997) (Figure 1.4). Soil degradation is a principal component of land degradation, due to adverse changes in the pedosphere. Among three principal types of soil degradation (e.g., physical, chemical, and biological), accelerated soil erosion is driven by the interactive effects of decline in soil structure and harsh climate. Soil erosion, in turn, starts a chain reaction that exacerbates all aspects of environmental degradation due to adverse changes in the atmosphere (emission of dust and radiatively active gases into the atmosphere), biosphere (reduction in species diversity and activity due to destruction of the habitat), hydrosphere (due to eutrophication of surface water, contamination of groundwater, and disturbance in components of the hydrologic cycle), and lithosphere (due to deep gullies, ravines, landslides, mass movement, and sand dune migration) (Figures 1.5–1.7).

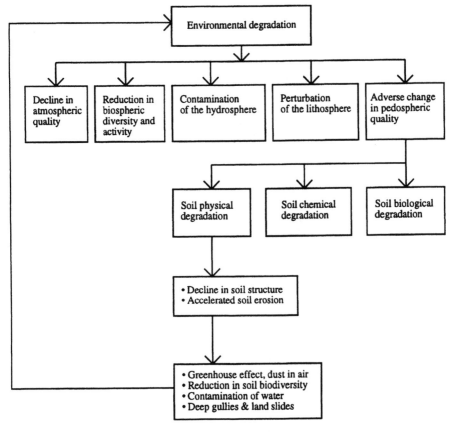

FIGURE 1.4 Soil degradation is linked to overall environmental degradation.

Accelerated soil erosion is the dominant type of soil degradation on a global scale. Land area affected by water erosion is estimated at 1094 million ha. Wind erosion is the second most important type of soil degradation on all continents. Land area affected by wind erosion is estimated at 548 million ha (Oldeman, 1994).

SOIL QUALITY

Accelerated soil erosion and other degradative processes influence agronomic productivity and environment through their impact on soil quality. Because of its global importance, therefore, there has been an enthusiastic response by the scientific community in the 1990s toward defining and assessing soil quality (Lal, 1993; National Research Council, 1993; Doran et al., 1994; Doran and Jones, 1996; Johnson et al., 1997; Gregorich and Carter, 1997). Soil quality refers to the productivity and environmental regulatory capacity of soil (Lal, 1997). It is a measure of the condition of soil relative to the requirement of one or more species and/or to any human need or purpose (Johnson et al., 1997).

(b)

FIGURE 1.5   Gully erosion is a form of pedospheric degradation. (a) Initiation of a gully in a pineapple field in Puerto Rico. (b) An aerial view of a severely gullied land in eastern Nigeria. (c) A gully head of an active gully in eastern Nigeria. (d) An active gully also expands laterally because of weak soil strength.

(d)

FIGURE 1.5 (continued)

FIGURE 1.6 Landslides and mass movement are severe pedospheric degradative processes on steeplands (landslides on high land in Malaysia).

Similar to soil degradation, there are also three types of soil quality: soil physical quality, soil chemical quality, and soil biological quality. Soil erosion adversely affects all three aspects of soil quality (Figure 1.8), especially soil physical quality, which are due to decline in soil structure and imbalance in soil–water regime. Erosion impacts on productivity are due to loss of available water capacity, decline in soil fertility, and deterioration of soil structure. There is a complex relationship between soil structure and erosion. While decline in soil structure (as evidenced by crusting and low infiltration rate) (Figure 1.9) leads to erosion, accelerated erosion exacerbates the problems of poor soil structure (Figure 1.10).

VANISHING LAND, SHRINKING FIELD, DECLINING SOIL QUALITY, AND FOOD INSECURITY

The problems of soil degradation and declining soil quality are confounded by high demographic pressure, low per capita arable land area, and resource-poor farmers who cannot afford to invest in soil restoration and erosion control measures. World grain harvested or the per capita arable land area is rapidly declining, especially in densely populated countries of Asia and Africa. It is also in these regions that soil erosion and erosion-induced soil degradation are severe problems. The same regions that are plagued by the problems of agrarian stagnation are also the regions that have the problem of perpetual food deficit and malnutrition and starvation.

(a)

(b)

FIGURE 1.7   Sand dunes are a severe form of pedospheric degradation in arid regions. (a) Sand dune in Syria. (b) Sand dunes in Tunisia.

The vicious cycle of soil erosion → low crop yields → poverty and malnutrition → low resource agriculture → more severe soil erosion and degradation must be broken through technological and policy interventions. The World Food Summit (FAO, 1996) adopted a resolution of combating drought and desertification to achieve sustainable food and agricultural production.

Understanding the complex and interactive relationship between soil erosion and soil quality will be a major step toward solving the problem of food insecurity and the decline in environment quality.

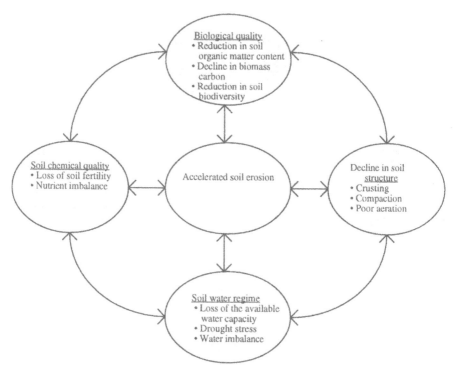

FIGURE 1.8   Accelerated soil erosion has adverse impacts on soil physical, chemical, and biological qualities. Decline in soil physical quality leads to adverse impact on soil structure and soil–water regime.

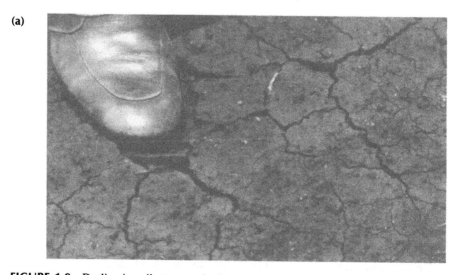

FIGURE 1.9   Decline in soil structure leads to crusting, runoff, and onset of erosion. (a) Crusting of an Oxisol in Brazil. (b) Crusting of an Alfisol in Nigeria impedes seedling emergence. (c) Crusting accentuates water runoff from an Alfisol in India.

(b)

(c)

FIGURE 1.9 (continued)

(a)

(b)

FIGURE 1.10   Severe soil erosion leads to formation of massive or structureless soil conditions. (a) Massive structure of a severely eroded soil in highlands of Rwanda. (b) Massive structure of a soil in transitional stage from rill to gully erosion.

REFERENCES

Brown, L.R. 1997. *The Agricultural Link: How Environmental Deterioration Could Disrupt Economic Progress.* Worldwatch Paper 136. Worldwatch Institute, Washington, D.C., p. 73.

Brown, L.R., M. Renner, and C. Flavin. 1997. *Vital Signs. The Environmental Trends That Are Shaping Our Future.* Worldwatch Institute, Washington, D.C., p. 165.

Doran, J.W. and A.J. Jones (eds.). 1996. *Methods of Assessing Soil Quality.* SSSA Special Publication No. 49, Soil Science Society of America, Madison, WI, p. 409.

Doran, J.W., D.C. Coleman, D.F. Bezdicek, and B.A. Stewart (eds.). 1994. *Defining Soil Quality for a Sustainable Environment.* SSSA Special Publication No. 35, Soil Science Society of America, Madison, WI, p. 244.

FAO. 1996. *World Food Summit. Rome Declaration on World Food Security and World Food Summit Plan of Action.* FAO, Rome, p. 43.

Gardner, G. 1996. *Shrinking Fields: Cropland Loss in a World of Eight Billion.* Worldwatch Paper 131. Worldwatch Institute, Washington, D.C., p. 55.

Gregorich, E.G. and M.R. Carter (eds.). 1997. *Soil Quality for Crop Production and Ecosystem Health.* Development in Soil Science 25. Elsevier, Amsterdam, 448 pp.

Johnson, D.L., S.H. Ambrose, T.J. Bassett, M.L. Boven, D.E. Crummey, J.S. Isaacson, D.N. Johnson, P. Lamb, M. Saul, and A.E. Winter-Nelson. 1977. Meanings of environmental terms. *J. Environ. Qual.* 26:581–589.

Lal, R. 1988. Soil degradation and the future of agriculture in sub-Saharan Africa. *J. Soil Water Conserv.* 43:444–451.

Lal, R. 1993. Tillage effects on soil degradation, soil resilience, soil quality and sustainability. *Soil Tillage Res.* 27:1–8.

Lal, R. 1997. Degradation and resilience of soils. *Philos. Trans. R. Soc. London Ser. B* 352:997–1010.

Lal, R. and B.A. Stewart. 1990. Soil degradation. A global threat. *Adv. Soil Sci.* 11:xiii–xvii.

Lal, R. and B.A. Stewart (eds.). 1994. *Soil Processes and Water Quality.* Lewis Publishers, Boca Raton, FL, 398 pp.

Lal, R., G.F. Hall, and F.P. Miller. 1989. Soil degradation. I. Basic processes. *Land Degrad. Rehabil.* 1:51–69.

Lal, R., J.M. Kimble, E. Levine, and B.A. Stewart (eds.). 1995a. *Soils and Global Change.* Lewis Publishers, Boca Raton, FL, p. 440.

Lal, R., J.M. Kimble, E. Levine, and B.A. Stewart (eds.). 1995b. *Soil Management for Mitigating the Greenhouse Effect.* Lewis Publishers, Boca Raton, FL, p. 385.

Lal, R., J.M. Kimble, R.F. Follett, and B.A. Stewart (eds.). 1997a. *Soil Processes and the Carbon Cycle.* CRC Press, Boca Raton, FL.

Lal, R., J.M. Kimble, R.F. Follett, and B.A. Stewart (eds.). 1997b. *Management of Carbon Sequestration in Soil.* CRC Press, Boca Raton, FL.

National Research Council. 1993. *Soil and Water Quality: An Agenda for Agriculture.* National Academy Press, Washington, D.C.

Oldeman, L.R. 1994. The global extent of soil degradation. In D.J. Greenland and I. Szabolcs (eds.). *Soil Resilience and Sustainable Land Use.* CAB International, Wallingford, U.K., pp. 99–118.

Pimentel, D., E.C. Terhune, R. Dyson-Hudson, S. Rochereau, R. Samis, E.A. Smith, D. Denman, D. Reifschneider, and M. Shepard. 1976. Land degradation: effects on food and energy resources. *Science* 194:145–155.

UNDP. 1995. *Human Development Report.* Oxford University Press, Oxford.

Uvin, P. 1994. The Hunger Report: 1993. Alan Shawn Feinstein World Hunger Program Brown University, Providence, RI.
Young, E.M. 1997. *World Hunger.* Routledge, London, p. 184

2 Determinants of Soil Quality and Health

John W. Doran, Alice J. Jones,
M.A. Arshad, and J.E. Gilley

INTRODUCTION

Interest in reducing soil erosion and maintaining soil quality has been stimulated by renewed awareness that soil is vital to both production of food and fiber and global ecosystem function. Soil quality or health can be broadly defined as the capacity of a living soil to function, within natural or managed ecosystem boundaries, to sustain plant and animal productivity, maintain or enhance water and air quality, and promote plant and animal health. Soil quality and health change over time due to natural events or human use. They are enhanced by management and land use decisions that weigh the multiple functions of soil and are impaired by decisions which focus only on single functions, such as crop productivity. Criteria for indicators of soil quality and health relate mainly to their utility in defining ecosystem processes and integrating physical, chemical, and biological properties; their sensitivity to management and climatic variations; and their accessibility and utility to specialists, producers, conservationists, and policymakers. Selection of useful indicators of soil quality and health, however, will depend upon identification of strategies for sustainable management of our natural resources. Although soils have an inherent quality as related to their physical, chemical, and biological properties within the constraints set by climate and ecosystem, the ultimate determinant of soil quality and health is the land manager.

SOIL: AN ESSENTIAL LINK IN THE CYCLE OF LIFE

Increasing human populations, decreasing resources, social and economic instability, and environmental degradation pose serious threats to the natural processes that sustain the global ecosphere and life on earth. Often, the thin layer of soil covering

1-57444-100-0/99/$0.00+$.50
© 1999 by Soil and Water Conservation Society

the surface of the earth represents the difference between survival and extinction for most land-based life (Glanz, 1995). Soil is a dynamic, living, nonrenewable resource whose condition is vital to both the production of food and fiber and to global balance and ecosystem function (Doran et al., 1996). The quality and health of soils determine agricultural sustainability (Acton and Gregorich, 1995; Papendick and Parr, 1992), environmental quality (Pierzynski et al., 1994), and, as a consequence of both, plant, animal, and human health (Haberern, 1992).

Soil is a dynamic, living, natural body that is vital to the function of terrestrial ecosystems and represents a unique balance between the living and the dead. The perception that soil is "living" seems beyond dispute given the observation that the number of living organisms in a teaspoon of fertile soil (10 g) can exceed 9 billion, 1.5 times the human population of the earth. Like water, soil is a vital natural resource essential to civilization, but unlike water, soil is nonrenewable on a human time scale. Soils form slowly, averaging 100–400 years/cm of topsoil, through the interaction of climate, topography, living organisms (microorganisms, animals, plants, and humans), and mineral parent material over time; thus the soil resource is essentially nonrenewable in human life spans (Jenny, 1980; Lal, 1994). Soils are composed of different sized inorganic mineral particles (sand, silt, and clay); reactive and stable forms of organic matter; myriad living organisms such as earthworms, insects, bacteria, fungi, algae, nematodes, etc.; water; and gases including O_2, CO_2, N_2, NO_x, and CH_4. The physical and chemical attributes of soil regulate soil biological activity and interchanges of molecules/ions between the solid, liquid, and gaseous phases, which influence nutrient cycling, plant growth, and decomposition of organic materials. The inorganic components of soil play a major role in retaining cations through ion exchange and nonpolar organic compounds and anions through sorption reactions. Essential parts of the global C, N, P, and S and water cycles occur in soil, and soil organic matter is a major terrestrial pool for C, N, P, and S; the cycling rate and availability of these elements are continually being altered by soil organisms in their constant search for food and energy sources.

The sun is the basis for most life on earth and provides radiant energy for heating the biosphere and for the photosynthetic conversion of carbon dioxide (CO_2) and water into food sources and oxygen for consumption by animals and other organisms. Most living organisms utilize oxygen to metabolize these food sources; capture their energy; and recycle heat, CO_2, and water to the environment to begin this cycle of "life" again.

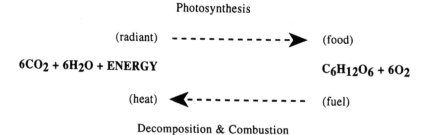

Photosynthesis

(radiant) - - - - - - - - - - - - ⟶ (food)

$6CO_2 + 6H_2O + ENERGY$ $C_6H_{12}O_6 + 6O_2$

(heat) ⟵ - - - - - - - - - - - - (fuel)

Decomposition & Combustion

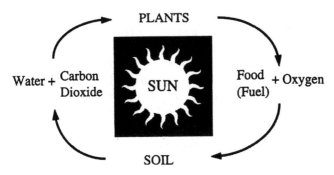

FIGURE 2.1 Simplified version of the "equation of life."

A simplified version of this "equation of life" is depicted in Figure 2.1.

The amount of CO_2 in the atmosphere is rather small and represents less than 0.04% of all gases in the atmosphere. If all the combustion and respiration processes on earth were halted, the plant life of the earth would consume all available CO_2 within a year or two. Thus, there is a fine balance between CO_2 production and utilization in the biosphere. Decomposition processes in soil play a predominant role in maintaining this balance, both in recycling C to the atmosphere as CO_2 and in the formation of soil organic matter, which serves as a sink for atmospheric C. These processes are brought about by a complex web of organisms in soil, each playing unique roles in the physical and chemical breakdown of organic plant and animal residues. The physiological diversity of this decomposer community, however, enables continued activity over a wide range of conditions, an essential attribute in a soil environment which is continually changing. Soils breathe and play a major role in transforming sunlight and stored energy and recycling matter through plants and animals. As such, "living" soils are vital to providing human food and fiber needs and in maintaining the ecosystems on which all life ultimately depends.

THREATS TO GLOBAL SUSTAINABILITY

Increasing human populations, decreasing resources, social instability, and environmental degradation threaten the natural processes that sustain the global ecosphere and life on earth (Costanza et al., 1992; Postel, 1994). Global climate change, depletion of the protective ozone layer, serious declines in species biodiversity, and degradation and loss of productive agricultural land are among the most pressing concerns associated with our technological search for a higher standard of living for an ever-growing human population. Past management of agriculture and other ecosystems to meet the needs of increasing populations has taxed the resiliency of soil and natural processes to maintain global balances of energy and matter (Bhagat, 1990; Sagan, 1992). The quality of many soils in North America and elsewhere has declined significantly since grasslands and forests were converted to arable agriculture and cultivation was initiated. Mechanical cultivation and the production of continuous row crops have resulted in physical soil loss and displacement through erosion, large decreases in soil organic matter content, and a concomitant release of organic C as carbon dioxide to the atmosphere (Houghton et al., 1983). Within the last decade, inventories of soil productive capacity indicate severe degradation on well over 10% of the earth's arable land as a result of soil erosion, atmospheric pollution, extensive soil cultivation, overgrazing, land clearing, salinization, and desertification (Oldeman, 1994). The quality of surface and subsurface water has been jeopardized in many parts of the world by intensive land management practices and the consequent imbalance of C, N, and water cycling in soil. At present, agriculture is considered the most widespread contributor to nonpoint source water pollution in the United States (CAST, 1992a; National Research Council, 1989). The major water contaminant in North America and Europe is nitrate–N, the principal sources of which are conversion of native to arable land use, animal manures, and fertilizers. Soil management practices such as tillage, cropping patterns, and pesticide and fertilizer use are known to influence water quality. However, these management practices can also influence atmospheric quality through changes in the soil's capacity to produce or consume important atmospheric gases such as carbon dioxide, nitrous oxide, and methane (CAST, 1992b; Rolston et al., 1993). The present threat of global climate change and ozone depletion, through elevated levels of atmospheric gases and altered hydrological cycles, necessitates a better understanding of the influence of land management on soil processes.

Present-day agriculture evolved as we sought to control nature to meet the food and fiber needs of an increasingly urbanized society (Quinn, 1993). With the development of modern chemistry during and after World War II, agriculturists often assumed a position of dominance in their struggle against a seemingly hostile natural environment, often failing to recognize the consequences of management approaches upon long-term productivity and environmental quality. Increased monoculture production of cash grain crops and greater reliance on chemical fertilizers and pesticides to maintain crop growth have resulted in two- to threefold increases in grain yields and on-farm labor efficiency (Avery, 1995; Brown et al., 1994; Power and Papendick, 1985). However, these management practices have also increased soil organic matter loss, soil erosion, and surface and groundwater contamination in the United States and elsewhere (Gliessman, 1984; Hallberg, 1987; Reganold et al., 1987). Motivations for shifting from input-intensive management to reduced external input farming include concern for protecting soil, human, and animal health from the potential hazards of pesticides; concern for protection of the environment and soil resources; and a need to lower production costs in the face of stagnant farm-gate receipts (Northwest Area Foundation, 1994; Soule and Piper, 1992; U.S. Department of Agriculture, 1980).

Developing *sustainable* agricultural management systems is complicated by the need to consider their utility to humans, their efficiency of resource use, and their ability to maintain a favorable balance with the environment that is favorable to both humans and most other species (Harwood, 1990). An Iowa farmer (Tom Franzen, Soil Health Conference, Decatur, Illinois, December 1994) recently shared a simple definition of agricultural sustainability:

Sustainable Agriculture — Sustains the People and Preserves the Land.

We are challenged to develop management systems which balance the needs and priorities for production of food and fiber with those for a safe and clean environment. Assessment of soil quality and health is invaluable in determining the sustainability of land management systems (Karlen et al., 1997). Some index of soil quality is needed to identify problem production areas, make realistic estimates of food production, monitor changes in sustainability and environmental quality as related to agricultural management, and assist government agencies in formulating and evaluating sustainable agricultural and land use policies (Granatstein and Bezdicek, 1992). In the United States, the importance of soil quality in maintaining a balance between environmental and production concerns was reflected by one conclusion of a National Academy of Science report: "Protecting soil quality, like protecting air and water quality, should be a fundamental goal of national environmental policy" (National Research Council, 1993). A recent call for development of a soil health index was stimulated by the perception that human health and welfare are associated with the quality and health of soils (Haberern, 1992). However, defining and assessing soil quality or soil health is complicated by the need to consider the multiple functions of soil in maintaining productivity and environmental well-being and to integrate the physical, chemical, and biological soil attributes which define those functions (Papendick and Parr, 1992; Rodale Institute, 1991). Most people recognize that maintaining the health and quality of soil should be a major goal of a "sustainable" society. The major question in many minds, however, is what constitutes a healthy or quality soil and how it might best be managed.

DETERMINING SOIL QUALITY AND HEALTH: THE USE OF INDICATORS

The quality of soil is largely defined by soil function and represents a composite of its physical, chemical, and biological properties that (1) provide a medium for plant growth and biological activity, (2) regulate and partition water flow and storage in the environment, and (3) serve as an environmental buffer in the formation and destruction of environmentally hazardous compounds (Larson and Pierce, 1994). Soil serves as a medium for plant growth by providing physical support, water, essential nutrients, and oxygen for roots. The suitability of soil for sustaining plant growth and biological activity is a function of physical properties (porosity, water-holding capacity, structure, and tilth) and chemical properties (nutrient-supplying ability, pH, salt content, etc.). Many of the soil's biological, physical, and chemical properties are a function of soil organic matter content (Rovira, 1993). Soil plays a key role in completing the cycling of major elements required by biological systems, decomposing organic wastes, and detoxifying certain hazardous compounds. The key role played by soils in recycling organic materials into carbon dioxide and water and degrading synthetic compounds foreign to the soil is brought about by microbial decomposition and chemical reactions. The ability of a soil to store and transmit water is a major factor regulating water availability to plants and transport of environmental pollutants to surface and groundwater.

Soil Quality versus Soil Health

Much like air or water, the quality of soil has a profound effect on the health and productivity of a given ecosystem and the environments related to it. However, unlike air or water, for which we have legislated quality standards, the definition and quantification of soil quality are complicated by the fact that soil is not directly consumed by humans and animals, as are air and water. Soil quality is often thought of as an abstract characteristic of soils which cannot be defined because it depends on external factors such as land use and soil management practices, ecosystem and environmental interactions, socioeconomic and political priorities, and so on. Perceptions of what constitutes a "good" soil vary depending on individual priorities for soil function and intended land use. However, to manage and maintain our soils in an acceptable state for future generations, soil quality and health must be defined, and the definition must be broad enough to encompass the many functions of soil (Karlen et al., 1997). The terms soil quality and soil health are often used interchangeably in the scientific literature and popular press, with scientists in general preferring soil quality and producers preferring soil health (Harris and Bezdicek, 1994). Some prefer the term soil health because it portrays soil as a living, dynamic organism that functions holistically rather than as an inanimate mixture of sand, silt, and clay. Others prefer the term soil quality and descriptors of its innate quantifiable physical, chemical, and biological characteristics. Efforts to define the concept of soil quality and soil health have produced a polarization of attitudes concerning these terms. On the one hand are those, typically speaking from outside agriculture, who view maintenance of soil health as an absolute moral imperative — critical to our very survival as a species. On the other hand is the attitude, perhaps ironically expressed most adamantly by academics, that the term is a misnomer — a viewpoint seated, in part, in fear that the concept requires value judgments which go beyond scientific or technical fact. The producers, and therefore society's management of the soil, are caught in the middle of these opposing views and the communication failures that result. In this chapter, the terms soil quality and soil health will be used synonymously. However, the term soil health is preferred in that it more clearly portrays the idea of soil as a living dynamic organism that functions in a holistic way depending on its condition or state, rather than as an inanimate object whose value depends on its innate characteristics and intended use. With consideration of the aforementioned factors, soil quality or soil health can be defined as the continued capacity of soil to function as a vital living system, within ecosystem and land use boundaries, to sustain biological productivity, promote the quality of air and water environments, and maintain plant, animal, and human health (Doran et al., 1996). Soil quality has been conceptualized as a three-legged stool, the function and balance of which require integration of its three major components — sustainable productivity, environmental quality, and plant and animal health (Figure 2.2). The challenge we face, however, is in quantitatively defining the state of soil quality or

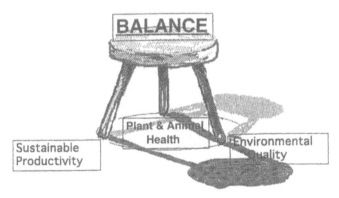

FIGURE 2.2   Soil quality or health as analogous to a three-legged stool, the major components (legs) of which must be in balance for optimum function or fitness.

health and its assessment using measurable properties or parameters. Unlike human health, the magnitude of critical indicators of soil quality and health ranges considerably over dimensions of time and space.

Assessing the health or quality of soil can be likened to a medical examination for humans, where certain measurements are taken as basic indicators of system function. Larson and Pierce (1991) proposed that a minimum data set be adopted for assessing the health of world soils and that standardized methodologies and procedures be established to assess changes in soil quality. The need for basic soil quality and health indicators is reflected in the question commonly posed by farmers, researchers, and conservationists: "What measurements should I make to evaluate the effects of management on the ability of soil to function now and in the future?" Too often, scientists confine their interests and efforts to the discipline with which they are most familiar. Microbiologists often limit their studies to soil microbial populations, having little or no regard for soil physical or chemical characteristics which define the limits of microbial activity or of other life forms. Approaches in assessing soil quality and health must be holistic, involving integration of all parts of the soil system, and not reductionistic in segregating and measuring only the function of individual parts. The indicators chosen must also be measurable by as many people as possible, especially managers of the land, and not limited to a select cadre of agricultural and environmental research scientists. These indicators should define the major ecological processes in soil and ensure that measurements made reflect conditions as they actually exist in the field under a given management system. They should relate to major ecosystem functions such as C and N cycling (Visser and Parkinson, 1992) and be components of computer models which emulate ecosystem function. Some indicators, such as soil bulk density, must be measured in the field so that laboratory results for soil organic matter and nutrient content can be converted to actual field conditions at time of sampling. Soil bulk density is also required for calculation of soil properties such as water-filled pore space which serves as an excellent integrator of soil physical, chemical, and biological soil properties and aeration-dependent microbial processes important to C and N cycling

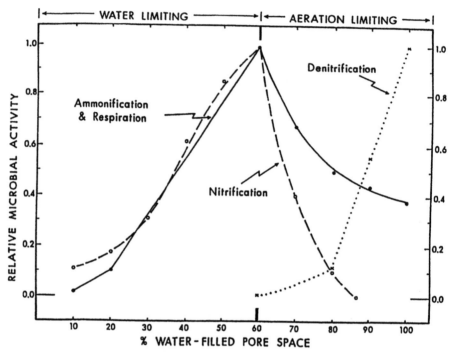

FIGURE 2.3 Relationship between soil water-filled pore space (WFPS) and the aerobic microbial processes of respiration and nitrification and anaerobic denitrification in carbon-amended soil. Divergence of relationship between ammonification/respiration and nitrification above 60–70% WFPS results from production of carbon dioxide and ammonification by anaerobic microbial fermentations. (After Linn and Doran, 1984.)

in soil (Figure 2.3). Many basic soil properties are useful in estimating other soil properties or attributes which are too difficult or expensive to measure directly. A listing of these basic indicators and input variables and the soil attributes they can be used to estimate is given in Table 2.1.

Starting with the minimum data set proposed by Larson and Pierce (1991), we developed a list of basic soil properties (Table 2.2) which meet the aforementioned requirements of indicators for screening soil quality and health. Appropriate use of such indicators will depend to a large extent on how well the relevance of these indicators is interpreted with respect to consideration of the ecosystem of which they are a part. Thus, interpretation of the meaning of soil biological indicators apart from soil physical and chemical properties is of little value and, with respect to assessment of soil quality and health, can actually be misleading.

VALUE OF QUALITATIVE/DESCRIPTIVE ASSESSMENTS

The concept of soil health is in many ways producer-generated and rooted in observational field experiences which translate into descriptive properties such as its

TABLE 2.1
A Limited Listing of Soil Attributes or Properties Which Can Be Estimated from Basic Input Variables Using Pedotransfer Functions or Simple Models[a]

Soil attribute or property	Basic input variables	Reference
CEC[b]	Organic C + clay type and content	Larson and Pierce, 1991, 1994
Water retention character (AWHC)	% sand, silt, clay, + organic C + BD	Gupta and Larson, 1979
Hydraulic conductivity	Soil texture	Larson and Pierce, 1991, 1994
Soil bulk density	Organic C + soil textural class	Rawls, 1983
Aerobic and anaerobic microbial activity	WFPS as calculated from BD and water content	Linn and Doran, 1984; Doran et al., 1990
C and N cycling	Soil respiration (soil temperature + WFPS)	Parkin et al., 1996
Plant/microbial activity or pollution potential	Soil pH + EC	Smith and Doran, 1996
Nitrate loss via denitrification/wet soils	Organic C, % sand, EC	Aulakh and Singh, 1996
Soil productivity	BD, AWHC, pH, EC, and aeration	Larson and Pierce, 1991, 1994
Rooting depth	BD, AWHC, pH	Larson and Pierce, 1991, 1994
Leaching potential	Soil texture, pH, organic C (hydraulic conductivity, CEC, depth)	Shea et al., 1992
Soil resistance/structure	Organic C, % clay, BD, and water content	da Silva and Kay, 1997
Soil erodibility factor in RUSLE	Organic C, soil texture, structure, and permeability	A.J. Jones, personal communication, June 1996

[a] AWHC = available water-holding capacity, BD = soil bulk density, CEC = cation exchange capacity, EC = soil electrical conductivity, and WFPS = water-filled pore space.

[b] Cation exchange capacity (cmol/kg) can be estimated by [(% C/0.58) × 200] + [% clay × (average exchange capacity of clay types)], where vermiculite = 150, montmorillonite = 100, illite = 30, and kaolinite = 8 cmol/kg (meq/100 g).

look, feel, resistance to tillage, smell, presence of biota, etc. Harris and Bezdicek (1994) conclude that farmer-derived descriptive properties for assessing soil health are valuable for (1) defining or describing soil quality/health in meaningful terms, (2) providing a descriptive property of soil quality/health, and (3) providing a foundation for developing and validating an analytical component of soil health based on quantifiable chemical, physical, and biological properties that can be used as a basis for management and policy decisions. Unfortunately, the potential contribution of indigenous farmer knowledge to management of soil quality/health throughout the world has not been fully utilized (Pawluk et al., 1992).

TABLE 2.2

Proposed Minimum Data Set of Soil Physical, Chemical, and Biological Indicators for Screening the Condition, Quality, and Health of Soil

Indicators of soil condition	Relationship to soil condition and function; rationale as a priority measurement
Physical	
Texture	Retention and transport of water and chemicals; needed for many process models; estimate of degree of erosion and field variability of soil types
Depth of soil, topsoil, and rooting[a]	Estimate of productivity potential and erosion; normalizes landscape and geographic variation
Soil bulk density and infiltration[a]	Indicators of compaction and potential for leaching, productivity, and erosivity; density needed to adjust soil analyses to field volume basis
Water-holding capacity (water retention character)	Related to water retention, transport, and erosivity; available water can be calculated from soil bulk density, texture, and soil organic matter
Chemical	
Soil organic matter (total organic C and N)	Defines soil fertility, stability, and erosion extent; use in process models and for site normalization
pH	Defines biological and chemical activity thresholds; essential to process modeling
Electrical conductivity	Defines plant and microbial activity thresholds, soil structural stability, and infiltration of added water; presently lacking in most process models; can be a practical estimator of soil nitrate and leachable salts
Extractable N, P, and K	Plant-available nutrients and potential for loss from soil; productivity and environment quality indicators
Biological	
Microbial biomass C and N	Microbial catalytic potential and repository for C and N; modeling; early warning of management effects on organic matter
Potentially mineralizable N (anaerobic incubation)	Soil productivity and N-supplying potential; process modeling; surrogate indicator of microbial biomass N
Soil respiration, water content, and temperature[a]	Measure of microbial activity (in some cases plants); process modeling; estimate of microbial biomass activity

[a] In field measurements for varying crop row and topographic positions and management conditions.

After Doran et al., 1996; Larson and Pierce, 1991; and Doran and Parkin, 1994, 1996.

Use of descriptive soil information is not commonly used in scientific literature dealing with characterization of soil quality/health. However, Arshad and Coen (1992) indicate that many soil attributes can be estimated by calibrating qualitative observations against measured values and recommend that qualitative (descriptive) information should be an essential part of soil quality monitoring programs. Visual and morphological observations in the field can be used by both producers and scientists to recognize degraded soil quality caused by (1) loss of organic matter, reduced aggregation, low conductivity, soil crusting, and sealing; (2) water erosion, as indicated by rills, gullies, stones on the surface, exposed roots, and uneven topsoil; (3) wind erosion as indicated by ripple marks, dunes, sand against plant stems, plant damage, dust in air, etc.; (4) salinization, as indicated by salt crust and salt-tolerant plants; (5) acidification and chemical degradation, as indicated by growth response of acid-tolerant and -intolerant plants and lack of fertilizer response; and (6) poor drainage and structural deterioration, as indicated by standing water and stunted or chlorotic plant stands.

Doran et al. (1994) stressed the importance of holistic management approaches which optimize the multiple functions of soil, conserve soil resources, and support strategies for promoting soil quality and health. They proposed use of the basic set of soil quality and health indicators (Table 2.2) to assess soil health in various agricultural management systems. However, while many of these key indicators are extremely useful to specialists (i.e., researchers, consultants, extension staff, and conservationists), many of them are beyond the expertise of the producer to measure (Hamblin, 1991). In response to this dilemma, Doran (1995) presented strategies for building soil quality and health which also included generic indicators which are measurable by and accessible to producers within the time constraints imposed by their normally hectic and unpredictable schedules (Table 2.3). Soil organic matter, crop appearance, and erosion were ranked by farmers in the northern U.S. Corn and Dairy Belt as the top three properties for describing soil health (Romig et al., 1995).

MANAGEMENT AS A DETERMINANT OF SOIL QUALITY AND HEALTH

Worldwide, declines in the quality of our soil resources have often been associated with nonscientific or unwise intensification of agriculture and soil cultivation. Plowing and mechanical cultivation, intensified use of fossil fuels, monoculture cropping, and bare fallowing have resulted in soil loss through erosion, large decreases in soil organic matter content, and increased release of carbon dioxide to the atmosphere. Intensive crop production, overgrazing, and deforestation have also resulted in accelerated loss of topsoil through wind and water erosion. Decreased organic matter content and the use of large tillage and harvesting equipment have resulted in decreased soil structure, tilth, water-holding capacity, and water infiltration and increased compaction. Development of saline and sodic soils after initiation of cultivation has resulted from both inefficient irrigation techniques and natural processes.

Early settlers in North America reaped immediate benefit from clearing and cultivating virgin prairie and forest soils which released their stored fertility to produce abundant crops. However, within the lifetimes of many of these early farmers, soil organic matter levels, native fertility, and production declined significantly and mineralization of soil nitrogen fell below that needed for sustained grain crop production (Campbell et al., 1976). The tremendous soil erosion which oc-

TABLE 2.3

Sustainable Management Strategies for Building Soil Quality and Health and Associated Indicators Which Are Assessable by Producers and Land Managers

Strategy	Indicators
Conserve soil organic matter *through* Maintaining C and N balance where inputs ≈ outputs	**Direction and change in soil organic matter with time** Visual or remote sensing by color or chemical analysis
Minimize soil erosion *through* Conservation tillage and increased cover (residue, cover crops, green fallow, etc.)	**Visual:** Gullies, rills, dust, etc. **Surface soil properties:** Depth of topsoil, soil organic matter/texture, infiltration rate, and percent cover
Substitute renewable for nonrenewable resources *through* Relying less on fossil fuels/petrochemicals, conservation tillage, and greater use of natural balance and diversity (crop rotation, legume cover crops, green fallow, etc.)	**Crop characteristics:** Yield, N content, color, rooting, soil physical state/compaction **Soil/water nitrate levels** **Input/output ratios** of costs and energy
Management which coexists with rather than dominates Nature *through* Optimizing productivity needs with environmental quality by synchronizing available N with crop needs during year	**Crop characteristics:** (yield, N content, color, vigor) Visual/remote sensing of vegetation and soil **Soil/water nitrate levels**

curred during the 1930s Dust Bowl was a graphic depiction of the need to consider both production and environment in developing agricultural management strategies. Emergence of the first reduced tillage and surface residue management systems in North America was catalyzed by the need to halt soil erosion and environmental degradation caused by changing weather conditions, the mechanization of agriculture, and extensive use of the moldboard plow.

The use of conservation and alternative management practices has done much to curtail soil and water degradation and enhance soil quality and health (Table 2.4). However, the specific effects of tillage, residue management, and cropping systems on soil productivity and environmental quality vary considerably across ranges in climate, seasonal weather patterns, soil type, and landscapes (Doran et al., 1993). Tillage and residue management practices act within the bounds set by climate, cropping systems, and soil type to determine soil environmental factors such as aeration, water content, temperature, and nutrient availability. These factors regulate biological activity, nutrient fluxes, and water balance, which ultimately determine

TABLE 2.4
Key Soil and Environmental Indicators[a] as Influenced by
Agricultural Management Practices

Soil or environmental indicator	General trend/change	Long-term agricultural practices affecting the indicator
Soil organic matter	Increase	Continuous cropping with well-managed crop residue, zero or minimum tillage, legume-based and other crop rotations, legume plowdown (green manure), cover crops, forages
	Decrease	Excessive tillage, summer fallow, crop residue removed or burned
Microbial biomass and biological diversity	Increase or decrease	Same as for soil organic matter
Soil aggregate stability	Increase	Conservation tillage, maintenance of crop residue, forages and legumes in crop rotations
	Decrease	Same as for soil organic matter decrease
Hydraulic conductivity	Increase	Reduced and zero tillage, maintenance of crop residue, forages and legumes in crop rotations — degree and extent of change vary with different practices
	Decrease	Same as for soil organic matter decrease
Soil depth/rooting volume	Increase	Conservation tillage and forage-based crop rotations should reduce erosion and allow soil-forming factors to maintain and rehabilitate topsoils
	Decrease	Excessive tillage, summer fallow cropping system, and crop residue removal or burning are the main agricultural practices that subject soils to serious wind and water erosion resulting in topsoil removal
Water quality	Positive or negative?	Data are lacking on how soil water quality is affected by different agricultural practices; in general, zero or minimum tillage, forage-based cropping systems, and maintenance of crop residue reduce surface runoff and soil loss to water streams; excessive use of herbicides and fertilizers may result in deterioration of water quality

[a] Additional indicators include soil pH, water-holding capacity, bulk density, and nutrient retention capacity. However, they are affected, to a large extent, by factors such as soil organic matter, aggregate size distribution and stability, etc. These factors are listed in the table.

Source: Arshad, 1996; unpublished data.

soil productivity and environmental quality. Although conservation tillage systems tend to induce cooler, wetter, more compact, and less aerobic soil environments (Doran and Linn, 1994; Mielke et al., 1986), the potential and magnitude of their effects on consequent biological activity, nutrient fluxes, and water balance are largely dependent on climate, landscape position, and soil characteristics. Thus the specific assessment of soil quality and health is critical to evaluation of the effectiveness of soil and crop management practices as determinants of productivity and environmental quality.

Soil quality (or health) is conceptualized as the major linkage between conservation management practices and achievement of the major goals of sustainable agriculture (Figure 2.4). As such, soil quality assessment is critical to sustainable land management (Dr. Fran Pierce, personal communication, April 1996). The use of conservation management practices, and in the United States the Conservation Reserve Program (CRP) in particular, has done much to curtail soil and water degradation and enhance soil quality and health. The CRP has removed 14.8 million ha (36.5 million acres), about 8% of all U.S. cropland, of highly erodible land from row crop production since 1985. This transition to soil-saving grasses and trees has reduced soil erosion losses by an estimated 22% or 635 million metric tons (700 million tons) of soil per year (U.S. Department of Agriculture, 1994). The total cost of this program, as represented by participation and management incentives to producers, is estimated to be $19.2 billion over the 10- to 15-year life of contracts. Maximum benefit to soil quality from CRP management, as related to increased organic matter levels and the structure and stability of soil, requires 3–10 years to accrue depending on soil, vegetation, and climatic factors (Lindstrom et al., 1994). These benefits to soil quality and sustainability, however, may be lost in a relatively short time period when CRP land is returned to production of grain crops (Gilley et al., 1997). Degradation of soil quality on CRP land returned to production of cash grain crops can be minimized through use of no-tillage management practices (Table 2.5).

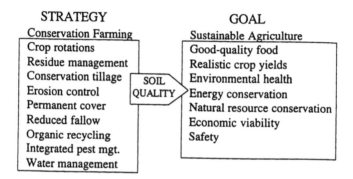

FIGURE 2.4 Soil quality links conservation farming to sustainable agriculture. (After Acton and Gregorich, 1995; Parr et al., 1992.)

TABLE 2.5
**Effect of Cropping and Tillage Management on Soil Organic Matter,
Water Infiltration, and Erodibility of Silty Clay Loam in Iowa Soils
After 7 Years in the Conservation Reserve Program**

Management practice	Crop	Soil organic matter (kg/ha, 30 cm)		Infiltration (min)[a]		Rainfall simulation[b]	
		C	N	First	Second	Runoff (mm)	Soil loss (tons/ha)
CRP	Grass/legume	80,200	5,730	0.3	16	10.1	0.01
No tillage	Corn	70,070	5,340	0.4	4	8.5	0.09
Plow, spring 1994	Soybean	59,340	4,660	0.1	28	35.4	3.14
Plow, fall 1993	Soybean	60,940	4,900	0.2	31	19.3	4.11

[a] Time for first and second 2.54 cm of water to infiltrate the soil.

[b] Runoff and soil loss after half an hour at a rainfall intensity of 95 mm/hr.

After Gilley et al., 1997.

Indicators of soil quality and health must not only identify the condition of the soil resource but also define the economic and environmental sustainability of land management practices. The theme of a recent conference in Australia ("Advances in Soil Quality: Science, Practice, and Policy," University of Ballarat, April 1996), "Soil Quality Is in the Hands of the Land Manager," highlighted the critical importance of the land manager in determining soil quality. A cotton grower at this conference shared his frustration with the direction soil quality indicators were taking: "I need help from scientists more with tools for management than with indicators of soil quality." Economic viability and survival are the primary factors motivating managers of the land, and while most appreciate the need for environmental conservation, the simple fact is that "it's hard to be green when you are in the red" (Ann Hamblin, Ballarat, Australia, April 1996). Gomez et al. (1996) presented a unique framework for determining the sustainability of hill country agriculture in the Philippines which employed indicators that considered the satisfaction of both farmer needs for productivity, profitability, stability, and viability and those needed for conservation of soil and water resources (Figure 2.5).

CONCLUSIONS

Soil and land management practices are primary determinants of soil quality and health. Producers and other managers of the land need practical tools and approaches to measuring the effects of management on soil quality and health which enable them to "fine-tune" and determine the sustainability of their production approaches. These tools may include some specific measures of soil quality that are useful to soil

Farmer Needs	Resource Conservation
Acceptable	Adequate
• Yields	• Soil organic matter
• Profits	• Soil depth
• Risk	• Soil cover

FIGURE 2.5 Measuring the sustainability of agricultural systems at the farm level requires consideration of both the farmer's needs and natural resource conservation. (After Gomez et al., 1996.)

specialists, as given in Table 2.2. However, they will likely involve more practical generic indicators such as water use efficiency; crop yield and growth characteristics; input costs; energy input/output ratios; soil loss from wind/water; soil structure; water storage and uptake; organic matter levels; and nutrient levels in soils, water, and farm products. On-farm assessment of soil quality and health will help producers evaluate the effects of management on agricultural sustainability and permit dialogue with researchers and conservationists in interpreting management effects. Strategies presented in Table 2.3 for building soil quality and health maximize the benefits of natural cycles, reduce dependence on nonrenewable resources, and help producers identify long-term goals for sustainability which also meet short-term needs for crop production. The generic indicators provided with these strategies are measurable by and assessable to producers, generally within the time constraints imposed by their normally hectic and unpredictable management schedules.

Assessing changes in sustainability poses a professional challenge to agriculture researchers, extension workers, and conservationists to develop standards for soil quality and health which are practical and useful to producers. Successful development of such tools and standards for assessment of soil health and sustainability, however, can only be accomplished through consultation and partnership with agricultural producers, who are the primary stewards of the land.

REFERENCES

Acton, D.F. and L.J. Gregorich. 1995. The Health of Our Soils: Toward Sustainable Agriculture in Canada. Agric. Agri-Food Can., Ottawa.

Arshad, M.A. 1996. Indicator of Soil Quality. Unpublished report. Agric. Canada, Beaverlodge.

Arshad, M.A. and G.M. Coen. 1992. Characterization of soil quality: physical and chemical criteria. Am. J. Altern. Agric. 7:12–16.

Aulakh, M.S. and B. Singh. 1996. Gaseous Losses of Nitrogen Through Denitrification from Soils Under Different Cropping Systems. 4th Annual Research Progress Report for Project FG-In-757 (IN-ARS-526). Department of Soils, Punjab Agricultural University, Ludhiana, India.

Avery, D.T. 1995. Saving the Planet with Pesticides and Plastic: The Environmental Triumph of High-Yielding Farming. Hudson Institute, Indianapolis, IN.

Bhagat, S.P. 1990. Creation in Crisis. Brethren Press, Elgin, IL, 173 pp.

Brown, L.R., H. Kane, and D.M. Roodman. 1994. Vital Signs 1994: The Trends That Are Shaping Our Future. Worldwatch Institute, W.W. Norton, New York, p. 27.

Campbell, C.A., E.A. Paul, and W.B. McGill. 1976. Effect of cultivation and cropping on the amounts and forms of soil N. In W.A. Rice (ed.). Proc. Western Can. Nitrogen Symp., Calgary, Alberta, Canada, January 20–21. Alberta Agriculture, Edmonton, pp. 9–101.

CAST. 1992a. Water Quality: Agriculture's Role. Task Force Report No. 120. Council for Agricultural Science and Technology, Ames, IA.

CAST. 1992b. Preparing U.S. Agriculture for Global Climate Change. Task Force Report No. 119. Council for Agricultural Science and Technology, Ames, IA.

Costanza, R., B.G. Norton, and B.D. Haskell. 1992. *Ecosystem Health: New Goals for Environmental Management.* Island Press, Washington, D.C.

da Silva, A.P. and B.D. Kay. 1997. Estimating the least limiting water range of soils from soil properties and management. *Soil Sci. Soc. Am. J.* 61:877–883.

Doran, J.W. 1995. Building quality soil. In Proc. 1995 Alberta Conservation Tillage Workshop on Opportunities and Challenges in Sustainable Agriculture, Red Deer, Alberta, Canada, February 23–25.

Doran, J.W. and D.M. Linn. 1994. Microbial ecology of conservation management systems. In J.L. Hatfield and B.A. Stewart (eds.). *Advances in Soil Science, Soil Biology: Effects on Soil Quality.* Lewis Publishers, Boca Raton, FL, pp. 1–27.

Doran, J.W. and T.B. Parkin. 1994. Defining and assessing soil quality. In J.W. Doran, D.C. Coleman, D.F. Bezdicek, and B.A. Stewart (eds.). *Defining Soil Quality for a Sustainable Environment.* SSSA Special Publication No. 35. Soil Science Society of America, Madison, WI, pp. 3–21.

Doran, J.W. and T.B. Parkin. 1996. Quantitative indicators of soil quality: a minimum data set. In J.W. Doran and A.J. Jones (eds.). *Methods for Assessing Soil Quality.* SSSA Special Publication No. 49. Soil Science Society of America, Madison, WI, pp. 25–27.

Doran, J.W., L.N. Mielke, and J.F. Power. 1990. Microbial activity as regulated by soil water-filled pore space. Symposium on ecology of soil microorganisms in the microhabitat environment III:94–99. In Trans. 14th Int. Congress of Soil Science, Kyoto, Japan, August 1990.

Doran, J.W., G.E Varvel, and J.B.L. Culley. 1993. Tillage and residue management effects on soil quality and sustainable land management. In Proc. Int. Workshop on Sustainable Land Management, Lethbridge, Alberta, Canada, June 15–24.

Doran, J.W., M. Sarrantonio, and R. Janke. 1994. Strategies to promote soil quality and soil health. In Proc. Organization for Economic Co-operation and Development (OECD) Int. Workshop on Management of the Soil Biota in Sustainable Farming Systems. Coop. Research Center for Soil & Land Management, Adelaide, South Australia, March 15–18.

Doran, J.W., M. Sarrantonio, and M. Liebig. 1996. Soil health and sustainability. In D.L. Sparks (ed.). *Advances in Agronomy,* Vol. 56. Academic Press, San Diego, pp. 1–54.

Gilley, J.E., J.W. Doran, D.L. Karlen, and T.C. Kaspar. 1997. Runoff, erosion, and soil quality characteristics of a former conservation reserve program site. *J. Soil Water Conserv.* 52(3):189–193.

Glanz, J.T. 1995. *Saving Our Soil: Solutions for Sustaining Earth's Vital Resource.* Johnson Books, Boulder, CO.

Gliessman, S.R. 1984. An agroecological approach to sustainable agriculture. In W. Jackson, W. Berry, and B. Colman (eds.). *Meeting the Expectations of the Land.* North Point Press, San Francisco, pp. 160–171.

Gomez, A.A., D.E. Swete Kelly, J.K. Seyers, and K.J. Coughlan. 1996. Measuring sustainability of agricultural systems at the farm level. In J.W. Doran and A.J. Jones (eds.). *Methods for Assessing Soil Quality.* SSSA Special Publication No. 49. Soil Science Society of America, Madison, WI, pp. 401–410.

Granatstein, D. and D.F. Bezdicek. 1992. The need for a soil quality index: local and regional perspectives. *Am. J. Altern. Agric.* 7:12–16.

Gupta, S.C. and W.E. Larson. 1979. Estimating soil water retention characteristics from particle size distribution, organic matter percent, and bulk density. *Water Resour. Res.* 15:1633–1635.

Haberern, J. 1992. Viewpoint: a soil health index. *J. Soil Water Conserv.* 47:6.

Hallberg, G.R. 1987. Agricultural chemicals in ground water: extent and implications. *Am. J. Altern. Agric.* II:3–15.

Hamblin, A. 1991. Environmental Indicators for Sustainable Agriculture. Report of a National Workshop. Publ. LWRRDC and GRDC, 96 pp.

Harris, R.F. and D.F. Bezdicek. 1994. Descriptive aspects of soil quality/health. In J.W. Doran, D.C. Coleman, D.F. Bezdicek, and B.A. Stewart (eds.). *Defining Soil Quality for a Sustainable Environment.* SSSA Special Publication No. 35. Soil Science Society of America, Madison, WI, pp. 22–35.

Harwood, R.R. 1990. A history of sustainable agriculture. In C.A. Edwards et al. (eds.). *Sustainable Agricultural Systems.* Soil and Water Conservation Society of America, Ankeny, IA, pp. 3–19.

Houghton, R.A., J.E. Hobbie, J.M. Melillo, B. Moore, B.J. Peterson, G.R. Shaver, and G.M. Woodwell. 1983. Changes in the carbon content of terrestrial biota and soils between 1860 and 1980: a net release of CO_2 to the atmosphere. *Ecol. Monogr.* 53:235–262.

Jenny, H. 1980. *The Soil Resource: Origin and Behavior.* Ecological Studies 37. Springer-Verlag, New York.

Karlen, D.L., M.J. Mausbach, J.W. Doran, R.G. Cline, R.F. Harris, and G.E. Schuman. 1997. Soil quality: a concept, definition, and framework for evaluation. *Soil Sci. Soc. Am. J.* 61:4–10.

Lal, R. 1994. Sustainable land use systems and soil resilience. In D.J. Greenland and I. Szabolcs (eds.). *Soil Resilience and Sustainable Land Use.* CAB International, Wallingford, Oxon, U.K., pp. 41–67

Larson, W.E. and F.J. Pierce. 1991. Conservation and enhancement of soil quality. In *Evaluation for Sustainable Land Management in the Developing World,* Vol. 2. IBSRAM Proc. No. 12(2). International Board for Soil Research and Management, Bangkok, Thailand.

Larson, W.E. and F.J. Pierce. 1994. The dynamics of soil quality as a measure of sustainable management. In J.W. Doran, D.C. Coleman, D.F. Bezdicek, and B.A. Stewart (eds.). *Defining Soil Quality for a Sustainable Environment.* SSSA Special Publication No. 35. Soil Science Society of America, Madison, WI, pp. 37–51.

Lindstrom, M.J., T.E. Schumacher, and M.L. Blecha. 1994. Management considerations for returning CRP lands to crop production. *J. Soil Water Conserv.* 49:420–425.

Linn, D.M. and J.W. Doran. 1984. Effect of water-filled pore space on carbon dioxide and nitrous oxide production in tilled and nontilled soils. *Soil Sci. Soc. Am. J.* 48: 1267–1272.

Mielke, L.N., J.W. Doran, and K.A. Richards. 1986. Physical environment near the surface of plowed and no-tillage soils. *Soil Tillage Res.* 7:355–366.

National Research Council. 1989. *Alternative Agriculture: Committee on the Role of Alternative Farming Methods in Modern Production Agriculture. Board on Agriculture, National Research Council.* National Academy Press, Washington, D.C..

National Research Council. 1993. *Soil and Water Quality: An Agenda for Agriculture; Committee on Long-Range Soil and Water Conservation. Board on Agriculture, National Research Council.* National Academy Press, Washington, D.C.

Northwest Area Foundation. 1994. *A Better Row to Hoe: The Economic, Environmental, and Social Impact of Sustainable Agriculture.* Northwest Area Foundation, St. Paul, MN.

Oldeman, L.R. 1994. The global extent of soil degradation. In D.J. Greenland and I. Szabolcs (eds.). *Soil Resilience and Sustainable Land Use.* CAB International, Wallingford, Oxon, U.K., pp. 99–118

Papendick, R.I. and J.F. Parr. 1992. Soil quality: the key to a sustainable agriculture. *Am. J. Altern. Agric.* 7:2–3.

Parkin, T.B., J.W. Doran, and E. Franco-Vizcaino. 1996. Field and laboratory tests of soil respiration. In J.W. Doran and A.J. Jones (eds.). *Methods for Assessing Soil Quality.* SSSA Special Publication No. 49. Soil Science Society of America, Madison, WI, chap. 14.

Parr, J.F., R.I. Papendick, S.B. Hornick, and R.E. Meyer. 1992. Soil quality: attributes and relationship to alternative and sustainable agriculture. *Am. J. Altern. Agric.* 7:5–11.

Pawluk, R.R., J.A. Sandor, and J.A. Tabor. 1992. The role of indigenous knowledge in agricultural development. *J. Soil Water Conserv.* 47:298–302.

Pierzynski, G.M., J.T. Sims, and G.F. Vance. 1994. *Soils and Environmental Quality.* Lewis Publishers, Boca Raton, FL.

Postel, S. 1994. Carrying capacity: earth's bottom line. In LR. Brown et al. (eds.). *State of the World, 1994.* W.W. Norton, New York, pp. 3–21.

Powell, D. and J. Pratley. 1991. Sustainability Kit Manual. Centre for Conservation Farming, School of Agriculture, Charles Sturt University-Riverina, Wagga Wagga, Australia.

Power, J.F. and R.I. Papendick. 1985. Organic sources of nutrients. In O.P. Engelstad (ed.). *Fertilizer Technology and Use,* 3rd ed. Soil Science Society of America, Madison, WI, pp. 503–520.

Quinn, D. 1993. *Ishmael.* Bantam Books, New York.

Rawls, W.J. 1983. Estimating soil bulk density from particle size analysis and organic matter content. *Soil Sci.* 135:123–125.

Reganold, J.P., L.F. Elliott, and Y.L. Unger. 1987. Long-term effects of organic and conventional farming on soil erosion. *Nature* 330:370–372.

Rodale Institute. 1991. *Conference Report and Abstracts, International Conference on the Assessment and Monitoring of Soil Quality.* Rodale Press, Emmaus, PA.

Rolston, D.E., L.A. Harper, A.R. Mosier, and J.M. Duxbury. 1993. *Agricultural Ecosystem Effects on Trace Gases and Global Climate Change.* Special Publication No. 55. American Society of Agronomy, Madison, WI, 206 pp.

Romig, D.E., M.J. Garlynd, R.F. Harris, and K. McSweeney. 1995. How farmers assess soil health and quality. *J. Soil Water Conserv.* 50:229–236.

Rovira, A.D. 1993. Sustainable farming systems in the cereal livestock areas of the Mediterranean region of Australia. Keynote paper at the Int. Conf. on Soil Management in Sustainable Agriculture, Wye College, U.K., September 1993.

Sagan, C. 1992. To avert a common danger. *Parade Magazine* March 1, pp. 10–14.

Shea, P.J., L.N. Mielke, and W.D. Nettleton. 1992. Estimation of Relative Pesticide Leaching in Nebraska Soils. Research Bulletin 313-D. Agric. Res. Div., IANR, University of Nebraska, Lincoln.

Smith, J.L. and J.W. Doran. 1996. Measurement and use of pH and electrical conductivity for soil quality analyses. In J.W. Doran and A.J. Jones (eds.). *Methods for Assessing Soil*

   Quality. SSSA Special Publication No. 49. Soil Science Society of America, Madi-
      son, WI, chap. 10.
Soule, J.D. and J.K. Piper 1992. *Farming in Nature's Image.* Island Press, Washington, D.C.
U.S. Department of Agriculture. 1980. Report and Recommendations on Organic Farming.
      U.S. Government Printing Office, Washington, D.C., 94 pp.
U.S. Department of Agriculture. 1994. Bringing idled land back into production. In Agricul-
      tural Research. USDA, Agricultural Research Service, Beltsville, MD, pp. 10–11.
Visser, S. and D. Parkinson 1992. Soil biological criteria as indicators of soil quality: soil
      microorganisms. *Am. J. Altern. Agric.* 7:33–37.

Section II

Types of Soil Quality

3 Erosion and Soil Chemical Properties

Darrell Norton, Issac Shainberg, Larry Cihacek, and J.H. Edwards

INTRODUCTION

Soil erosion by water involves complex interactions between rainwater and soil, as influenced by soil properties and soil condition at the time of interaction. Depending on the constituents making up the soil and their relative proportions, soils may behave differently than rainfall and have different amounts of runoff and erosion. In addition to the permanent soil properties, soil conditions as determined by antecedent moisture content, duration of time at this moisture content (aging), and the rate of prewetting also determine amount of runoff and erosion. In this chapter we summarize how soil chemical properties directly affect production of runoff and affect erosion processes and how these same properties may interact with time, leading to soil condition which further affects the interaction between soil and rainfall. Inherent chemical properties such as mineralogy and charge density of clays are not easily altered and are important in determining soil physical properties such as aggregation, density, water-holding capacity, porosity, etc. However, in this review, the effect of chemical properties which are easily altered (e.g., composition of soil solution and exchange phase, pH, Eh, organic matter content) is discussed. Many problem soils have adverse physical conditions due to their inherent chemical properties. Other chemical properties of the soil are easily modified and can be changed by amending soil. Amending chemical constituents may change the physical condition and soil erodibility by affecting processes important to erosion, including dispersion and flocculation, cohesion, hard-setting, self-mulching, slaking, swelling, and surface sealing. An understanding of how chemical properties combined with soil condition (e.g., antecedent moisture content, aging time, prewetting rate, etc.) affect erosion processes will lead to strategies to amend soil chemistry to favorably affect soil quality and reduce erosion.

1-57444-100-0/99/$0.00+$.50
© 1999 by Soil and Water Conservation Society

Soil erosion by water occurs when precipitation exceeds the infiltration rate of the soil, producing runoff containing soil. Infiltration rate is controlled by the saturated hydraulic conductivity of a thin layer at the very soil surface where the raindrops impact the soil. Typically when soil quality is reduced, the infiltration rate decreases. This is largely due to changes in soil chemical properties that result in structural instability, causing clay dispersion and development of a thin surface seal (Norton et al., 1993). This thin surface seal develops on many soils with a wide range of composition and texture. Many processes can occur at the soil–air–water interface, depending on the soil composition and condition at the time rainfall occurs. The detachment of soil or aggregates by raindrops or flowing water is largely physical; however, the condition the soil is in and its behavior with respect to rainwater are largely physiochemical and dependent on the chemistry of the soil and soil solution.

Soil chemical quality is also important for plant nutrition and growth and providing the essential plant nutrients. Soil chemistry and its effect on plant growth are well known and have been the subject of study in edaphology for hundreds of years. In fact, humans have been amending the soil for fertility purposes for thousands of years. But soil chemical quality also affects the physical quality and the environment for the plant; therefore, physical soil quality cannot be completely separated from the chemical aspects. Those processes that restrict water exchange also affect air exchange for plant roots that may limit yield. The chemistry effect on soil physical quality and its effect on plant production received little study. Poor structural stability in the field causes runoff in some places on the landscape and flooding in others. Both may limit yield in the respective landscape areas. The chemistry of the eroding soils also affects the water quality of the runoff. Runoff production and erosion affect both the on-site quality of the soil and the off-site quality of surface waters and sediment.

The ability of the soil to remain open and have a high saturated hydraulic conductivity is controlled by the stability of aggregates, which is affected by type of saturating cations, surface charge balances, and the electrical conductivity of rainwater. Soil composition is difficult and unpractical to change, and if the component can be changed, such as organic matter, the change is very slow. Soil chemical composition can be changed much quicker and therefore should be considered as an amendable practice to be used to improve soil chemical and physical quality. Understanding how soil chemical properties affect structural instability will result in management practices and strategies to amend the soil to improve its quality and control erosion while potentially improving yield.

SOIL COMPOSITION

Soil composition is due to the naturally occurring materials making up the soil that have been weathered and altered to secondary minerals as well as organic materials that have been added over time imparting good physical and chemical quality. Soil composition varies considerably both vertically and spatially on the landscape. Soil composition may impart either favorable or unfavorable soil quality conditions

TABLE 3.1
Common Soil Clay Minerals and General Size

Mineral	Diameter (μm)	Surface area (m²/g)	Planar:edge	CEC[a] units
Smectite	0.03	600–800	50	80–120
Vermiculite	0.03	400–800	50	100–150
HIV	0.1–1	80–150	4–6	10–40
Illite	0.3–1	60–200	8	15–40
Kaolinite	0.3–2	5–40	4	3–15
Hematite	0.02–0.05	50–120		
Gibbsite	0.1	10–30		
Allophane	0.003–0.005	1000		

[a] CEC = cation exchange capacity.

depending on the relative amounts and properties of its constituents. The composition is difficult to change without large and extensive modification of the soil. The most easily changed component is soil organic matter. As topsoil is eroded, the soil organic carbon generally is reduced, and for soils with an argillic horizon, the clay composition of the exposed horizon is increased. Organic matter additions have been shown to increase soil organic carbon of eroded soil (Langdale et al., 1992) (Table 3.1), but the improvement is slow.

INORGANIC COMPONENTS

Skeletal

Skeletal soil components are the backbone of the soil and impart considerable mass but are largely chemically inert. The main skeletal components are quartz and feldspar grains. Because of their low solubilities, they tend to be stable in the silt and sand fractions of soils. These minerals do weather in soil environments but at a very slow rate as compared to the processes taking place at the time scale of raindrop impact. However, the weathering of skeletal components, in particular sodium plagioclase feldspars, does affect the chemical composition of the soil colloids. Also, the chemical composition of the soil colloids is affected by dust storms that carry in salt from desert areas.

Colloidal

Soil colloids (<2 mm), because of their large surface area and the surface charge component, provide most of the nutrient-holding capacity of the soil in addition to soil organic matter. They may also be thought of as the plasma that holds the skeletal grains into aggregates. Depending on the type of clay mineral in the soil and the amount of organic matter, the cation exchange capacity (CEC) can vary considerably (Table 3.1). Typically, soils high in smectitic clays will have a high CEC and those with kaolinite a low CEC. Likewise, organic matter will impart greater CEC and increased buffering capacity to a soil. As erosion of the soil progresses, the CEC may remain similar because of the offsetting factors of lowering of organic matter and increasing clay content for smectitic soils. However, for soils with kaolinitic or oxidic subsoils, the loss in organic matter can result in a considerable lowering of the CEC.

Adsorbed Ions

Sodium has long been known for its dispersive effects in soils, and its presence has been known for the deterioration of soil structure and poor soil quality. Soil classification systems have recognized this adverse behavior and provide groupings for these soils such as solonetz, natrargids, etc. Even soils with very low exchangeable sodium percentages can have clay dispersion, surface sealing, and low final infiltration rates (Gal et al., 1984).

Exchangeable Mg can also cause soil structural deterioration, and some classification systems recognize a "magnesium solonetz." Keren (1989) found that even for smectitic soils with no Na in Israel the infiltration rate was always lower for those with Mg as compared to Ca on the exchange complex. He attributed this difference to the larger radii of hydration of Mg compared to Ca. Apparently, the effect of Mg on lowering the infiltration rate is only measurable in a Ca/Mg system because of the overriding dispersive effect of increasing exchangeable sodium percentage (Shainberg, 1992). In addition, the ancillary effects of Mg on lowering the Ca/Mg ratio can also increase the dispersive effect on given levels of Na (Paliwal and Ghandi, 1976: Alperovitch et al., 1981; Seelig et al., 1990).

The effect of K on dispersion and seal formation has not been extensively studied. Levy and van der Watt (1990) showed that when K was the counterion to Ca, increasing the exchangeable potassium percentage decreased the final infiltration rate. They concluded that K had an intermediate effect between Ca and Na on the final infiltration rate of soils exposed to rain.

PROPERTIES OF RAINWATER AND SOIL SOLUTIONS

Rainwater, even acid rain, is relatively devoid of electrolytes as compared to the soil solution. When bare soil is exposed to rain, the impact energy of the drop compacts the soil and provides the mechanical energy for enhanced dispersion and surface sealing. Surface sealing causes a low final infiltration rate for many soils even in the absence of Na on the exchange complex (Norton et al., 1993). This combined effect is even more pronounced in the presence of Mg as compared to Ca on the exchange complex (Keren, 1989; Norton and Dontsova, 1998).

Because the dispersive effect of the low electrolyte concentration of rainwater is largely responsible for surface sealing on many soils, addition of low amounts of an electrolyte source to the soil surface has been successful as a means of improving infiltration rates (Agassi et al., 1981; Gal et al., 1984; Miller, 1987). Likewise, polymers that promote flocculation, such as polyacrylamide, have been successful in reducing surface sealing (Shainberg, 1992). In a micromorphological study of surface sealing, Norton (1987) determined that only the properties of the very few millimeters of the surface seal controlled the final infiltration rate. This led to the development of amending only this zone with low amounts of either electrolytes or polymers for erosion control (Norton, 1995). This thin zone of interactions was verified by Zhang et al. (1997).

AMENDABLE CHEMICAL PROPERTIES

Soil pH

Acid soil in this chapter is defined as soil having a pH below 7. Soil acidity is the dominating factor of acid soils, which influences almost every aspect of the soil system. Total soil acidity is divided into active acidity and potential acidity. Active acidity is a measure of H^+ concentration or activity in the soil solution, which is represented by soil pH. Soil pH is defined as the negative logarithm of the H^+ activity in soil solution. Active acidity, which is presumably in equilibrium with potential acidity, is also referred to as the intensity factor of soil acidity. Potential acidity is the fraction of total acidity which needs to be neutralized in order to achieve a selected soil pH. It is well reflected by the liming requirement and therefore is often referred to as the capacity factor. Potential acidity can be further separated into exchangeable acidity and titratable acidity. Exchange acidity is the acidity that can be replaced from exchange sites with neutral salts. Titratable acidity is the fraction that can be neutralized by titration at a selected pH, such as that produced by Al hydrolysis and dissociation of H^+ from organic and clay mineral compounds.

Formation of Soil Acidity

Soil acidity is determined by soil components and the nature and extent of chemical and biological reactions in the soil systems and is largely controlled by the degree of weathering. Weathering and dissolution of parent materials by hydrolysis of CO_2, followed by leaching of basic cations (Na, Ca, and Mg) with bicarbonate, is the dominant soil acidification process in nature. Weathering of acid parent materials or oxidation of acid minerals produces additional soil acidity. Organic matter or humus contains reactive carboxylic and phenolic groups which act as weak acids and dissociate H^+ through deprotonation. It is also a potential source of stronger acid HNO_3, produced by nitrification of organic nitrogen. Inorganic soil components such as aluminosilicate clay minerals and sesquioxides produce titratable acidity through deprotonation from hydroxyl groups at the clay surfaces and edges. Aluminum and iron hydrolyses are often major sources of soil acidity for highly weathered soils. Acid chemical fertilizers can also produce a considerable amount of acid by nitrification of ammonium fertilizer materials. In addition, removal of basic cations by plants, redox reactions, and acid rain may also cause soil acidification to a certain degree.

Significance of Soil pH

The pH of acid soils is ordinarily between 4 and 7. It is seldom below 4 unless free acids are present due to oxidation of sulfides. As an intensity factor of soil acidity, soil pH is a master indicator of the soil system and influences many of the chemical and biological processes occurring in soils, though it produces no hint about the total soil acidity. The pH is a good indicator of Al toxicity, which is generally the most limiting factor for crop production in acid soils. Severe problems are normally encountered when pH is below 5 because the most toxic species of Al^{3+} predominates in soil solution. No serious problems exist at pH > 5.5, particularly in highly weathered soils. The pH also indicates the solubilities of most chemical elements in soils and their associated availability, deficiency, or toxicity to plant growth. At low pH, Al and Mn become more soluble and can be toxic to plants. Solubilities of most micronutrients except for Mo decrease as pH increases, and deficiency symptoms are often seen when pH is greater than 7. The degree of Al hydrolysis is controlled by soil pH, which further affects total soil acidity and the level of Al toxicity to plants. Soil pH also has a significant impact on soil microbiological activity. It has been shown that the rates of mineralization and nitrification increase with the increase of pH up to 7 (Alexander, 1980).

The pH-dependent characteristics of elemental solubility and chemical reactions influence chemical compositions and concentrations of exchangeable and soil solution cations. As pH increases, the Al saturation percentage normally decreases, while basic metal cation saturation increases. This has a significant effect on clay flocculation, aggregate stability, and soil surface sealing. A decrease in exchangeable Al^{3+} due to an increase in soil pH tends to induce clay dispersion when electrolyte concentration in soil solution is low. For variable-charge soils, negative charges at the clay edges increase through deprotonation or dissociation of H^+ from hydroxyl groups as soil pH increases (charge reversal). This charge reversal reduces edge-to-face flocculation and therefore promotes clay dispersion (van Olphen, 1977). Enhanced clay dispersion after liming is more likely to occur in highly weathered, acid soils where Al^{3+} is the dominant exchangeable cation due to replacement of Al^{3+} with Ca^{2+}.

Saturating Ions and Exchange Capacity

The CEC of a soil is largely determined by the total negative charge of soils. Total negative charge is composed of permanent charge and variable or pH-dependent charge. Permanent charge, which is independent of environmental conditions such as soil pH under which it is characterized (Figure 3.1), is formed from isomorphous substitution of higher valency cations by lower valency cations within the clay lattice. This type of charge varies greatly with type of clay minerals and changes with the degree of weathering, with 2:2 and 2:1 clay > 1:1 clay > R_2O_3. Negative variable charge is caused by deprotonation from carboxylic and phenolic groups of organic matter or hydroxyl groups of clay and sesquioxides. This type of charge is dependent upon soil conditions such as soil pH, electrolyte concentration, and type

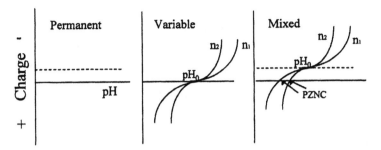

FIGURE 3.1   Surface charge versus pH for permanent-, variable-, and mixed-charge systems at two electrolyte concentrations. $n_2 > n_1$. (After Sumner, 1992).

of cation (Figure 3.1). Soil pH at which net variable charge vanishes is called pH_0 (Uehara and Gillman, 1981). The variable charge is negative above pH_0 and becomes positive below pH_0. Organic matter has a pH_0 around 3, and pH_0 of sesquioxides is between 8 and 9. In the pH range between 4 and 7, organic matter is negatively charged, while sesquioxides are positively charged. On a weight basis, organic matter by far has more charges than either layer silicates or sesquioxides or complexes of both. In addition, the negative charge produced by specific adsorption or exchange of ligands at the clay edges can be significant under certain circumstances. It also should be kept in mind that soil is usually a mixture of various types of clay minerals and organic matter so that charge interactions and intercompensations may exist and should be considered when interpreting charge-related experimental data.

As CEC is largely dependent on total negative charge of soils, all factors that alter total negative charge affect CEC as well. Therefore, the CEC of a variable-charge soil is also dependent on soil pH and the electrolyte concentration under which it is determined. Due to the pH-dependent nature of CEC, effective CEC, which is measured at soil pH by unbuffered neutral salt solutions such as KCl or $CaCl_2$, is preferred for soils with variable charges, such as highly weathered acid soils, because it reflects more intrinsic cation-holding capacity of the soil. In comparison, the other widely used methods, such as $BaCl_2$–triethanolamine at pH 8.2 and NH_4OAc at pH 7, tend to overestimate cation-holding capacity. However, the CECs determined with these buffered methods can be used as standard references because the experimental conditions are well controlled in these methods. The percent ratio of effective CEC to that at pH 8.2 varies considerably with soil constituents and organic matter content, ranging from 30 to 70% (Thomas and Hargrove, 1984). For soils with dominantly permanent charge, such as those with predominant layer silicates, the ratio is close to 1, while for soils which are high in organic matter and sesquioxides, the difference can be substantial. It has been shown that the pH-dependent CECs of acid surface soils increase slowly with pH from pH 3 to 5 and rather rapidly afterward (Pratt and Bair, 1962), and soil organic matter is the major contributor to pH-dependent CEC (Helling et al., 1964). CEC is determined by total negative charge, but the relationship is not always 1:1. At very low pH, exchange of Al^{3+} by neutral salt solution is often less than 100% because Al^{3+} is held tightly on adsorption sites due to large charge-to-size ratio. More importantly, complexation of Al^{3+} with organic matter, which makes it unexchangeable, counters some charges. As pH increases, Al^{3+} hydrolyzes, and the resultant hydroxyl–Al that is not exchangeable by neutral salt (Chernov, 1964) still counters some charges. However, more sites are expected to be liberated as Al hydrolysis continues. This may explain Pratt and Bair's (1962) observation that rapid increase of CEC occurred at pH > 5. In addition, the negative charge, which contributes to charge deficit in the electric double layer due to charge repelling (covered later), is not accounted for by CEC measures.

Percent base saturation with metal cation Ca^{2+}, Mg^{2+}, Na^+, and K^+ is another indicator of soil acidity and has significant implications for soil physical properties as well. Since the removal of these basic cations with bicarbonate from CO_2 hydrolysis is the major process causing soil acidification in nature, percent base saturation is inversely related to the degree of soil acidification. In contrast, percent saturation of Al^{3+}, which is the major source of exchangeable acidity, increases as acidity increases. CEC at pH 8.2 (or percent base saturation at pH 7) was found to be around 80% for less weathered soils with predominant 2:1 and 2:2 clay minerals. The percent base saturation was between 30 and 50% for organic matter as well as highly weathered soils with predominant kaolinite and oxides (Mehlich, 1942). Percent base saturation on an effective CEC basis may be more meaningful, but the problem of using effective CEC lies in the difficulties in obtaining reliable data due to large experimental variations.

The Ca^{2+}, Mg^{2+}, and Al^{3+} ions are the dominant cations exchangeable by neutral salts in acid soils. At pH < 5, Al^{3+} typically becomes predominant. Exchangeable H^+ is relatively small due to low affinity of adsorption sites to a lower valent cation (Coleman et al., 1959) and because H-clay is chemically not stable and the clay dissolves and releases Al. This is supported by the fact that hydrogen-saturated montmorillonite is still dispersive and shows considerable swelling. Chemical composition of adsorbed and/or dissolved exchangeable cations influences not only chemical properties but also physical properties of an acid soil. It dictates clay behavior (dispersion or flocculation), which further influences aggregate stability, formation of surface seals, water infiltration, and soil erosion. Chemical composition of adsorbed cations affects clay behavior more than that of the soil solution (Miller et al., 1990). The flocculation power of a cation is related positively to its valence but negatively to its size. In general, the flocculation power of metal cations follows the order $Al^{3+} > Ca^{2+} > Mg^{2+} > K^+ > Na^+$. At pH < 5, predominant Al^{3+} keeps clay flocculated and aggregates stable. Shifting from Al^{3+} to Ca^{2+} domination can cause a degree of clay dispersion for some soils, depending on the ionic strength of the soil solution. This will reduce saturated hydraulic conductivity and increase erosion due to clogging of conductive pores by dispersed clay particles (Norton et al., 1993). On the other hand, substitution of monovalent cations and/or Mg^{2+} by Ca^{2+} promotes clay flocculation and, therefore, increases water infiltration. Clay flocculation can be further enhanced by elevated electrolyte concentration after liming or adding gypsum. Contrarily, the deterioration of soil physical properties is usually observed when Ca^{2+} is substituted by Mg^{2+} (Norton and Dontsova, 1998). Therefore, for achieving better soil physical properties, calcitic limestones appear superior to dolomitic limestones.

Clay Flocculation and Dispersion

Clay minerals are the most active components in a soil due to the large surface area, small size, and charge phenomena. Soil physical and chemical properties are largely dependent on the clay characteristics and behavior (flocculation or dispersion). Since clay may act as a cementing agent, clay dispersion or flocculation has a tremendous effect on soil physical properties such as aggregation and structural stability. Clay flocculation promotes soil aggregation and structural stability, while clay dispersion often leads to the deterioration or loss of soil structure. The behavior of clay particles can be understood with the Gouy–Chapman electric double-layer theory.

The electric double layer refers to a clay surface charge and a compensating counterion charge (Figure 3.2). The clay surface charge, normally negative in most soils, is composed of permanent charge and/or variable charge. The counterion charge is a diffuse cloud of ions that are electrostatically attracted by the oppositely charged surface and have a tendency to diffuse away from the surface to the bulk solution. Simultaneously, counterions, which have the same sign as the surface charge, are repelled by the surface charge and tend to diffuse back into the counterion cloud. Based on the electrostatic and diffusion theories (Boltzmann's theorem and Poisson's equation), concentration distributions of positive and negative ions in the double layer can be estimated. A schematic illustration of the computed ion distributions at two electrolyte concentrations for both permanent- and variable-charge surfaces is presented in Figure 3.3. The n_+ and n_- are cation and anion concentration distributions (number of ions per unit volume), and σ_+ and σ_- are total cation excess and anion deficit, respectively. Symbols with the prime sign stand for the corresponding values at great distance from the clay surface. The total surface charge ($\sigma = \sigma_+ + \sigma_-$) and the charge ratio of σ_-/σ_+ in response to the changes in electrolyte concentration are also given in Figure 3.3. It clearly illustrates that the double layer

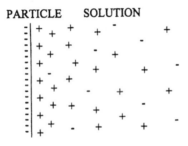

FIGURE 3.2 Distribution of diffuse double-layer charges according to Gouy–Chapman theory. (After van Olphen, 1977.)

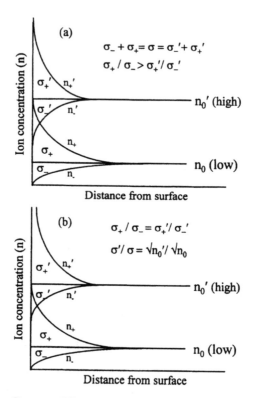

FIGURE 3.3 Gouy–Chapman diffuse double-layer charge or ion distribution at two electrolyte concentrations for (a) constant surface charge and (b) constant surface potential. (After van Olphen, 1977.)

is compressed toward the clay surface when electrolyte concentration is elevated. A similar response is also exhibited when a lower valency cation is replaced by a greater valency cation. The double-layer compression or expansion dictates clay behavior.

When two particles approach each other, their diffuse counterion clouds begin to overlap or repel each other. Based on the electric double-layer theory, this repulsive potential or repulsive energy (V_r) decreases quasi-exponentially with increasing particle separation. As shown in Figure 3.4, curve V_r (namely, repulsive potential curve) is compressed toward the particle surface when the counterion concentration is increased. An attractive energy (V_a), which works against the repulsive energy, is derived from the van der Waals attractive forces that are the sum of all the attractive forces between every atom of one particle and every atom of another particle. The attractive potential curve decreases rapidly with particle separation and is inversely proportional to the second power of the distance from the particle surface.

The summation of the repulsive and attractive potential curves at each distance (termed net potential curve, V_n) determines the nature of the colloidal suspension.

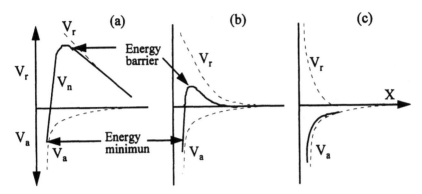

FIGURE 3.4 Particle separation as a result of attractive and repulsive energies at (a) low, (b) medium, and (c) high electrolyte concentration. V_r = diffused double-layer repulsion energy, V_a = van der Waals attraction energy, and V_n = net potential curve. (After van Olphen, 1977.)

The V_n is suppressed as electrolyte is increased due to the reduction in the repulsive potential (Figure 3.4). This results in a reduction in the energy barrier. When a critical concentration, at which the energy barrier disappears, is reached, rapid coagulation takes place. At this concentration (termed CFC), the flocculation rate depends on the frequency of particle collisions. It should be pointed out that theoretically flocculation could occur at any concentration. However, the flocculation rate becomes extremely slow when the energy barrier is large because the chances for clay particles to overcome the energy barrier and to be "trapped" in the energy minimum are very small, provided Brownian motion is the major means to cause particle encounters.

Under the conditions where the energy barrier vanishes, the CFC can be derived as:

$$CFC = \frac{\lambda \{\tanh[ze\psi_0 /(4kT)]\}^4}{A^2 z^6} \tag{3.1}$$

where λ = a constant number, A = Hamaker constant, z = valence of the counterions, e = electronic charge, ψ_0 = electrical potential at the clay surface, k = Boltzmann constant, and T = absolute temperature.

From Equation 3.1, the dependence of CFC on $1/z^6$, known as the Schulze–Hardy rule, indicates that z has a dominant effect on flocculation. To obtain qualitative, comparative information on the effects of valence and electrolyte concentration on the double-layer thickness, an "effective thickness" (δ), which is defined as the center of gravity of the space charge in the double layer, can be computed as:

$$\delta = \left(\frac{\epsilon kT}{8\pi z^2 e^2 c N_A} \right)^{1/2} \tag{3.2}$$

where ε = dielectric constant of the solvent, c = the molar concentration, and N_A = Avogadro constant.

The δ from Equation 3.2 can only be used to represent the symbolic length of the double-layer extension, since the double layer actually extends to infinity from the clay surface. However, Equation 3.2 depicts the impact of c and z on clay behavior. Swelling pressure (p_s), which is the repelling force between the two overlapped double layers, can be computed as:

$$ p_s = 2N_A ckT \left[\cosh\left(\frac{ze\psi_d}{kT} \right) - 1 \right] \tag{3.3} $$

where ψ_d is the potential midway between the plates. The swelling pressure is considered to be identical to the osmotic pressure generated midway between the two plates. The dependence of p_s on z and c is not obvious in Equation 3.3 because the embedded function of ψ_d decreases exponentially with z and the square root of c. As a matter of fact, p_s decreases as z and c increase. This relationship has significant implications in soil mechanics and civil engineering applications.

The reverse process or clay dispersion is invoked when electrolyte concentration decreases to below CFC. This process is of great importance in preserving soil aggregate and structural stability. Spontaneous dispersion often takes place for soils with predominant swelling-type clays or those high in exchangeable sodium percentage (ESP) when electrolyte concentration in the soil solution is below the CFC. Mechanically induced dispersion (with energy input, e.g., raindrop impact) is the dominant process for the soils with nonswelling clay minerals and low in ESP. Gradual loss of clay particles deteriorates soil structure and sometimes leads to severe piping or tunneling. Clay dispersion also plays a significant role in surface seal formation. The low electrolyte concentration of rainwater, coupled with the mechanical raindrop impact, results in severe clay dispersion at the soil surface. Dispersed clay particles, which then clog conducting pores near the soil surface or settle on the soil surface during water infiltration, lead to formation of a less permeable surface seal. This thin layer greatly increases runoff and erosion due to its extremely low hydraulic conductivity.

EFFECTS OF CHEMICAL COMPOSITION AND pH ON FLOCCULATION

Chemical composition and concentration on exchangeable sites and in soil solution greatly influence the clay behavior in a soil. The flocculating power of cations, depending on their valences and sizes, follows the sequence $Al^{3+} > Ca^{2+} > Mg^{2+} > K^+ > Na^+$. The Al^{3+} has the greatest flocculating power, while Na^+ often acts as a dispersing agent in soils. Therefore, soil ESP is a good indicator of the degree of clay dispersion. Miller et al. (1990) reported that CFC increases linearly with ESP for three soil clays. This implies that electrolyte concentration necessary for flocculation increases as Na$^+$ concentration increases. Soils that are high in ESP, such as sodic soils, often have poor aggregation and structure and very low hydraulic conductivity. Many studies have shown that ESP has a tremendous effect on surface seal and crust formation. The enhanced clay dispersion with Na$^+$ is because replacement of higher valency cations with Na$^+$ causes electric double-layer expansion, which results in an increase in repulsive potential between clay particles.

Exchangeable Mg^{2+} and K$^+$ may also promote clay dispersion and cause soil structural deterioration compared with Ca^{2+} (Shainberg et al., 1988; Keren, 1989). Keren (1989) found that soils showed a higher water infiltration rate when treated with Ca salt than when treated with Mg salt. This is because of the greater hydration radii of Mg (compared with Ca^{2+}), which may shift the diffuse ion cloud away from the surface. Levy and van der Watt (1990) reported that an increase in K$^+$ exchangeable percentage in the K–Ca system enhanced clay dispersion and decreased hydraulic conductivity.

For variable-charge soils, soil pH is an important factor influencing soil clay behaviors. Since negative pH-dependent charge increases as soil pH increases, the increased total negative charge, of course, will result in an increase in repulsive potential. As a result, clay becomes more dispersive as pH increases. Reichert and Norton (1996) showed that surface application of fluidized bed combustion bottom ash (containing 23% CaO and 73% anhydrite) on highly weathered soils caused enhanced clay dispersion due to elevated pH, which further resulted in greater runoff and soil loss rates for sesquioxcidic soils. For acid soils when Al^{3+} is the dominant exchangeable cation (normally pH < 5), liming can cause severe clay dispersion and soil structural deterioration. As pH increases, Al^{3+} is replaced by exchangeable Ca^{2+} and then hydrolyzed. The lower flocculation power of Ca^{2+} compared to Al^{3+}, coupled with increased negative charges resulting from elevated soil pH, would enhance clay dispersion.

The beneficial effects of liming on soil physical properties also result from the elevated ionic strength in soil solution and the domination of Ca^{2+} on exchangeable sites. As discussed earlier, Ca^{2+} has the second highest flocculating power following Al^{3+}, which is the dominant species only at very low pH. Increasing exchangeable calcium percentage by replacing monovalent cations or Mg^{2+} inhibits clay dispersion and therefore promotes aggregation and soil structural stability. Increased exchangeable calcium percentage and ionic strength increases soil resistance to aggregate breakdown (Zhang and Miller, 1996). This preserves soil aggregates from disintegration by raindrop impact and reduces surface sealing and crusting (Miller and Radcliffe, 1992; Agassi et al., 1981; Shainberg, 1992). It also increases hydraulic conductivity (Chiang et al., 1987), while reducing surface runoff and erosion (Norton et al., 1993; Zhang and Miller, 1996; Miller, 1987; Shainberg et al., 1989). For highly weathered soils, which are often lower in cation-holding capacity and electrolyte concentration, increased ionic strength usually plays a dominant role in flocculating clays, but for some soils, replacement of Al^{3+} by Ca^{2+} and/or increased pH-dependent negative charge can enhance clay dispersion and deteriorate soil structure. A thorough knowledge of the soil chemistry and mineralogy is required to determine the optimum liming materials for improving soil quality.

SOIL CONDITION

The condition of the soil may change considerably over a short time and may be considered as a time-dependent soil property. Soil condition is affected by antecedent moisture content, the length of time the soil is at a given moisture content (aging) prior to the rainstorm, the wetting rate of the rain (e.g., the initial wetting of the soil was completed by low-intensity rain prior to the intense storm), etc. Aging of the soil, in terms of hours and days, increases the cohesive forces between soil particles and stabilizes soil aggregates (Kemper et al., 1987; Shainberg et al., 1996) so that runoff and erosion are reduced significantly. Similarly, slow prewetting prevents aggregate breakdown and the soil surface is much less susceptible to crusting and runoff. Soil condition is affected by the composition but is largely a function of the state that the materials comprising the soil are in at a particular time. Soil condition may vary depending on soil composition (texture, clay mineralogy), soil pH, saturating cations, moisture content, organic content, near-surface hydraulic gradients, etc. Soil erodibility is related to soil condition and composition and can vary considerably as a soil is eroded (Gabbard et al., 1997).

RELATION OF SOIL QUALITY TO EROSION PROCESSES

Soil erosion processes are typically divided into rill and interrill flow processes. Detachment and transport of soil particles and aggregates can occur in both regimes. The main difference is that interrill detachment and transport is aided by raindrop impact whereas rill processes are turbulent and generally not affected by raindrops. Typically, transport of particles is unaffected by raindrops when flow depths are greater than three raindrop diameters (Kinnell, 1991).

Interrill Processes

Raindrop impact, as stated earlier, is the dominant process in detachment in interrill areas. The interaction between the raindrops and the soil typically leads to a reduction in surface roughness elements and depressional storage. The development of the surface seal with low roughness and depressional storage promotes runoff and transport by rain-impacted flows (Eltz and Norton, 1997). However, the increase in strength of the seal results in a reduction in the material detached and partially offsets the increased runoff and transport capacity (Norton, 1987).

Slaking of aggregates occurs when soil at low antecedent moisture is wetted by rainfall (Reichert et al., 1994). This fast wetting causes entrapment of air and spatial swelling in the internal structure of the aggregates, which leads to breakdown of the aggregates. This leads to a rapid reduction in infiltration rate and is accentuated by the chemical interactions of the soil and the rainwater. These processes, coupled with the chemical and physical dispersion of clays, lead to smaller particles and aggregates which can be more easily transported by the shallow interrill flows. Compaction of the surface also occurs and leads to a fusing or welding together of aggregates, further developing the surface seal. A reduction of the chemical quality of the soil which makes it more susceptible to interrill processes, such as a reduction in organic matter or aggregate stability, tends to increase the erosion risk or soil erodibility (Langdale et al., 1992).

Rill Processes

While interrill processes are important for runoff production, breakdown of aggregates, and clay dispersion, rill processes generally lead to the actual removal of soil from a hillslope or field. Similar interactions with rainwater occur in rills, but to a lesser extent since the water in rills has already interacted with the soil and contains electrolytes. However, swelling, slaking, and breakdown of unstable aggregates are very important processes in rills because these processes lead to a plentiful supply of small transportable aggregates or primary particles. Any process that leads to increased detachment or reduction in the size of the materials will increase the erosion rate and rill erodibility of a soil. Unlike interrill erosion studies, where it is important to simulate the quality of the rainwater, Flanagan et al. (1997) found that the quality of water used does not affect measured results in rill erosion studies. This indicates that processes active in rills are largely physical rather than chemical in nature.

DEGRADATION OF SOIL CHEMICAL PROPERTIES AND EROSION

Soil degradation is the subject of much research and debate. Degraded soil infers that it has less quality than undegraded "virgin" soil. Typically, as the soil is disturbed, organic matter is reduced and aggregates become less stable, and the soil more susceptible to erosion (Darmody and Norton, 1994). As erosion of a soil occurs, it is often chemically degraded because of the selective removal of organic matter and changes in accompanying soil chemical factors. Cihacek and Swan (1994) observed differences among soils in the effect of erosion on soil chemical properties that were related to surface and subsoil properties caused by differences in parent material, climate, vegetation, and time during the soil-forming process. Eroded soil can cause adverse environmental problems off-site due to eutrophication of water bodies. In addition to clays, electrolytes that are released from the soil are removed, and these typically are nutrients needed by plants. The coupling of less plant cover to protect the soil from raindrop impact and exposure of high-clay subsoils increases the potential for surface sealing, runoff, and erosion, further degrading the soil.

Erosion not only degrades the soil profile but also reduces the productive capacity of the entire landscape. In the worst case, this involves total land destruction and abandonment (Lal, 1990). In many parts of the world, as erosion of topsoil progresses, less fertile and acid subsoils are exposed, resulting in a degradation of the chemical status of the topsoil. This results in a dramatic degradation in the productive potential of these soils, and production can be maintained only with large inputs of chemical fertilizers. Resource-poor farmers are not able to buy fertilizers, and the result is less food production. The reduction in plant canopy cover due to exposure of infertile subsoils or rocks leads to increased wind and water erosion and a further accentuation of the problem.

SUMMARY AND CONCLUSIONS

Soil quality is a rather arbitrary term. What may provide favorable soil quality for one application may not be suitable or desirable for another. For soils where water erosion is a potential problem, the chemistry of the soil is very important in explaining its behavior to application of rainwater. The surface physical chemistry effects are quite important in explaining the aggregate stability and soil surface sealing characteristics. These effects also are important in determining the condition of the soil at the time of its interaction with rainwater. The low electrolyte content of rainwater and the mechanical beating of the drops generally cause a lowering of the infiltration rate and runoff production containing sediment. The beating of raindrops can be controlled with mulching and residues; however, the physical chemical processes still occur, leading to runoff production underneath residues. These processes include dispersion, flocculation, cohesion, hard-setting, self-mulching, slaking, swelling, and sealing.

Soil composition cannot be easily changed, but the soil chemical quality often has been. Conventional tillage systems generally lower the amount of carbon and nitrogen in the soil. Conversion to conservation tillage can also restore some of these losses and improve soil quality. Soil chemical quality has been largely expressed as the amount of available N, P, and K in a soil in addition to some meso- and micronutrients, and these have been the major soil amendments for plant production. Soil pH is important for plant nutrition and in the dispersion/flocculation processes related to erosion. The type of saturating cations on the exchange complex can have a considerable effect on soil structure and erosion. Using gypsum and other soil amendments can improve soil structure and reduce erosion. More research is needed to determine the optimum percentage of Ca, Mg, and K in the exchange complex to have favorable soil structure and control erosion with and without the presence of an electrolyte source such as gypsum. A thorough understanding of how soil physical chemical processes affect soil quality and erosion will lead to better management strategies that may improve yields through better water distribution on the landscape while economically controlling erosion through the use of soil amendments.

REFERENCES

Agassi, M.J., I. Shainberg, and J. Morin. 1981. Effect of electrolyte concentration and soil sodicity on infiltration rate and crust formation. *Soil Sci. Soc. Am. J.* 45:848–851.

Alexander, M. 1980. Effect of acidity on microorganisms and microbial processes in soil. In T. Hutchinson and M. Harvas (eds.). *Effects of Acid Precipitation on Terrestrial Ecosystems.* Plenum Publishing, New York, pp. 363–364.

Alperovitch, N., I. Shainberg, and R. Keren. 1981. Specific effect of magnesium on the hydraulic conductivity of soils. *J. Soil Sci.* 32:543–554.

Chernov, V.A. 1964. *The Nature of Soil Acidity* (English translation). Soil Science Society of America, Madison, WI.

Chiang, S.C., D.E. Radcliffe, W.P. Miller, and I.D. Newman. 1987. Hydraulic conductivity of three southeastern soils as affected by sodium, electrolyte concentration and pH. *Soil Sci. Soc. Am. J.* 51:1293–1299.

Cihacek, L.J. and J.B. Swan. 1994. Effects of erosion on soil chemical properties in the north central region of the United States. *J. Soil Water Conserv.* 49:259–265.

Coleman, N.T., S.B. Weed, and R.J. McCracken. 1959. Cation-exchange capacity and exchangeable cations in Piedmont soils of North Carolina. *Soil Sci. Soc. Am. Proc.* 23:146–149.

Darmody, R.G. and L.D. Norton. 1994. Structural degradation of a prairie soil from long-term management. In A.J. Ringrose-Voase and G.S. Humphreys (eds.). *Soil Micromorphology: Studies in Management and Genesis.* Developments in Soil Science 22. Elsevier, Amsterdam, pp. 641–650.

Eltz, F.L. and L.D. Norton. 1997. Soil surface roughness changes as affected by tillage and rainfall erosivity. *Soil Sci. Soc. Am. J.* (in press).

Flanagan, D.C., L.D. Norton, and I. Shainberg. 1997. Effect of water chemistry and soil amendments on a silt loam soil. I. Infiltration and runoff. *Trans. ASAE* (in press).

Gabbard, D.S., C. Huang, and L.D. Norton. 1997. Landscape position, surface hydraulic gradients and erosion processes. *Earth Surf. Process. Landforms* (in press).

Gal, M., L. Arcan, I. Shainberg, and R. Keren. 1984. The effect of exchangeable Na and phosphogypsum on the structure of soil crust — SEM observations. *Soil Sci. Soc. Am. J.* 48:872–878.

Helling, C.S., G. Chesters, and R.B. Corey. 1964. Contributions of organic matter and clay to soil cation-exchange capacity as affected by the pH of the saturating solution. *Soil Sci. Soc. Am. Proc.* 28:517.

Kemper, W.D., R.C. Rosenau, and A.R. Dexter. 1987. Cohesion development in disrupted soils as affected by clay and organic matter content and temperature. *Soil Sci. Soc. Am. J.* 51:860–867.

Keren, R. 1989. Water-drop kinetic energy effect on water infiltration in calcium and magnesium soils. *Soil Sci. Soc. Am. J.* 53:1624–1628.

Kinnell, P.I.A. 1991. The effect of flow depth on sediment transport induced by raindrops impacting shallow flows. *Trans. ASAE* 34:161–168.

Lal, R. 1990. Soil erosion and land degradation: the global risks. In R. Lal and B.A. Stewart (eds.). *Soil Degradation.* Advances in Soil Science, Vol. II, Springer-Verlag, New York, pp. 129–172.

Langdale, G.W., L.T. West, R.R. Bruce, W.P. Miller, and A.W. Thomas. 1992. Restoration of eroded soil with conservation tillage. *Soil Technol.* 5:81–90.

Levy, G.J. and H.V.H. van der Watt. 1990. Effect of exchangeable potassium on the hydraulic conductivity and infiltration rate of some South African soils. *Soil Sci.* 149:69–77.

Mehlich, A. 1942. Base saturation and pH in relation to soil type. *Soil Sci. Soc. Am. Proc.* 6:150–154.

Miller, W.P. 1987. Infiltration and soil loss of three gypsum-amended Ultisols under simulated rainfall. *Soil Sci. Soc. Am. J.* 51:1314–1320.

Miller, W.P. and D.E. Radcliffe. 1992. Soil crusting in the southeastern United States. In M.E. Sumner and B.A. Stewart (eds.). *Soil Crusting: Chemical and Physical Processes.* Lewis Publishers, Boca Raton, FL, pp. 233–266.

Miller, W.P., H. Frenkel, and K.D. Newman. 1990. Flocculation concentration and sodium/calcium exchange of kaolinitic soil clays. *Soil Sci. Soc. Am. J.* 54:346–351.

Norton, L.D. 1987. Micromorphological study of surface seals developed under simulated rainfall. *Geoderma* 40:127–140.

Norton, L.D. 1995. Mineralogy of high calcium/sulfur-containing coal combustion by-products and their effect on soil surface sealing. In D.L. Karlen, R.J. Wright, and W.O. Kemper (eds.). *Agricultural Utilization of Urban and Industrial By-Products*. Am. Soc. of Agronomy Special Publication No. 58. American Society of Agronomy, Madison, WI, pp. 87–106.

Norton, L.D. and K. Dontsova. 1998. Use of soil amendments to prevent soil surface sealing and control erosion. In Proc. IXth Int. Soil Conservation Organization (accepted for publication).

Norton, L.D., I. Shainberg, and K.W. King. 1993. Utilization of gypsiferous amendments to reduce surface sealing in some humid soils in the eastern USA. In J.W.A. Poesen and M.A. Nearing (eds.). *Soil Sealing and Crusting*. Catena Supplement 24. Catena Verlag, Cremlinge, Germany, pp. 77–92.

Paliwal, K.V. and A.P. Ghandi. 1976. Effect of salinity, SAR, Ca:Mg ratio in irrigation water, and soil texture on the predictability of exchangeable sodium percentage. *Soil Sci.* 122:85–90.

Pratt, P.F. and F.L. Bair. 1962. Cation-exchange properties of some acid soils of California. *Hilgardia* 33:689.

Reichert, J.M. and L.D. Norton. 1996. Fluidized bed combustion bottom-ash effects on infiltration and erosion of variable-charge soils. *Soil Sci. Soc. Am. J.* 60:275–282.

Reichert, J.M., L.D. Norton, and C. Huang. 1994. Sealing, amendment, and rain intensity effects on erosion of high-clay soils. *Soil Sci. Soc. Am. J.* 58:1199–1205.

Seelig, B.D., J.L. Richardson, and W.T. Barker. 1990. Characteristics and taxonomy of sodic soils as a function of landform position. *Soil Sci. Soc. Am. J.* 54:1690–1697.

Shainberg, I. 1992. Chemical and mineralogical components of crusting. In M.E. Sumner and B.A. Stewart (eds.). *Soil Crusting: Chemical and Physical Processes*. Lewis Publishers, Boca Raton, FL, pp. 33–53.

Shainberg, I., N. Alperovitch, and R. Keren. 1988. Effect of magnesium on the hydraulic conductivity of sodic smectites–sand mixtures. *Clays Clay Miner.* 36:432.

Shainberg, I., M.E. Sumner, W.P. Miller, M.P.W. Farina, M.A. Pavan, and M.V. Fey. 1989. Use of gypsum on soils: a review. In B.A. Steward (ed.). *Advances in Soil Science*, Vol. 9. Springer-Verlag, New York, pp. 1–112.

Shainberg, I., D. Goldstein, and G.J. Levy. 1996. Rill erosion dependence on soil water content, aging, and temperature. *Soil Sci. Soc. Am. J.* 60:916–922.

Sumner, M.E. 1992. The electrical double layer and clay dispersion. In M.E. Stewart and B.A. Stewart (eds.). *Soil Crusting: Chemical and Physical Processes*. Lewis Publishers, Boca Raton, FL, pp. 1–31.

Thomas, G.W. and W.L. Hargrove. 1984. The chemistry of soil acidity. In F. Adams (ed.). *Soil Acidity and Liming*. Agron. Monogr. No. 12, ASA, CSSA, and SSSA, Madison, WI, 380 pp.

Uehara, G. and G. Gillman. 1981. *The Mineralogy, Chemistry, and Physics of Tropical Soils with Variable Charge Clays*. Westview Press, Boulder, CO.

van Olphen, H. 1977. *An Introduction to Clay Colloid Chemistry*. John Wiley & Sons, New York, 318 pp.

Zhang, X.C. and W.P. Miller. 1996. Physical and chemical crusting processes affecting runoff and erosion in furrows. *Soil Sci. Soc. Am. J.* 60:860–865.

Zhang, X.C., L.D. Norton, and M.A. Nearing. 1997. Chemical transfer from soil solution to surface runoff. *Water Resour. Res.* 33:809–815.

4 Impact of Soil Organisms and Organic Matter on Soil Structure

D.E. Stott, A.C. Kennedy, and
C.A. Cambardella

INTRODUCTION

Historically, soil quality has been equated solely with the soil's ability to promote the growth of plants. However, within the last several years, it has come to include the soil's capacity to protect watersheds by regulating infiltration and partitioning of water and to prevent water and air pollution by buffering potential pollutants (National Research Council, 1993). The final determination of a soil's relative quality depends upon its use. Thus, the water infiltration and retention characteristics for a good quality soil used as a building site will be quite different from a good quality agricultural soil. In this chapter, we will concentrate on agricultural uses of soil.

To regulate infiltration and partition water flow, a good quality agricultural soil will need to exhibit several characteristics. These include a structure that allows water to infiltrate, the capability of retaining beneficial amounts of water, a low tendency to crust or form a surface seal, and the ability to resist erosive forces. Many of these characteristics are strongly influenced by the soil microbial population. A major activity of the soil microbial community is the decomposition and transformation of organic residues into soil organic matter (SOM). SOM greatly influences soil physical characteristics such as structure, surface crusting and seal formation, water retention characteristics, and erodibility. There is evidence that management strategies influence not only the amount of SOM in the soil, but also how SOM is distributed in the various organic matter fractions. The dynamic nature of soil biological communities, microbial and macrofaunal, makes them a sensitive indicator for assessing soil quality alterations due to changing management practices (Kennedy and Papendick, 1995). Soil populations may provide advanced evidence of subtle changes in the soil before changes in soil physical and chemical properties become apparent.

SOIL QUALITY INDICATORS

Over the last decade, scientists have worked on developing a set of basic soil characteristics that serve as key soil quality indicators (Doran and Parkin, 1996; Parr et al., 1992). These indicators need to be sensitive to changes in both management and climate. Soil characteristics that change within only a few weeks or months in response to the changing seasons, shifting weather patterns, and plant growth cycles are not appropriate soil quality indicators. Characteristics that begin to show change only after 5 or more years are not helpful indicators, often showing progressive soil degradation only after much of the productive topsoil is lost. The best soil quality indicators are those characteristics that show significant change between 1 and 3 years, with 5 years being an upper limit of usefulness.

Soil quality indicators are usually classified as physical, chemical, or biological indicators (Doran and Parkin, 1996). Physical indicators include soil texture, depths of soil, topsoil and rooting, bulk density, infiltration rate, and water retention characteristics. The chemical indicators are total organic C and N, pH, electrical conductivity, and extractable N, P, and K. The basic biological indicators are microbial biomass C and N content, potentially mineralizable N, soil respiration, water content, and soil temperature. The latter two indicators are measured in order to meaningfully interpret other data. Two useful calculations are the ratio of biomass C to total soil organic C and the ratio of soil respiration to microbial biomass respiration (Rice et al., 1996).

The divisions between physical, chemical, and biological factors are not clear-cut, as will be discussed in this chapter. Soil physical characteristics are affected by the biological and chemical condition of the soil. In turn, the soil biological activity is impacted by the soil's chemistry and physical structure. Likewise, the chemistry is impacted by other soil factors.

Soil quality itself is the product of a combination of soil-degradative and soil-conserving processes (Parr et al., 1992). Soil-degradative processes may involve soil erosion, nutrient runoff, acidification, compaction, crusting, organic matter loss, salinization, nutrient depletion, and accumulation of toxins or heavy metals. Management schemes that help conserve the soil are conservation tillage practices, crop rotations, residue management schemes that leave the bulk of the crop residues on or near the surface, fertilizers, organic amendments, water conservation techniques, terracing, contour farming, improved drainage, and better management systems that match cultivar to the soil and climate conditions. Enhancing soil quality may be our first line of defense against water and air pollution.

BENEFICIAL ACTIVITIES OF SOIL ORGANISMS

Soil is a living system. It is the domain of a profusion of microorganisms and small animals that are closely associated with the soil organic fraction. The organic fraction represents a carbon, energy, and nutrient source for most soil organisms. The numbers, kinds, and activities of these organisms are influenced by the type and amount of food material available, including plant and animal residues and soil humus. Soil characteristics such as soil texture, pH, moisture, aeration, salt concentrations, and available nutrients will impact soil biological activity.

Although in some soils a few organisms may parasitize or injure plant roots, or infest animals or humans, the vast majority perform beneficial functions which are important for soil, plants, and animals, including humans. Through decomposition of organic residues and soil humus, soil organisms release C, N, and other nutrients, making them available for new generations of living things. As plant and animal residues are degraded by soil microbes, part of the organic material, along with organics synthesized by soil organisms, is transformed into soil humus (Melillo et al., 1989; Stott et al., 1983a; Verma and Martin, 1976; Voroney et al., 1989). Toxic organic substances from plants, as well as most applied organic amendments and pesticides, are either destroyed, utilized as a carbon and energy source, or linked into newly forming soil humus through the action of soil microbes (Bollag et al., 1980; Stott et al., 1983b; Wolf and Martin, 1975). Soil aggregation or tilth is improved through the activity of microorganisms during organic residue decomposition. Microbial activity solubilizes inorganic nutrients and either oxidizes or reduces N and S compounds, depending upon environmental conditions. Specific groups of soil bacteria play a key role in fixing atmospheric nitrogen and then transforming it into a plant-useable form.

RESIDUES AND RESIDUE DECAY

Plant and animal residue decay is one of the chief functions of soil microorganisms. Soil microbes not only derive food and energy from the residues, but create new biomass and transform some of the residue components into soil organic matter. Residues and residue decay impact soil erodibility through the formation of biomass, metabolic by-products, and SOM. Together these fractions can bind soil particles together into soil aggregates, thereby increasing soil resistance to erosion. In addition, surface residues protect the soil mechanically by absorbing the wind and water impact energies. As residue decays, the surface cover protection is reduced and the soil surface's exposure to erosive forces is increased.

Residue decomposition rates are controlled primarily by environmental factors in the field (Martin and Haider, 1986). Especially important are the residue water content and temperature (Orchard, 1983; Parr and Papendick, 1978). Considerable information exists on the effects of soil water potential on the growth and activity of individual microbial species. Microbial tolerance to low water potentials varies widely. The maximum tolerance of the bacterial genus *Vibrio* has been reported as –20 MPa (1 MPa = 10 bars) water potential and *Micrococcus* as –25 MPa (Rose, 1976). Ascomycetes and yeasts have tolerances down to –65 MPa (Leistner and Rodel, 1976). When relating soil water potential to microbial activity, –0.33 MPa or –1/3 bar is considered optimal (Sommers et al., 1981). While soil water potential is a useful indicator of microbial activity when water is limiting, soil water content may be a more useful measurement where water availability is not limiting (Skopp et al., 1990). As water-filled pore space increases above 60%, O_2 becomes limiting and microbial activity decreases (Linn and Doran, 1984b). In loams and silt loams, 60% water-filled pore space is roughly equivalent to a water potential of -0.33 MPa, which approximates the "field capacity" of a soil.

Many of the soil bacteria frequently found in temperate regions are psychrotrophic and generally grow over a wide temperature range (Baross and Morita, 1978; Biederbeck and Campbell, 1971; Stott et al., 1986). Common soil bacterial genera, such as *Pseudomonas, Bacillus,* and *Clostridium,* can grow at soil temperatures from below 0 to 40°C (Druce and Thomas, 1970). Wiant (1967) determined that microbial activity followed Van't Hoff and Arrhenius laws (activity doubles for every 10°C rise in temperature) at temperatures below 40°C; Reiners (1968) confirmed this down to 0°C. However, these relationships must be viewed with caution as dramatic changes occur in the soil microbial population's makeup and activity as environmental conditions fluctuate. As temperatures approach 0°C for long periods of time, psychrotrophic and psychrophyllic organisms become more active in the soil environment, causing the system to no longer respond to Van't Hoff and Arrhenius laws. Stott et al. (1986) determined that during the initial stages of decomposition, significant decay still occurred at 0°C, as long as there was sufficient liquid water present. As the soil solution contains salts, the freezing point is depressed to -1 to -2°C.

Another factor impacting residue decay rates is the position of the residue within the soil profile. Tillage operations displace the aboveground biomass from the surface to within the A_p horizon. The amount of residue buried depends on the type and configuration of the tillage implement. An implement such as the moldboard plow inverts the residue, shifting the majority of it to the bottom of the plow layer. A chisel/disk system will mix the residues throughout the plow layer. Conservation tillage systems will leave a majority of residues on or near the soil surface. Each succeeding tillage pass not only turns some of the residue under, but also can, in a few cases, return some of the previously buried residue to the surface (Hill and Stott, 1997; Johnson, 1988).

Position of the residue will impact the rate of decay, primarily through changes in the temperature and water conditions. Residues that remain standing after harvest are exposed to greater fluctuations in air temperature and dry quickly when wetted by precipitation, limiting the amount of water available for microbial activity. Residues left flat on the surface decay at a faster rate than standing residues, primarily at the residue soil interface, where sufficient water is present to sustain activity. Buried residues will decay fastest, as the soil temperature fluctuates less than air temperatures, and there are longer time periods when the soil water content will maintain microbial activity.

Chemical composition of residue also controls the decay rate. The C-to-N ratio is frequently stated in the literature as the primary indicator of how chemical composition will impact decomposition. This is true only in the broadest sense, when comparing legumes and nonlegumes. Legumes, with a lower C-to-N ratio, also decay more rapidly than other types of plant residue. When determining rates among similar plant species, the C-to-N ratio is a very poor predictor of decay. A field study

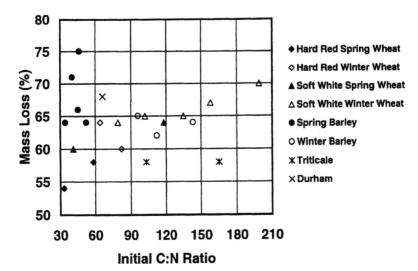

FIGURE 4.1 Relation of the C-to-N ratio of small grain residues versus mass lost during 1 year of field exposure. (From Smith, J.H. and R.E. Peckenpaugh. 1986. *Soil Sci. Soc. Am. J.* 50:928–932. With permission.)

by Smith and Peckenpaugh (1986) showed that wheat and barley (*Hordeum vulgare* L.) residues, with similar initial C-to-N ratios, may have widely varying decomposition values. Rates may vary up to 20% over a 1-year period (Figure 4.1). Stott (unpublished data) has shown the same trend in corn (*Zea mays* L.) and soybean varieties. A lab study using the leaves, stems, and roots from cotton (*Gossypium hirsutum*), sorghum (*Sorghum bicolor* [L.] Moench), and peanut (*Arachis hypogaea*) showed that 96% of the variation in the initial decay rate was explained by the proportion of N, sugars, hemicelluloses, and lignin in the residue as well as the residue particle size (Diack, 1994).

In no-till management systems, earthworms and other fauna increase in number and activity, increasing the rate of residue loss from the surface through ingestion and movement of the residues to within burrows or middens. While microarthropods and other insects play a major role in residue decay in other ecosystems, their role appears to be limited in agroecosystems where insecticides are used routinely. Stroo et al. (1989) noted that the presence of *Collembola* did not alter the rate of wheat (*Triticum aestivum*) residue decay under "normal" Pacific Northwest conditions. Another little studied factor is the impact of microbial diversity on the rate of decay, as well as on the by-products produced from the decay process. Differences in by-products may have a profound impact on soil structure development. Changes in the amount or type of carbohydrate produced would alter the degree of macroaggregate formation and stabilization.

The main factors impacting residue decay rates have been included in a mechanistic model for winter wheat residues, published by Stroo et al. (1989). The model simulates decay under constant environmental factors using C and N dynamics and then determines the impact of the temperature and water content conditions, calculating the fraction of an optimum "decomposition day" occurring in a 4-hr period. The model incorporates mechanistic relationships developed in the laboratory and a series of field studies across climatic zones (Bristow et al., 1986; Stott et al., 1986, 1990). After the theoretical crop residue decomposition model was developed, a simplified model for use as a component in a residue management computer-based decision support system (RESMAN) was produced and expanded for use with many crops (Stott, 1991; Stott and Rogers, 1990; Stott et al., 1988, 1995). The goal for developing RESMAN was to incorporate residue decomposition knowledge with site- and situation-specific tillage management, including residue burial rates. Inputs for the expert system needed to be relatively simple, readily available to a wide variety of users, and processed quickly; thus the system was unable to deal with the complexity of a true research model. The details of the model can be found in Stott et al. (1995) and Elliott and Stott (1997).

Since its release in 1990, RESMAN has been widely used by industry personnel, extension and soil conservation advisors, and private consultants to develop crop residue management strategies for soil conservation. Due to its wide acceptance, the theory and equations used in RESMAN were incorporated into several erosion models being developed by the USDA, including RUSLE (Revised Universal Soil Loss Equation) (Renard et al., 1997), RWEQ (Revised Wind Erosion Equation), WEPP (Water Erosion Prediction Project Model) (Stott et al., 1995), and WEPS (Wind Erosion Prediction System). RUSLE was implemented nationwide in 1996 by the USDA Natural Resource Conservation Service. RWEQ is expected to be implemented in 1998, while WEPP and WEPS, utilizing new erosion prediction technologies, are expected to be implemented by the end of the decade.

AGGREGATION

Soil structure is defined as the size and arrangement of particles and pores in soils. A variety of pores sizes within the soil is necessary for the storage of water, transmission of water and air through the soil, and root growth. Aggregates are considered the primary unit of soil structure. It is their size, shape, and stability that control pore size distribution within the soil (Table 4.1). Edwards and Bremner (1967) stated that in a well-structured soil, the majority of macroaggregates should range in diameter between 1 and 10 mm.

This arrangement of particles must remain stable against a variety of disruptive forces. Dry aggregate stability is more important in areas where wind erosion processes are dominant. Water, either directly by rainfall or as surface runoff, is the main cause of aggregate breakdown in most soils. Water-stable aggregation, therefore, becomes key in maintaining soil structure, especially on the soil surface.

A major cause of water erosion, reduction in hydraulic conductivity, and reduced water infiltration is the formation of a crust and subsequent surface sealing. The first step in crust formation and surface sealing is the breakdown of surface aggregates (Le Bissonnais, 1990). There are four main mechanisms for aggregate breakdown:

TABLE 4.1
Relation Between Aggregate Diameters and the Soil Pores
and the Associated Function for Water Availability

Aggregate size (μm diam.)	Aggregate type	Pore size range (μm)	Water availability
<2.0	Micro	<0.2	Contains bound or nonplant-available water
2–250	Micro	0.2–25	Plant-available water
250–1000	Macro	25–100	Capillary conduction of water; involved in maintaining soil aeration
>1 mm	Macro	>100	Rapid drainage of water; important for soil aeration and root growth

Information from Oades, 1984.

slaking, differential swelling, mechanical, and physicochemical dispersion (Le Bissonnais, 1996). Slaking is due to the compression of entrapped air during rapid wetting of aggregates. The effect is on the volume of air inside the aggregate, the rate of wetting, and the shear strength of the wet aggregate (Le Bissonnais, 1996; Nearing and Bradford, 1985). Aggregate microcracking from differential swelling and shrinkage depends on the same properties as slaking and produces microaggregates similar to those produced from slaking. The two processes, however, involve different types of soil in terms of clay content. Differential swelling breakdown increases as clay content increases, while slaking decreases as clay contents increase (Le Bissonnais, 1996). Mechanical breakdown occurs as a result of the kinetic energy of raindrops (Bradford and Huang, 1992) and plays a significant role on wet soils because aggregates are weaker when soil is wet. Raindrop impact detaches soil fragments and displaces previously detached particles. Particles resulting from raindrop detachment are usually small, being either primary particles or small microaggregates less that 100 μm in diameter. Physicochemical dispersion results when the attractive forces between colloidal particles are reduced during wetting.

Stability or dispersion depends on cation size and valence. Polyvalent cations cause flocculation, whereas monovalent cations cause dispersion. In surface soils, dispersion depends mainly on the exchangeable sodium percentage of the soil. The dispersion process produces primary particles rather than microaggregates. It also results in rapid crusting, slow infiltration rates, and greater mobility of particles in water (Ben-Hur et al., 1992).

Biotic factors play an important role in aggregate formation in medium to coarse textured soils, with clay contents of 35% or less (Oades, 1993). Earthworms contribute to the biotic aggregate-forming factors through the formation of casts and biopores. Roots also contribute to the formation of biopores. Roots push through the soil, forming channels, and when the root dies and decays, the root channel is left behind as a large biopore. Microarthropods, dominated by mites and *Collembola*, contribute to soil aggregate formation via the production of fecal pellets. These pellets are a mixture of plant debris and humic material, usually less than 1 mm in diameter. Fungi quickly colonize these pellets, utilizing portions of the organic matter and transforming parts into relatively stable SOM. In addition, fungal hyphae contribute to aggregate stabilization.

A theory of aggregate stabilization, published by Tisdall and Oades (1982), is based primarily on information from Alfisols. Later work by Elliott (1986) supported the theory. The model proposed four stages of aggregation. First, water-stable aggregates less than 2 μm in diameter are made up of permanent inorganic binding agents sorbed onto clay surfaces. These binding agents act as a cement, holding the clay platelets together through electrostatic bonding and flocculation. The binding agents consist of amorphous aluminosilicates, oxides, and condensed humic substances. Management will have no impact on this set of binding agents. Other water-stable particles less than 2 μm in diameter are not aggregates, but consist of clay plates held together by flocculation. Second, water-stable aggregates between 2 and 20 μm in diameter consist of particles less than 2 μm in diameter bonded together so strongly that they are not impacted by management practices. These persistent binding agents consist of humic material and soil polysaccharides associated with di- and trivalent metal cations, as well as degradation-resistant fragments of roots, fungal hyphae, and bacterial cells. The organic matter is thought to be the center of the aggregate, with fine clay particles sorbed onto the surfaces. Third, the 20- to 200-μm-diameter aggregates, including young macroaggregates, are bound together by temporary binding agents consisting of roots, root hairs, and fungal hyphae, especially vesicular–arbuscular mycorrhizal hyphae (Figure 4.2). These agents persist for months to years and are affected by soil management. The fourth stage consists of the largest macroaggregates that range from 200 to 2000 μm in diameter. They are held together by transient binding agents, the most important of which are microbial and plant polysaccharides. These agents are produced when plant and animal residues are added to the soil and subsequently degraded by the microbial population or are associated with the roots and microbial biomass within the rhizosphere. The transient binding agents are produced rapidly but degrade quickly, lasting for days to weeks in the soil, and are highly responsive to changes in soil management.

Oades and Waters (1991) expanded the concept of aggregate hierarchy to include Mollisols. They also noted in the same study that the hierarchical nature of aggregate formation and stability does not hold true for all soil types. While aggregates from Alfisols and Mollisols appear to break down in steps, with macroaggregates greater that 250 μm in diameter breaking into 20- to 250-μm-diameter microaggregates before disintegrating into particles less than 20 μm in diameter. Oxisol macroaggregates, while stable to rapid wetting treatments, break down immediately into the small particles less than 20 μm when subjected to vigorous disruption treatments. While the aggregate binding agents in Alfisols and Mollisols are dominated by organic materials, the dominant binding agents in Oxisols are oxides, apparently inhibiting the expression of aggregate hierarchy. Other data in the literature suggest that while organic materials are at the core of aggregates in Alfisols and Mollisols, the core of the Oxisol aggregate consists of aluminum and iron oxides. In Oxisols,

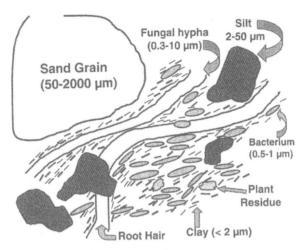

FIGURE 4.2 Representative diagram of microorganisms and organic material binding soil particles. (Based on Lynch and Bragg, 1985.)

the organic matter is probably only involved in binding the larger macroaggregates together, probably as an exterior layer.

Soil biota play a major role in stabilizing micro- and macroaggregates. There are three requirements for optimum stabilization to occur: (1) primary producers must contribute photosynthetic input into soil (e.g., plants), (2) the form and distribution of the photosynthetic products must be in a readily available form for decomposition (i.e., surface or buried residues, large or fine roots, and the type of root exudates produced), and (3) good soil conditions must be present for soil organisms to flourish and decompose organic litter.

ORGANIC MATTER IMPACT ON SOIL STRUCTURE

Soil organic matter is not homogeneous. It encompasses the living microbial biomass within the soil, as well as the dead plant matter and partially decayed and transformed plant and animal residues. It also includes the soil humus, which is that portion of the SOM that has been transformed into a relatively stable form by microorganisms. Organic residues are the plant and animal materials that have been added to the soil and are still recognizable as to their origin.

SOM also has many beneficial properties. It serves as a plant nutrient storehouse, improves soil structure and water-holding capacity, and reduces the toxicity of toxic substances within the soil. SOM supports a greater and more varied microbial population, favoring biological control of plant diseases. It aids in micronutrient element nutrition of plants and in the solubilization of plant nutrient elements from insoluble minerals. SOM has a high adsorptive or exchange capacity for plant nutrient elements, aiding in soil fertility. The dark color that organic matter imparts to the topsoil favors heat adsorption for earlier spring planting.

Organic matter content is a critical soil property, along with textural characteristics, for determining soil erodibility (Wischmeier and Smith, 1978; Wischmeier and Mannering, 1969). Erodibility is a soil's susceptibility to erosive forces such as wind and water (Soil Science Society of America, 1997). The primary role of SOM in reducing soil erodibility is by stabilizing the surface aggregates. This leads to a reduction in crust formation and surface sealing and increases water infiltration rates (Le Bissonnais, 1990).

As discussed in the previous section, microorganisms and SOM are intimately involved in the process of aggregate formation and stabilization. In soils stabilized by organic matter, the degree of stabilization depends on the amount of organic material, especially carbohydrates added to the soil, and the microbial biomass and length of the fungal hyphae present in the soil (Lynch, 1984; Tisdall, 1991). For this reason, soils under pasture have more stable macroaggregation than cropland, no-till managed soils are more stable than tilled, and soil cropped every year is more stable than land that lies fallow every other year (Tisdall, 1994). An experiment done on a silty loam Aeric Haplaquept, under continuous silage corn, a 4-year corn–barley–barley–wheat rotation, or a continuous hay system showed that slaking-resistant macroaggregates are enriched in and probably stabilized by recently deposited organic matter (Angers and Giroux, 1996). Organic matter surrounding an aggregate protects the aggregate from slaking by reducing the wetting rate (Dinel et al., 1991; Zhang and Hartge, 1992). Also, the more humified the material, the slower the water infiltrated into the aggregate, reducing the amount of slaking that occurred.

The degree of soil aggregation and the distribution of organic matter within the soil matrix will impact the length of time required to decompose organic residues (Golchin et al., 1994). Particulate organic matter in the soil, either pieces of root or aboveground biomass that has entered the soil, is initially colonized by soil microbes and adsorb mineral particles. The plant fragments, encrusted with clay platelets, become the center of the water-stable aggregates, especially in the 100- to 200-μm size range, and thus are protected from rapid decomposition. During the initial stages of decay of these organic cores, when the organic materials are rich in carbohydrates and chemically attractive to microbes, aggregates show high stability because the polysaccharide gums and other metabolites produced by the microorganisms during the decay process permeate the surrounding layer of clay minerals and thus stabilize the whole aggregate. As decomposition proceeds within the aggregates, the more labile portion of the interior organic residues, such as proteins and carbohydrates, is consumed by the microbes, and the carbon enters the soil organic pool as microbial biomass or exuded metabolites. Over time, the central core of organic matter is transformed from a recognizable piece of residue to more resistant organic structures that are high in aromatic and alkyl C and low in O-containing moieties (Golchin et al., 1994; Stott and Martin, 1990). Eventually, decomposition proceeds to the point where the microbial products that serve as binding agents are themselves degraded, weakening the aggregate, in a process often referred to as mellowing.

There is evidence that management strategies influence not only the amount of SOM present in the soil, but how it is distributed in various fractions: humic, fulvic, or humin; light or heavy; or particulate versus soluble. New research indicates that the relative distribution of SOM among the fractions impacts soil structure, crusting, and erodibility. While 50–80% of the SOM is in the form of humic substances, from 10 to 30% is carbohydrates, mostly in the form of polysaccharides (Stott and Martin, 1990). Golchin et al. (1995) showed that total organic matter was not closely correlated with aggregate stability, suggesting only part of the soil C was involved in aggregation, specifically the carbohydrate fraction. After 10 years of continuous barley, Arshad et al. (1990) found significant changes in the SOM quantity and quality. The whole soil and the C-enriched fractions from the no-till management were higher in C and N, carbohydrates, and amino acids than the conventionally managed soil. In another study, Arshad and Schnitzer (1987) compared two Boralf soils, one that only slightly crusts (Debolt) and one that exhibits severe crusting (Demmitt). They noted that the Debolt topsoil is richer in total C, total N, carbohydrates, and proteinaceous materials than the Demmitt topsoil. The humic acid extracted from the Debolt soil is more aliphatic and contains fewer CO_2H groups than the Demmitt. They concluded that SOM quality appears to play a role in crusting.

Management has a profound impact on the soil. Choice of crop plays an important role in structural stabilization and enhancing soil quality. Monocots are superior to dicots in stabilizing aggregates and grasses are better than soils (Oades, 1993). For instance, Baldock and Kay (1987) noted that as bromegrass (*Bromus inermis* Leyss) becomes more dominant in a cropping history, the size and stability of soil aggregates increase, and the rate of structural deterioration induced by grain corn production under a conventional tillage system was faster than the rate of structural enhancement promoted by bromegrass production.

It had long been noted that soil cropped to soybean is more erodible than the same soil cropped to corn (McCracken et al., 1985; Morachan et al., 1972). This same phenomenon has been noted in other grain–legume systems. Macroaggregates are easily disrupted by wetting or low energy inputs, while microaggregates are more stable, requiring higher energy inputs before disruption will occur (Golchin et al., 1994).

When management shifts from a conventionally tilled system to a no-till system, there are many alterations that occur within the soil that impact the microbial population. In a conventional system, the plant residues are buried, being either inverted to the bottom of the plow layer or evenly distributed throughout the plow layer, depending on the tillage implements used. In a no-till system, the residues are concentrated on the surface of the soil. As plant residues are a primary food and energy source for much of the microbial population, the zone of high microbial activity in a no-till system resides in the surface 1 to 2 cm of soil, while it is more evenly distributed throughout the plow layer in a conventionally tilled system (Doran, 1980a,b; Linn and Doran, 1984a). Due to the placement of the residues, there can be a change in the moisture status of the soil. Savabi and Stott (1994) noted that the presence of residues on the soil surface or within the profile impacted the amount of rainfall infiltration, increasing with increasing residue mass. As the water content of the soil approaches optimum, about 60% water-filled pore space or field capacity in loam soils, microbial activity increases (Linn and Doran, 1984b).

Soil texture has an impact on microbial activity. In the warm humid climates of Texas, the portion of soil organic C present as microbial biomass C was increased as the clay content increased (Franzluebbers et al., 1996).

In soils irrigated with saline water, C and N increased in the microbial biomass, due to decreases in the rate of C and N mineralization (Sarig et al., 1993). The same study noted a decrease in soil carbohydrate contents and aggregate stability. While there was structural deterioration of soil due to wetting a dry soil, there was additional degradation in soil irrigated with saline water due to decreases in the carbohydrate content, as well as the chemical and physical interactions between the soil particles and the solution salts.

MICROBIAL INDICATORS OF SOIL QUALITY

The dynamic nature of soil biological communities, microbial and macrofaunal, makes them a sensitive indicator for assessing alterations in soil quality due to changing management practices (Kennedy and Papendick, 1995). Soil populations may provide advanced evidence of subtle changes in the soil before changes in soil physical and chemical properties become apparent. Management practices on the land result in changes in soil physical and chemical properties, altering the soil environment that supports the growth of the microbial population.

For a detailed look at microbial activity as it impacts soil quality, research scientists use a number of indicators beyond the basic set discussed earlier. The basic set includes total organic C and N, microbial biomass C and N, potentially mineralizable N, and soil respiration (Turco et al., 1994). Two useful ratios include biomass C to total organic C and the soil respiration rate compared to the total biomass. Imbalances in these two ratios may be an early indicator that the soil biology is responding to changes in the soil condition, and these changes may soon be reflected in the physical and chemical properties of the soil. To understand various nutrient cycles and levels of microbial activity, investigators have looked at various enzymatic activities in the soil, including dehydrogenase and fluorescein diacetate hydrolysis, both indicators of general microbial activity; phosphatase, involved in the P cycle; arginase, involved in protein hydrolysis; arylsulfatase, part of the S cycle; and β-glucosidase, involved in the C cycle. Microbial community fingerprinting is also developing as a possible indicator of soil quality and utilizes such methods as substrate utilization, fatty acid analysis, and nucleic acid analysis (Kennedy and Smith, 1995).

While several methods to assess the status of the soil microbial communities are now available, interpretation can be challenging. The research community is continuing to increase the understanding of the relation between assay results and status of the soil system. This leads to knowledge of how management, soil biological communities, and soil quality interact and change, thus allowing for more informed management decisions.

SUMMARY

Soil quality has moved beyond being equated solely with productivity to inclusion of the soil's capacity to partition water and maintain environmental quality. A final determination of a soil's relative quality is dependent upon its use. Concentrating on agricultural uses, we discussed the impact of the soil microbial communities on the soil quality aspect of water partitioning, especially soil erodibility, crusting and surface sealing, infiltration rate, and aggregate stability. SOM greatly influences these soil physical characteristics. A major activity of the soil microbial community is the decomposition and transformation of organic residues (plant and animal) into SOM. There is evidence that management strategies influence not only the amount of SOM present in the soil, but how it is distributed in various fractions: humic, fulvic, or humin; light or heavy; or particulate versus soluble. As microbial biomass and activity as well as SOM content increase in a soil, soil erodibility and crusting tend to decrease, while the infiltration rate and aggregate stability increase. New research indicates that the relative distribution of SOM among the fractions impacts soil structure, crusting, and erodibility. The dynamic nature of soil biological communities, microbial and macrofaunal, makes them a sensitive indicator for assessing alterations in soil quality due to changing management practices. Soil populations may provide advanced evidence of subtle changes in the soil before changes in soil physical and chemical properties become apparent. Several methods to assess the status of the soil microbial communities are now available, but interpretation can be challenging. The research community is continuing to increase the understanding of the relation between assay results and status of the soil system. This leads to knowledge of how management, soil biological communities, and soil quality interact and change, thus allowing for more informed management decisions.

REFERENCES

Angers, D.A. and M. Giroux. 1996. Recently deposited organic matter in soil water-stable aggregates. *Soil Sci. Soc. Am. J.* 60:1547–1551.

Arshad, M.A. and M. Schnitzer. 1987. Characteristics of the organic matter in a slightly and in a severely crusted soil. *Z. Pflanzenernähr. Bodenkde.* 150:412–416.

Arshad, M.A., M. Schnitzer, D.A. Angers, and J.A. Ripmeester. 1990. Effects of till vs no-till on the quality of soil organic matter. *Soil Biol. Biochem.* 22:595–599.

Baldock, J.A. and B.D. Kay. 1987. Influence of cropping history and chemical treatments on the water-stable aggregation of a silt loam soil. *Can. J. Soil Sci.* 67:501–511.

Baross, J.A. and R.Y. Morita. 1978. Microbial life at low temperatures: ecological aspects. In D.J. Kushner (ed.). *Microbial Life in Extreme Environments.* Academic Press, New York, pp. 9–71.

Ben-Hur, M., R. Stern, A.J. van der Merwe, and I. Shainberg. 1992. Slope and gypsum effects on infiltration and erodibility of dispersive and nondispersive soils. *Soil Sci. Soc. Am. J.* 56:1571–1576.

Biederbeck, V.O. and C.A. Campbell. 1971. Influence of simulated fall and spring conditions on the soil system. I. Effect on soil microflora. *Soil Sci. Soc. Am. Proc.* 35:474–479.

Bollag, J.-M., S.-Y. Liu, and R.D. Minard. 1980. Cross-coupling of phenolic humus constituents and 2,4-dichlorophenol. *Soil Sci. Soc. Am. J.* 44:52–56.

Bradford, J.M. and C. Huang. 1992. Mechanisms of crust formation: physical components. In M.E. Sumner and B.A. Stewart (eds.). *Soil Crusting: Physical and Chemical Processes.* Lewis Publishers, Boca Raton, FL, pp. 55–72.

Bristow, K.L., G.S. Campbell, R.I. Papendick, and L.F. Elliott. 1986. Simulation of heat and moisture transfer through a surface residue–soil system. *Agric. For. Meterol.* 36: 193–214.

Diack, M. 1994. Surface Residue and Root Decomposition of Cotton, Peanut and Sorghum. M.S. thesis. Purdue University, West Lafayette, IN, 166 pp.

Dinel, H., G.R. Mehuys, and M. Lévesque. 1991. Influence of humic and fibric materials on the aggregation and aggregate stability of a lacustrine silty clay. *Soil Sci.* 151:146–157.

Doran, J.W. 1980a. Microbial changes associated with residue management with reduced tillage. *Soil Sci. Soc. Am. J.* 44:518–524.

Doran, J.W. 1980b. Soil microbial and biochemical changes associated with reduced tillage. *Soil Sci. Soc. Am. J.* 44:765–771.

Doran, J.W. and T.B. Parkin. 1996. Quantitative indicators of soil quality: a minimum data set. In J.W. Doran and A.J. Jones (eds.). *Methods for Assessing Soil Quality.* Soil Science Society of America, Madison, WI, pp. 25–38.

Druce, R.G. and S.B. Thomas. 1970. An ecological study of psychrotrophic bacteria of soil, water, grass, and hay. *J. Appl. Bacteriol.* 33:420–425.

Edwards, A.P. and J.M. Bremner. 1967. Microaggregates in soils. *J. Soil Sci.* 18:64–73.

Elliott, E.T. 1986. Aggregate structure and carbon, nitrogen, and phosphorus in native and cultivated soils. *Soil Sci. Soc. Am. J.* 50:627–633.

Elliott, L.F. and D.E. Stott. 1997. The influence of no-till cropping systems on microbial relationships. *Adv. Agron.* 60 (in press).

Franzluebbers, A.J., R.L. Haney, F.M. Hons, and D.A. Zuberer. 1996. Active fractions of organic matter in soils with different texture. *Soil Biol. Biochem.* 28:1367–1372.

Golchin, A., J.M. Oades, J.O. Skjemstad, and P. Clarke. 1994. Soil structure and carbon cycling. *Aust. J. Soil Res.* 32:1043–1068.

Golchin, A., P. Clarke, J.M. Oades, and J.O. Skjemstad. 1995. The effects of cultivation on the composition of organic matter and structural stability of soils. *Aust. J. Soil Res.* 33:975–993.

Hill, P.R. and D.E. Stott. 1997. Operational effects of a combination chisel plow on corn residue. *Soil Sci. Soc. Am. J.* (submitted).

Johnson, R.R. 1988. Soil engaging tool effects on surface residue and roughness with chisel-type implements. *Soil Sci. Soc. Am. J.* 52:237–243.

Kennedy, A.C. and R.I. Papendick. 1995. Microbial characteristics of soil quality. *J. Soil Water Conserv.* 50:243–247.

Kennedy, A.C. and K.L. Smith. 1995. Soil microbial diversity and the sustainability of agricultural soils. *Plant Soil* 170:75–86.

Le Bissonnais, Y. 1990. Experimental study and modelling of soil surface crusting processes. In R.B. Bryan (ed.). *Soil Erosion: Experiments and Models.* Catena Verlag, Cremlingen-Destedt, pp. 13–28.

Le Bissonnais, Y. 1996. Aggregate stability and assessment of soil crustability and erodibility. I. Theory and methodology. *Eur. J. Soil Sci.* 47:425–437.

Leistner, L. and W. Rodel. 1976. Inhibition of microorganisms in food by water activity. In F.A. Skinner and W.B. Hugo (eds.). *Inhibition and Inactivation of Vegetative Microbes.* Academic Press, New York, pp. 219–237.

Linn, D.M. and J.W. Doran. 1984a. Aerobic and anaerobic microbial populations in no-till and plowed soils. *Soil Sci. Soc. Am. J.* 48:794–799.

Linn, D.M. and J.W. Doran. 1984b. Effect of water-filled pore space on carbon dioxide and nitrous oxide production in tilled and nontilled soils. *Soil Sci. Soc. Am. J.* 48: 1267–1272.

Lynch, J.M. 1984. Interactions between biological processes, cultivation and soil structure. *Plant Soil* 76:307–318.

Lynch, J.M. and E. Bragg. 1985. Microorganisms and soil aggregate stability. In B.A. Stewart (ed.). *Advances in Soil Science,* Vol. 2. Springer-Verlag, New York, pp. 133–171.

Martin, J.P. and K. Haider. 1986. Influence of mineral colloids on turnover rates of soil organic carbon. In P.M. Huang and M. Schnitzer (eds.). *Interactions of Soil Minerals with Natural Organics and Microbes.* SSSA Special Publication No. 17. Soil Science Society of America, Madison, WI, pp. 283–304.

McCracken, D.V., W.C. Moldenhauer, and J.M. Laflen. 1985. The impact of soybeans on soil physical properties and soil erodibility. In R. Shibles (ed.). *World Soybean Research Conference III: Proceedings.* Westview Press, Boulder, CO, pp. 988–994.

Melillo, J.M., J.D. Aber, A.E. Linkins, A. Ricca, B. Fry, and K.J. Nadelhoffer. 1989. Carbon and nitrogen dynamics along the decay continuum: plant litter to soil organic matter. *Plant Soil* 115:189–198.

Morachan, Y.B., W.C. Moldenhauer, and W.E. Larson. 1972. Effects of increasing amounts of organic residues on continuous corn. I. Yields and soil physical properties. *Agron. J.* 64:199–203.

National Research Council, Board of Agriculture. 1993. *Soil and Water Quality: An Agenda for Agriculture.* National Academy Press. Washington, D.C., 516 pp.

Nearing, M.A. and J.M. Bradford. 1985. Single waterdrop splash detachment and mechanical properties of soils. *Soil Sci. Soc. Am. J.* 49:547–552.

Oades, J.M. 1984. Soil organic matter and structural stability: mechanisms and implications for management. *Plant Soil* 76:319–337.

Oades, J.M. 1993. The role of biology in the formation, stabilization and degradation of soil structure. *Geoderma* 56:377–400.

Oades, J.M. and A.G. Waters. 1991. Aggregate hierarchy in soils. *Aust. J. Soil Res.* 29:815–828.

Orchard, V.A. 1983. Relationship between soil respiration and soil moisture. *Soil Biol. Biochem.* 15:447–453.

Parr, J.F. and R.I. Papendick. 1978. Factors affecting the decomposition of crop residues by microorganisms. In W.R. Oshwald (ed.). *Crop Residue Management Systems.* ASA, CSSA, SSSA, Madison, WI, pp. 101–129.

Parr, J.F., R.I. Papendick, S.B. Hornick, and R.E. Meyer. 1992. Soil quality: attributes and relationship to alternative and sustainable agriculture. *Am. J. Altern. Agric.* 7:2–3.

Reiners, W.A. 1968. Carbon dioxide evolution from the floor of three Minnesota forests. *Ecology* 44:471–483.

Renard, K.G., G.R. Foster, G.A. Weesies, D.K. McCool, and D.C. Yoder (eds.). 1997. Predicting Soil Erosion by Water: A Guide to Conservation Planning with the Revised Universal Soil Loss Equation (RUSLE). USDA Agriculture Handbook 703. 384 pp.

Rice, C.W., T.B. Moorman, and M. Beare. 1996. Role of microbial biomass carbon and nitrogen in soil quality. In J.W. Doran and A.J. Jones (eds.). *Methods for Assessing Soil Quality.* Soil Science Society of America, Madison, WI, pp. 203–215.

Rose, A.H. 1976. Osmotic stress and microbial survival. *Soc. Gen. Microbiol. Symp.* 26: 155–182.

Sarig, S., E.B. Roberson, and M.K. Firestone. 1993. Microbial activity–soil structure: response to saline water irrigation. *Soil Biol. Biochem.* 25:693–697.

Savabi, M.R. and D.E. Stott. 1994. Effect of rainfall interception by plant residues on the soil water balance. *Trans. ASAE* 37:1093–1098.

Skopp, J., M.D. Jawson, and J.W. Doran. 1990. Steady-state aerobic microbial activity as a function of soil water content. *Soil Sci. Soc. Am. J.* 54:1619–1625.

Smith, J.H. and R.E. Peckenpaugh. 1986. Straw decomposition in irrigated soil: comparison of twenty three cereal straws. *Soil Sci. Soc. Am. J.* 50:928–932.

Soil Science Society of America. 1997. *Glossary of Soil Science Terms. 1996.* SSSA, Madison, WI, 134 pp.

Sommers, L.E., C.M. Gilmour, R.E. Wildung, and S.M. Beck. 1981. The effect of water potential on decomposition processes in soil. In J.F. Parr, W.R. Gardner, and L.F. Elliott (eds.). *Water Potential Relations in Soil Microbiology.* Soil Science Society of America, Madison, WI, pp. 97–118.

Stott, D.E. 1991. RESMAN: a tool for soil conservation education. *J. Soil Water Conserv.* 46:332–333.

Stott, D.E. and J.P. Martin. 1990. Synthesis and degradation of natural and synthetic humic material in soils. In P. MacCarthy, C.E. Clapp, R.L. Malcolm, and P.R. Bloom (eds.). *Humic Substances in Soils and Crop Sciences: Selected Readings.* ASA–SSSA, Madison, WI, pp. 37–63.

Stott, D.E. and J.B. Rogers. 1990. RESMAN: A Residue Management Decision Support Program. Public Domain Software. NSERL Publication 5.

Stott, D.E., G. Kassim, W.M. Jarrell, J.P. Martin, and K. Haider. 1983a. Stabilization and incorporation into biomass of specific plant carbons during biodegradation in soil. *Plant Soil* 70:15–26.

Stott, D.E., J.P. Martin, D.D. Focht, and K. Haider. 1983b. Biodegradation, stabilization in humus, and incorporation into soil biomass of 2,4-D and chlorocatechol carbons. *Soil Sci. Soc. Am. J.* 47:66–70.

Stott, D.E., L.F. Elliott, R.E. Papendick, and G.S. Campbell. 1986. Low temperature of low water potential effects on the microbial decomposition of wheat residue. *Soil Biol. Biochem.* 18:577–582.

Stott, D.E., B.L. Stuart, and J.R. Barrett. 1988. Residue Management Decision Support System. Microfiche Collect. Paper No. 88-7541. American Society of Agricultural Engineers, St. Joseph, MI.

Stott, D.E., H.F. Stroo, L.F. Elliott, R.I. Papendick, and P.W. Unger. 1990. Wheat residues loss from fields under no-till management. *Soil Sci. Soc. Am. J.* 54:92–98.

Stott, D.E., E.E. Alberts, and M.A. Weltz. 1995. Plant residue decomposition and management. In D.C. Flanagan and M.A. Nearing (eds.). USDA Water Erosion Prediction Project (WEPP): Hillslope Profile and Watershed Model Documentation. National Soil Erosion Research Laboratory Report No. 10. West Lafayette, IN, pp. 9.1–9.16.

Stroo, H.F., K.L. Bristow, L.F. Elliott, R.I. Papendick, and G.S. Campbell. 1989. Predicting rates of wheat residue decomposition. *Soil Sci. Soc. Am. J.* 53:91–99.

Tisdall, J.M. 1991. Fungal hyphae and structural stability of soil. *Aust. J. Soil Res.* 29:729–743.

Tisdall, J.M. 1994. Possible role of soil microorganisms in aggregation in soils. *Plant Soil* 159:115–121.

Tisdall, J.M. and J.M. Oades. 1982. Organic matter and water-stable aggregates in soil. *Soil Sci.* 33:141–163.

Turco, R.F., A.C. Kennedy, and M.D. Jawson. 1994. Microbial indicators of soil quality. In J.W. Doran, D.C. Coleman, D.F. Bezdicek, and B.A. Stewart (eds.). *Defining Soil Quality for a Sustainable Environment.* Soil Science Society of America, Madison, WI, pp. 73–90.

Verma, L. and J.P. Martin. 1976. Decomposition of algal cells and components and their stabilization through complexing with model humic acid-type phenolic polymers. *Soil Biol. Biochem.* 8:85–90.

Voroney, R.P., E.A. Paul, and D.W. Anderson. 1989. Decomposition of wheat straw and stabilization of microbial products. *Can. J. Soil Sci.* 69:63–77.

Wiant, H.V. 1967. Influence of temperature on rate of soil respiration. *J. For.* 65:489–490.

Wischmeier, W.H. and J.V. Mannering. 1969. Relation of soil properties to its erodibility. *Soil Sci. Soc. Am. Proc.* 33:131–137.

Wischmeier, W.H. and D.D. Smith. 1978. Predicting Rainfall Erosion Losses — A Guide to Conservation Planning. U.S. Department of Agriculture, Washington, D.C., 58 pp.

Wolf, D.C. and J.P. Martin. 1975. Microbial decomposition of ring-^{14}C atrazine, cyanuric acid, and 2-chloro-4,6-diamino-s-triazine. *J. Environ. Qual.* 4:134–139.

Zhang, H. and K.E. Hartge. 1992. Zur auswirkung organischer substanz verschiedener humifizierungsgrade auf die aggregatstabilität durch reduziferung der benetzbarkeit (Effect of differently humified organic matter on aggregate stability by reducing aggregate wettability). *Z. Pflanzenernähr. Bodenkde.* 155:143–149.

5 Erosion Impact on Soil Quality and Properties and Model Estimates of Leaching Potential

B. Lowery, G.L. Hart, J.M. Bradford, K-J.S. Kung, and C. Huang

INTRODUCTION

Scientists, farmers, and others have come to accept the fact that during the process of erosion by water and wind, the fertile topsoil is lost. The many changes in soil physical properties as a result of erosion, however, are often not considered by many researchers. It has been shown that, in addition to the loss of fertility, the process of erosion alters important soil physical properties; this alteration is often in an unfavorable manner. As topsoil is eroded, total soil organic matter is significantly reduced because, in most soils, it is primarily located in the upper horizon. We think this loss of organic matter and associated degradation of soil physical properties reduces soil quality and could have significant effects on pesticide leaching potential. To test this hypothesis, we used the computer simulation model LEACHP to predict the effect of decreased soil organic matter and associated changes in key soil physical properties caused by erosion on the leaching potential of atrazine [6-chloro-N-ethyl-N'-(1-methylethyl)-1,3,5,-triazine-2,4-diamine] and atrazine metabolite deethylatrazine [6-chloro-N'-(1-methylethyl)-1,3,5,-triazine-2,4-diamine] in several soils. Much of the data used in this simulation were compiled by North Central-174 Regional Research Committee. Simulated model results indeed show deeper movement of pesticide with increasing erosion and lower organic matter and greater bulk density. The greatest simulated chemical leaching occurred in Sparta sand (mesic, uncoated, Typic Quartzipsamments). This soil is low in organic matter, and erosion by wind of the surface soil will significantly decrease its capability to retain certain chemicals. The depth to which a chemical leached increased with increasing amount of rainfall. Simulation showed both atrazine and deethylatrazine moved to greater depth in 1993 compared to 1991. The rainfall for 1993 for this 210-day period was 133.5 mm greater than in 1991.

EROSION IMPACT ON SELECTED SOIL PHYSICAL PROPERTIES AND QUALITY

Soil physical and chemical properties may be altered dramatically by erosion. As aggregate stability decreases or becomes destroyed, pore size distribution of a soil is reduced and the availability of water to plants and microorganisms is reduced. Thus, what has been accepted as overall good soil physical quality is reduced with erosion, and this has a negative impact on biological productivity, including reduced plant production. The fertile topsoil is lost and soil productivity is reduced by erosion. The sorting action of either water or wind erosion removes a large proportion of the clay and humus (organic matter) from the soil, leaving behind less productive coarse sand, gravel, and in some cases even stone (Troeh et al., 1980).

The most chemically and physically reactive portion of soil is the organic and clay fractions. Sheet, and to a certain extent rill, erosion by water and wind is a selective and destructive process because the surface layers of soil are lost (McDaniel and Hajek, 1985; White et al., 1985; Thomas et al., 1989). What makes erosion such a destructive process is that this upper part of most soils is rich in organic matter. Thus, erosion removes one of the key reactive fractions of soil (Lowery et al., 1995). Loss of topsoil will result in a reduction in soil organic matter and, thus, a degradation in soil physical properties. Lowery et al. (1995) noted that as erosion becomes more severe, the composition of lower horizons increasingly determines the physical properties of a given soil. This change has been attributed to unfavorable changes in surface soil physical and biological properties, which implies a decline in soil quality (Lowery and Larson, 1995). As previously mentioned, this decline is most likely caused by a loss of organic-enriched topsoil. Thus, one means of improving soil quality is by addition of organic material to replace that lost by erosion (Touray, 1994).

In most cases, erosion removes one or more of the reactive fractions (organic and clay) of soil. It has been shown that increasing soil erosion results in a decrease in organic matter content and an increase in bulk density and clay content of the surface soil horizon (Lowery et al., 1995). Examples of reductions in organic matter with erosion are shown in Table 5.1. There is a reduction in organic matter when the Egan (fine-silty, mixed, mesic Udic Haplustolls), Marlette (fine-loamy, mixed, mesic Glossoboric Hapludalfs), Sharpsburg (fine-montmorillonitic, mesic Typic Argiudolls), and Ves (fine-loamy, mixed, mesic Udic Haplustolls) soils are eroded and shift from slight erosion to moderate erosion phase (Table 5.1). For the latter three soils, there is a further reduction in organic matter as they move from moderate

TABLE 5.1
Changes in Selected Soil Physical and Chemical Properties
of Ap Horizon with Erosion

		Erosion phase		
Soil	Properties	Slight	Moderate	Severe
Dubuque	Bulk density (mg/m^3)	1.34	1.36	1.37
	Organic carbon (%)	1.48	1.68	2.06
	Silt (%)	82.6	77.8	77.4
	Clay (%)	12.6	16.3	17.2
Egan	Bulk density (mg/m^3)	1.21	1.18	1.22
	Organic carbon (%)	2.9	2.6	2.6
	Silt (%)	59.1	58.9	39.7
	Clay (%)	2.9	2.6	2.6
Marlette	Bulk density (mg/m^3)	1.37	1.48	1.50
	Organic carbon (%)	1.1	0.9	0.7
	Silt (%)	34	30	29
	Clay (%)	8	14	16
Sharpsburg	Bulk density (mg/m^3)	1.44	1.50	1.55
	Organic carbon (%)	2.01	1.64	1.38
	Silt (%)	69	66	61
	Clay (%)	29	32	37
Ves	Bulk density (mg/m^3)	1.22	1.30	1.33
	Organic carbon (%)	2.1	1.8	1.4
	Silt (%)	34.5	30.5	33.4
	Clay (%)	22.6	26.4	19.8

Data collected by the North Central-174 Committee.

to severe. In the case of Dubuque, there is an unexplained increase in organic matter as it is eroded.

Reduction in organic matter leads to less stable aggregates and soil crust formation in many soils, especially those with a large amount of silt and little organic matter at or near the soil surface. The crust in turn results in reduced infiltration and air permeability (McSweeney et al., 1988) and increased water runoff. Associated with soil crusting is the rearrangement of fines in the soil such that pores are clogged by particle migration. These changes in organic matter and soil physical properties produce a chain of undesirable reactions, including reduction in water storage and further erosion as runoff increases (Swan et al., 1987; Andraski and Lowery, 1992). In the case of extreme erosion, water storage capacity is reduced for soils with decreasing porosity in the subsoil. Soil bulk density often increases with erosion (Frye et al., 1982). Good soil structure and aggregation, which are considered desirable traits of good soil, are often reduced with erosion because of a loss of organic matter.

WATER-HOLDING CAPACITY OF SOIL

With adequate fertility and a good level of management, the potential of a soil to produce corn (*Zea mays* L.) is largely determined by the capacity of the soil to store and supply water to plants (Leeper et al., 1974). Given that water is the limiting factor for corn production, it is not surprising that soil quality and crop production are reduced with erosion. Reduction in crop yield with increasing erosion is well documented (Langdale et al., 1979; Alberts and Spomer, 1987; Swan et al., 1987; Andraski and Lowery, 1992). Andraski and Lowery (1992) found that the total quantity of plant-extractable water stored in the upper 1 m of slightly eroded Dubuque silt loam soil was 181 mm, which is 7% more than that for moderately eroded Dubuque soil. In this study, the moderately eroded soil had 14% more water stored (169 mm) than the severely eroded soil (159 mm). Swan et al. (1987) found a significant interaction effect on corn yield between climate and soil water-holding capacity. Corn yield increased as depth to a clay residuum increased. The clay residuum has a low plant-available water capacity; thus, as the depth to the clay increases, the amount of plant-available water increases, along with yield potential. Swan et al. (1987) suggested that in some landscapes, such as the driftless region of southwest Wisconsin, the depth of soil is a function of soil erosion (i.e., soils shallow to clay have been eroded). We are assuming that in general eroded soils are of low quality because the organic-enriched layers have been removed and the water storage capacity reduced.

MANAGEMENT TO IMPROVE SOIL PHYSICAL PROPERTIES

Addition of organic matter from crop residue or waste products such as animal manure can increase the organic matter of eroded soil (Touray, 1994). Chang et al. (1991) found an increase in organic matter in a clay loam soil from 2 to 4.5% with the application of cattle manure over 11 years. The depth of increase in organic matter ranged from 30 cm with no irrigation to 60 cm with irrigation. It has been shown that a continuous application of manure is better than a single application in that numerous applications resulted in higher soil organic matter content (Mathers and Stewart, 1984). However, Wei et al. (1985) reported a 13% increase in organic matter 5 year after a single application of 112 mg ha^{-1} of dry sewage sludge to a clay loam soil. Although this soil did not show signs of severe erosion, the texture of its surface is typical of an undesirable (low quality) subsurface horizon of many key agricultural soils (such as Dubuque silt loam) that would be exposed following severe erosion. Wei et al. (1985) found significant improvement in many soil physical properties, including bulk density, infiltration, and aggregate stability with increasing organic matter. They applied 0, 11.2, 22.4, 44.8, and 112.0 mg ha^{-1} dry dewatered sludge in 1977 and a yearly application from 1977 to 1982, which totals 134.4 mg ha^{-1}. They then made numerous measurements of soil physical properties in 1982, showing increased infiltration, hydraulic conductivity of saturated soil, and infiltration and decreased bulk density with higher levels of sludge.

SOIL EROSION BY WATER AND ITS IMPACT ON WATER QUALITY

Soil erosion results in reduced soil quality, and the deposition of eroded materials adversely affects water quality when transported sediment is deposited in a receiving body of water. In the absence of direct soil removal, runoff water laced with pesticides and/or nutrients also causes degradation of water quality. Nelson and Ehni (1976) considered pollution from agriculture to be the greatest potential threat to our aquatic environment. They reported that erosion of soil by water produces 4 billion tons of sediment yearly, with three-fourths coming from agricultural land. They suggested that sediment is the major source of nonpoint pollution to surface water.

The literature is replete with examples of water quality degradation from soil erosion by rainfall and runoff resulting in deposition of sediments in surface water. However, examples of the impact of wind erosion on water quality are less abundant, and we have chosen to focus more of our discussion on the impact of wind erosion on water quality.

WIND EROSION AND ITS IMPACT ON WATER QUALITY

Aerial deposition of particles containing agricultural-related chemicals into surface water is a potential cause of extensive water contamination/degradation. If these particles originate from agricultural fields as a result of wind erosion, there is also a potential for degradation of soil quality since this is loss of the organic-enriched topsoil. Many agricultural chemicals such as pesticides and some fertilizers are applied to the surface of soils in anticipation that they will bind to soil particles or are moved into the soil profile by rain or irrigation. In some areas, these particles are eroded by wind before they are displaced into the soil profile, particularly in the spring, when fields are being tilled and planted and the majority of pesticides and fertilizers are applied.

Wind erosion in spring and early summer is especially significant on muck and sandy soils (Behm, 1986). However, any soil can be eroded by wind during the winter in areas where soil freezes and surface desiccation occurs. There is a potential for wind erosion from freeze-dried soil in many parts of the United States, especially in northern regions when there is little or no snow cover. Even when there is significant snow cover, there is often drifting of snow mixed with wind-blown soil particles. Snow and soil drifts become trapped in or near road ditches and other contributing zones for surface water in rural areas. This material will run off during spring snow melt and potentially serve as a source of water pollution.

Wind erosion caused enormous destruction of large sections of land and other property in the United States in the mid-1930s. This was caused by a widespread drought and poor soil management. In some cases, erosion degraded soil quality and rendered large sections of land useless for many years (Troeh et al., 1980). Erosion by wind on a large scale has not occurred since the 1930s, but on local scales wind erosion occurs yearly (Troeh et al., 1980). It is a well-known fact that in many local areas of the United States (and other countries), wind erosion causes damage to crops and the deposition of the eroded materials is a source of property damage. However, an often overlooked result of wind erosion is deposition of soil particles in water courses. Wind removes the nutrient-rich and pesticide-laced surface soil particles from farmland. These particles become entrained in the air and are often deposited in surface waters such as streams, rivers, and lakes or low landscape positions that serve as discharge or recharge zones for groundwater.

Water contamination related to wind erosion has not been well documented. We found little published data on water contamination by wind erosion, although there is a wealth of data on wind erosion (Troeh et al., 1980) and a developing database on atmospheric deposition of pesticides (Glotfelty et al., 1990; Nations and Hallberg, 1992). However, in most studies of aerial pesticide deposition, researchers have primarily measured pesticides in wet (precipitation) deposition (and some dry deposition) and have assumed that the pesticide originated from pesticide drift and/or volatilization during and after application. There are very few published studies of transport and deposition (in surface water) of soil particles containing pesticides and nutrients originating from wind erosion.

There is a need for better understanding of all sources of water contamination, and deposition from wind erosion has not been adequately addressed to date. In Wisconsin, about 2 million ha of sandy and organic (muck) farmland is planted to some type of intense row crop annually during the spring. Some of these cropping systems require large quantities of pesticides. Behm (1986) reported 16.8 mT ha yr^{-1} of average annual soil loss as a result of wind erosion in Portage County (central Wisconsin). Given this amount of soil loss and an average early season pesticide concentration of 4 mg kg^{-1} in the top few millimeters of the soil, we calculated 4.35 kg of pesticide lost from an average field (65 ha).

There has been some speculation about the effect of wind erosion on environmental quality (mainly air), but little research has actually been conducted. In their discussion and conclusions, Hagen and Woodruff (1975) noted that wind erosion is relatively unimportant as a long-range mover of sediment. However, they suggest that the pollution hazards posed by erodible soils containing large amounts of pesticides and fertilizers need further research. After calculating potential wind erosion using the wind erosion equation, Wilson (1975) concluded that pollution from wind erosion is certainly a legitimate environmental concern.

As previously noted, limited data are available on soil pesticide/agricultural chemical-laced aerosols, but there have been some interesting studies. Gaynor and MacTavish (1981) reported that an early spring windstorm 8 days after application of granular simazine (6-chlor-N,N'-diethyl-1,3,5-triazine-2,4-diamine) to a sandy loam soil reduced the amount of simazine in the treated area to 57% of that applied. The herbicide was deposited up to 2.5 m downwind of the area of application at concentrations phytotoxic to susceptible crops. Risebrough et al. (1968) found that concentrations of chlorinated hydrocarbons in airborne dust carried by the trade winds from the European–African land areas to Barbados ranged from <1 to 165 ppm. They suggested that the amount of pesticides contributed to the tropical Atlantic by the trade winds is comparable to that carried to the sea by major river systems. Eolian-related distribution of pesticides was noted by Glotfelty et al. (1989) while they were conducting a study on the rate of volatilization of alachlor [2-chloro-*N*-(2,6-diethylphenyl)-*N*-(methoxymethyl)acetamide], atrazine, simazine, and toxaphene (chlorinated camphene). Glotfelty et al. (1989) reported daily volatilization patterns of atrazine and simazine which indicated that some wind erosion of wettable powder formulation occurred as the surface soil dried. However, they suggest that the transport they observed was small. They recorded wind speeds ranging from 0.5 to 5.5 m s^{-1}. In our extensive literature review, these were the most comprehensive reports we found that link wind erosion and pesticide movement to air/water quality. However, we did not find any references to the effect of erosion on reducing soil organic matter and the resulting increased leaching potential of agricultural chemicals.

SOIL EROSION IMPACTS LEACHING OF AGRICULTURAL CHEMICALS AND WATER QUALITY

It has been shown that in most cases soil erosion reduces soil quality because organic matter, water-holding capacity, soil depth, etc. are reduced and soil chemistry is altered (Lowery et al., 1995). Thus, we have focused the remainder of this chapter on the impact of these changes in soil physical and chemical properties on leaching potential of the herbicide atrazine, which has received considerable attention in recent years. The main objective of the remaining sections of this chapter is to predict the potential impact of soil erosion on water quality. We used LEACHP to estimate the impact of organic matter reduction and changes in bulk density, hydraulic conductivity, and soil depth caused by erosion on leaching potential of atrazine and its metabolite deethylatrazine. We conducted model simulations using data from the North Central-174 Regional Research (NC-174) Committee and other sources for three levels of erosion.

METHODS FOR MODEL SIMULATION

Simulated leaching of atrazine and deethylatrazine (DEA) in three soils was done with the LEACHP program of LEACHM (Wagenet and Hutson, 1989; LEACHM Ver. 3 documentation manual). Data for soils eroded by water were obtained from studies conducted by NC-174 Committee (Tables 5.1 and 5.2). Data for a soil affected by wind erosion were composed from various sources and are reported in Table 5.3.

Soils used in this study were Dubuque, Sparta, and Ves. Selected physical properties for the Dubuque and Ves soils, which are affected by water erosion, were collected by the members of the NC-174 Committee (Table 5.1). The NC-174 Committee has compiled numerous data on the impact of soil erosion on soil productivity and related properties for selected benchmark and other key soils (Lowery et al., 1995; Mokma et al., 1996; Cihacek and Swan, 1994; Shaffer et al., 1994; Schumacher et al., 1994). Soil physical properties, as a function of erosion, for the three soils used in model simulations are shown in Table 5.2. Data for Sparta

TABLE 5.2
Soil Properties for Three Levels of Erosion of Three Soils
Influenced by Water Erosion

Soil	Erosion level	Depth (cm)	Horizon	Organic carbon (%)	Bulk density (mg/m³)	Hydraulic conductivity at saturation (mm/day)	Silt (%)	Clay (%)
Dubuque	Slight	0–40	Ap	1.48	1.34	10,022	82	13
		40–66	Bt	0.37	1.39	437	66	32
		66–86	Bt1	0.37	1.39	437	66	32
		86–114	Bt2	0.28	1.39	437	66	32
		114–138	2Bt1	0.32	1.27	2.87	41	54
	Moderate	0–20	Ap	1.68	1.36	3,733	78	16
		20–43	Bt	0.46	1.50	437	69	29
		43–68	Bt1	0.46	1.50	437	69	29
		68–89	Bt2	0.23	1.50	437	69	29
		89–110	Bt3	0.11	1.50	437	69	29
		110–140	2Bt1	0.32	1.22	2.45	52	45
	Severe	0–17	Ap	2.06	1.37	4,294	77	17
		17–38	Bt	0.70	1.40	437	64	33
		38–50	Bt1	0.70	1.40	437	64	33
		50–65	Bt2	0.31	1.40	37	64	33
		65–112	2Bt1	0.53	1.16	3.92	56	40
Ves	Slight	0–25	Ap	2.00	1.22	372	30	23
		25–38	AB	1.30	1.32	372	31	24
		38–56	Bw	0.80	1.41	372	28	26
		56–97	Bw2	0.50	1.43	372	32	23
		97–152	C	0.30	1.57	372	38	20
	Moderate	0–20	Ap	2.10	1.30	206	29	26
		20–46	Bw	0.90	1.41	206	30	22
		46–69	BC	0.40	1.53	206	34	19
		69–152	C	0.30	1.60	206	36	16
	Severe	0–20	Ap	1.20	1.33	374	32	20
		20–64	C1	0.40	1.58	374	34	18
		64–120	C2	0.30	1.58	374	34	18

Data collected by members of the North Central-174 Committee.

sand were obtained from various sources (see Table 5.3). Data for the Ves and Dubuques soils were obtained from actual data of soils that had undergone erosion and are now considered to be in the slight, moderate, or severe erosion class. Data for Sparta sand were compiled from a slightly eroded soil and simulations were made assuming moderate (10 cm topsoil loss) or severe (20 cm topsoil loss) erosion (Lowery et al., 1995; Mokma et al., 1996).

The Sparta sand from which the data were compiled is located along the Lower Wisconsin River Valley (LWRV) and the Central Sands Area of central Wisconsin

TABLE 5.3
Selected Properties of Sparta Sand

Horizon	Depth (cm)	Sand (%)	Clay (%)	OC[a] (%)	Kd atrazine[b] (L kg^{-1})	Ksat[c] (m s^{-1})	Water[d] content FMC (%)
Ap	0–23	96	1.6	0.39	0.58	7.5×10^{-5}	11.1
A	23–33	96	1.6	0.56	0.60		11.9
AB	33–43	96	1.5	0.34	0.35		10.3
Bw1	43–54	96	1.0	0.27	0.30	9.1×10^{-5}	8.5
Bw2	54–66	97	1.2	0.15	0.12		7.2
Bw3	66–84	99	0.6	0.08	0.10	23×10^{-5}	4.8
BC	84–135	99	0.2	0.04	0.08		4.9
C	135–160	99	0.2	0.01	0.025		3.7

[a] OC = organic carbon content.

[b] Atrazine batch adsorption coefficient (Seybold et al., 1994).

[c] Ksat = water-saturated hydraulic conductivity (Wang et al., 1991).

[d] Volumetric soil water content at field moisture capacity (Hart et al., 1994). FMC = field moisture capacity.

From Fermanich, K.J., W.L. Bland, B. Lowery, and K. McSweeney. 1996. *J. Environ. Qual.* 25: 1291–1299.

(CS). The soil along the LWRV is about 95% sand and is cropped under irrigation to corn, vegetables used for canning, and potato (*Tuberosum solanum*). Several of the investigators have observed significant wind erosion events from fields in the LWRV. The herbicide atrazine is frequently found in the groundwater in this area; thus, this is an ideal soil for this evaluation (Postle, 1989).

Simulations of atrazine and DEA leaching were made for all soils under slight, moderate, and severe erosion indexes for observed rainfall patterns during 1991 and 1993. These two years are representative of the extreme differences in rainfall pattern and amount during the growing season in the Midwest. All simulations represented a 210-day growing period for corn (Table 5.4).

To test the importance of various parameters attributed to leaching of atrazine and DEA as determined by the computer simulation, we performed a multiple regression analysis. Significant predictors affecting the depth of atrazine leaching were determined according to the significance (*p*-value) of the coefficient for each predictor variable when fitted to a stepwise regression equation (Draper and Smith, 1981).

MODEL SIMULATION RESULTS AND DISCUSSION

Simulations of atrazine and DEA leaching were made for Dubuque and Ves soils using data from sites described as being slightly, moderately, and severely eroded and for Sparta sand with projected slight, moderate, and severe erosion levels for

TABLE 5.4
Results of Regression Analysis for Significant
Predictor Variables Affecting Depth of
Atrazine Leaching in the Three Soils Tested

Predictor variable	Significance level (p-value)[a]
Intercept	0.0743
Drainage	0.0153
Erosion level	0.9471
Organic carbon	0.0184
Rainfall	0.0001
Soil	0.0673

Note: Predictor variables considered were drainage (mm), erosion level (1
= slight, 2 = moderate, 3 = severe), organic carbon in Ap horizon
(%), rainfall (mm), and soil (Dubuque, Sparta, Ves).

[a] Significance determined from type III sum of squares analysis.

1991 and 1993 (data not presented for Dubuque and Ves in 1991) rainfall over 210 days. As shown in Figures 5.1–5.8, there is an increase in the simulated leaching of these chemicals with increasing erosion with Sparta sand and Ves soils. This increased leaching is believed to be strongly related to the reduction in organic matter with increasing erosion. Organic carbon had a significant effect on depth of leaching (Table 5.4). The apparent lack of increased leaching with erosion for the Dubuque soil is related to reported increase in organic matter with erosion for this soil (Lowery et al., 1995). As expected, there is also more leaching with increasing rainfall. Rainfall in 1993 for this 210-day period was 133.5 mm greater than that for 1991. Rainfall was found to be the most sensitive predictor of atrazine leaching. However, for both years, the predicted leaching of DEA was greater than that of atrazine.

As expected, drainage had a significant effect on atrazine leaching (Table 5.4). This is expected, as rainfall was significant and drainage would be related to water input (i.e., rainfall). Simulated leaching of atrazine and DEA for all erosion levels and soil was not below the root zone for 1991. However, leaching was well below the root zone and near the groundwater table for Sparta soil in 1993. This soil has a very low water-holding capacity, especially in the lower profile (Hart et al., 1994). Thus, increased leaching is expected with increasing levels of erosion. Depth to the water table in this soil ranges from 3 to 10 m in the LWRV (Hart et al., 1994) and from 2.5 to 4 m at the data selection site for this simulation (Fermanich et al., 1996). There is a trend of perched water table in the Dubuque soil during fall and spring (B.J. Andraski, 1985, personal communication). Thus, in the case of Sparta sand, there is a high potential of possible contamination of groundwater with increasing erosion. This is of great concern because there is already considerable evidence of groundwater contamination beneath this soil where little erosion has occurred (WDATCP/WDNR, 1989; Fermanich et al., 1996).

FIGURE 5.1 Simulated profile distribution of atrazine for slight, moderate, and severe erosion levels of Dubuque soil for 1993 growing season.

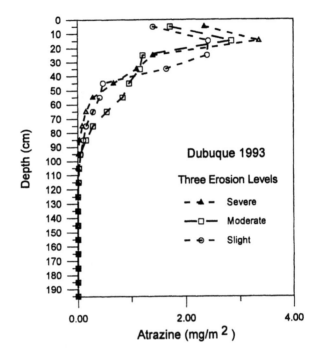

FIGURE 5.2 Simulated profile distribution of DEA for slight, moderate, and severe erosion levels of Dubuque soil for 1993 growing season.

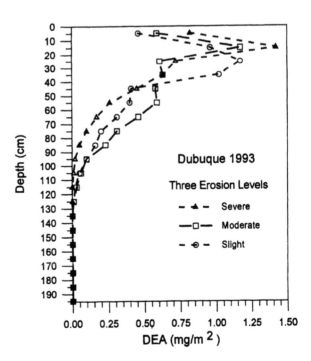

FIGURE 5.3 Simulated
profile distribution of
atrazine for slight, moderate,
and severe erosion levels of
Ves soil for 1993 growing
season.

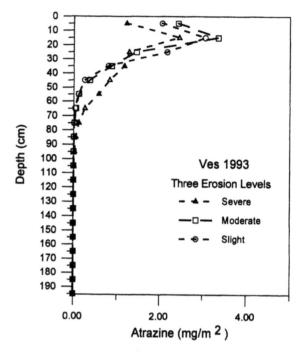

FIGURE 5.4 Simulated
profile distribution of DEA
for slight, moderate, and
severe erosion levels of Ves
soil for 1993 growing
season.

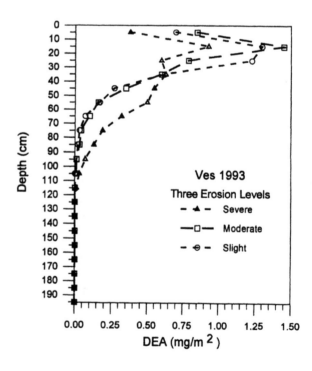

FIGURE 5.5   Simulated
profile distribution of
atrazine for slight, moderate, and severe erosion
levels of Sparta soil for
1991 growing season.

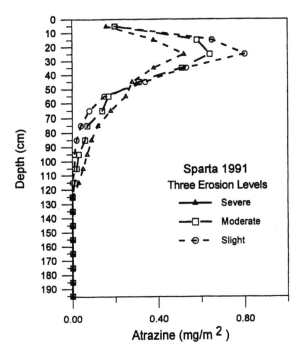

FIGURE 5.6   Simulated
profile distribution of DEA
for slight, moderate, and
severe erosion levels of
Sparta soil for 1991
growing season.

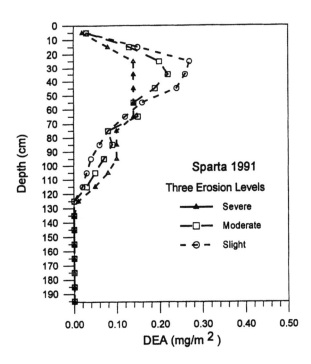

FIGURE 5.7 Simulated
profile distribution of
atrazine for slight, moderate,
and severe erosion levels of
Sparta soil for 1993 growing
season.

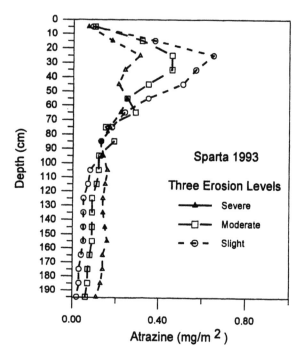

FIGURE 5.8 Simulated
profile distribution of DEA
for slight, moderate, and
severe erosion levels of
Sparta soil for 1993 growing
season.

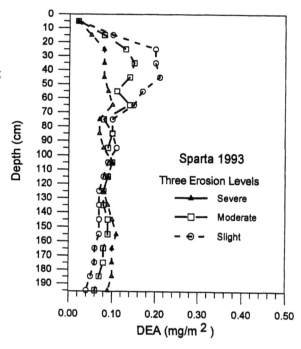

It appears from Table 5.4 that atrazine leaching is independent of erosion level because erosion level did not appear as a significant predictor of leaching in this simulation. However, it should be noted that organic carbon and fine soil particles are lost from the soil horizon as a result of erosion processes. We have shown in Table 5.1 that soil organic carbon content is decreased with increasing erosion. Therefore, if not directly, erosion level indirectly contributes to leaching potential. A test of these three soils showed a significance approaching 95% (Table 5.4).

SUMMARY AND CONCLUSIONS

Changes in soil physical and chemical properties with increasing soil erosion range from slight to extreme. The change that will have the greatest effect on soil physical properties and leaching of chemicals such as atrazine is the decrease in organic carbon. Based on our computer simulations of atrazine and DEA leaching, there is a potential for increased leaching of these compounds with increasing soil erosion for most soils. We think this is due to a reduction in organic matter with increasing erosion. Thus, it seems possible to reduce the potential of pesticide leaching with improved soil quality (i.e., increased organic matter in the surface horizon), made possible by additions of manure soil amendments.

REFERENCES

Alberts, E.E. and R.G. Spomer. 1987. Corn grain yield response to topsoil depth on deep loess soil. *Trans. ASAE* 30:977–981.

Andraski, B.J. and B. Lowery. 1992. Erosion effects on soil water storage, plant water uptake and corn growth. *Soil Sci. Soc. Am. J.* 56:1911–1919.

Behm, D.H. 1986. Assessment of Erosive Wind Frequency for Portage County, Wisconsin. M.S. thesis. University of of Wisconsin-Stevens Point, 168 pp.

Chang, C., T.G. Sommerfeldt, and T. Entz. 1991. Soil chemistry after eleven annual applications of cattle feedlot manure. *J. Environ. Qual.* 20:475–480.

Cihacek, L.J. and J.B. Swan. 1994. Effects of erosion on soil chemical properties in the north central region of the United States. *J. Soil Water Conserv.* 49:259–265.

Draper, N.R. and H. Smith. 1981. *Applied Regression Analysis.* John Wiley & Sons, New York, 709 pp.

Fermanich, K.J., W.L. Bland, B. Lowery, and K. McSweeney. 1996. Irrigation and tillage effects on atrazine and atrazine metabolite leaching from a sandy soil. *J. Environ. Qual.* 25:1291–1299.

Frye, W.W., S.A. Ebelhar, L.W. Murdock, and R.L. Blevins. 1982. Soil erosion effects on properties and productivity of two Kentucky soils. *Soil Sci. Soc. Am. J.* 46:1051–1055.

Gaynor, J.D. and D.C. MacTavish. 1981. Movement of granular simazine by wind erosion. *HortScience* 16:756–757.

Glotfelty, D.E., M.M. Leach, J. Jersey, and A.W. Taylor. 1989. Volatilization and wind erosion of soil surface applied atrazine, simazine, alachlor, and toxaphene. *J. Agric. Food Chem.* 37:546–551.

Glotfelty, D.E., G.H. Williams, H.P. Freeman, and M.M. Leech. 1990. Regional atmospheric transport and deposition of pesticides in Maryland. In D.A. Kurtz (ed.). *Long Range Transport of Pesticides.* Lewis Publishers, Boca Raton, FL, pp. 199–221.

Hagen, L.J. and N.P. Woodruff. 1975. Particulate loads caused by wind erosion in the Great Plains. *J. Air Pollut. Control Assoc.* 25:860–861.

Hart, G.L., B. Lowery, K.J. Fermanich, and K. McSweeney. 1994. *In-situ* characterization of hydrologic properties of Sparta sand: relation to solute movement. *Geoderma* 64: 41–55.

Langdale, G.W., J.E. Box, Jr., R.A. Leonard, A.P. Barnett, and W.G. Fleming. 1979. Corn yield reduction on eroded southern Piedmont soils. *J. Soil Water Conserv.* 34: 226–228.

Leeper, R.A., E.C.A. Runge, and W.M. Walker. 1974. Effect of plant-available stored soil moisture on corn yields. I. Constant climatic conditions. *Agron. J.* 66:723–727.

Lowery, B. and W.E. Larson. 1995. Symposium: erosion impact on soil productivity — preamble. *Soil Sci. Soc. Am. J.* 59:647–648.

Lowery, B., J. Swan, T. Schumacher, and A. Jones. 1995. Physical properties of selected soils by erosion class. *J. Soil Water Conserv.* 50:306–311.

Mathers, A.C. and B.A. Stewart. 1984. Manure effects on crop yields and soil properties. *Trans. ASAE* 27:1022–1026.

McDaniel, T. and B.F. Hajek. 1985 Soil erosion effects on crop productivity and soil properties in Alabama. In B.A. Stewart and R.F. Follett (eds.). *Erosion and Soil Productivity.* American Society of Agricultural Engineers, St. Joseph, MI, pp. 48–58.

McSweeney, K., A.A. Heshaam, G.S. LeMasters, and L.D. Norton. 1988. Micromorphological and selected physical properties of bare and residue covered soil surfaces. *Soil Tillage Res.* 12:301–322.

Mokma, D.L., T.E. Fenton, and K.R. Olson. 1996. Effect of erosion on morphology and classification of soils in the North Central United States. *J. Soil Water Conserv.* 51:171–175.

Nations, B.K. and G.R. Hallberg. 1992. Pesticides in Iowa precipitation. *J. Environ. Qual.* 21:486–492.

Nelson, W.C. and R.J. Ehni. 1976. Land use and nonpoint pollution in the Sheyenne Valley. *Water Resour. Res.* 1:381–390.

Postle, J.K. 1989. Results of the WDATCP Groundwater Monitoring Project for Pesticides. Memo Report. Wisconsin Department of Agriculture, Trade and Consumer Protection, Madison.

Risebrough, R.W., R.J. Huggett, J.J. Griffin, and E.D. Goldberg. 1968. Pesticides: trans-Atlantic movements in the northeast trades. *Science* 159:1233–1236.

Schumacher, T.E., M.J. Lindstrom, D.L. Mokma, and W.W. Nelson. 1994. Corn yield: erosion relationships of representative loess and till soils in the North Central United States. *J. Soil Water Conserv.* 49:77–81.

Seybold, C.A., K. McSweeney, and B. Lowery. 1994. Atrazine adsorption in sandy soils of Wisconsin. *J. Environ. Qual.* 23:1291–1297.

Shaffer, M.J., T.E. Schumacher, and C.L. Ego. 1994. Long-term effects of erosion and climate interactions on corn yield. *J. Soil Water Conserv.* 49:272–275.

Swan, J.B., M.J. Shaffer, W.H. Paulson, and A.E. Peterson. 1987. Simulating the effects of soil depth and climatic factors on corn yield. *Soil Sci. Soc. Am. J.* 51:1023–1032.

Thomas, A.W., R.L. Carter, and J.R. Carreker. 1989. Soil and water nutrient losses from Tifton loamy sand. *Trans. ASAE* 11:677–679.

Touray, K.S. 1994. Erosion and Organic Amendment Effects on the Physical Properties and Productivity of a Dubuque Silt Loam Soil. Ph.D. thesis. University of Wisconsin, Madison, 157 pp.

Troeh, F.R., J.A. Hobbs, and R.L. Donohue. 1980. *Soil and Water Conservation for Productivity and Environmental Protection.* Prentice-Hall, Englewood Cliffs, NJ, 718 pp.

Wagenet, R.J. and J.L. Hutson. 1989. *LEACHM: Leaching Estimation and Chemistry Model: A Process Based Model of Water and Solute Movement Transformations, Plant Uptake and Chemical Reactions in the Unsaturated Zone.* Continuum Vol. 2. Water Resources Institute, Cornell University, Ithaca, NY.

Wang, D., K. McSweeney, J.M. Norman, and B. Lowery. 1991. Preferential flow in soils with ant burrows. In T.J. Gish and A. Shirmohammadi (eds.). *Preferential Flow. Proceedings of the National Symposium.* American Society of Agricultural Engineers, St. Joseph, MI, pp. 183–191.

WDATCP/WDNR. 1989. Compilation of Private Water Supply Well Sampling in the LWRV. Memo Report. Wisconsin Department of Agriculture, Trade, and Consumer Protection and Wisconsin Department of Natural Resources, Madison.

Wei, Q.F., B. Lowery, and A.E. Peterson. 1985. Effect of sludge application on physical properties of a silty clay loam soil. *J. Environ. Qual.* 14:178–180.

White, W.P., R.R. Bruce, A.W. Thomas, G.W. Langdale, and H.F. Perkins. 1985. Characterizing productivity of eroded soils in the southern Piedmont. In B.A. Stewart and R.F. Follett (eds.). *Erosion and Soil Productivity.* American Society of Agricultural Engineers, St. Joseph, MI, pp. 83–95.

Wilson, L. 1975. Application of the wind erosion equation in air pollution surveys. *J. Soil Water Conserv.* 30:215–219.

Section III

Soil Quality Management

6 Effects of Long-Term Cropping on Organic Matter Content of Soils: Implications for Soil Quality*

T.E. Fenton, J.R. Brown, and M.J. Mausbach

INTRODUCTION

The importance of soil organic matter (SOM) as a soil constituent and its contributions to plant growth have long been recognized. The amount of organic matter in a soil is dependent upon a complex set of interactions of physical, chemical, and biological processes. Decline in the organic matter content of soils when virgin soils are converted to cultivated soils has been shown with many long-term experiments. Most of the data show an initial rapid decline, then a period of a lower rate of decline, followed by a quasi-equilibrium under the same use and management. Changes in land use and/or management disturb the quasi-equilibrium, and the system undergoes adjustment to a new organic matter level dictated by the impact of residues returned to the soil. More complete data sets, including depth distribution and bulk density measurements, are needed to quantify changes in the total amount of organic matter in soils. SOM is an important component of the current concept of soil quality, the definition of which is still evolving. Additional parameters for characterizing SOM need to be developed and evaluated. The summation of soil properties present in soil profiles and their distribution on the landscape should also be included in the concept of soil quality.

* Paper No. J-17822 of the Iowa Agriculture and Home Economics Experiment Station, Ames, Iowa, Project No. 2666, and supported by Hatch Act and State of Iowa.

95

Organic matter has long been recognized as an important factor in soil processes and crop growth. Waksman (1942) listed the following direct and indirect effects:

1. A source of nutrients for crops, especially N and P
2. Affects the physical condition of the soil, especially moisture-holding and buffering capacities
3. Supplies certain specific elements and compounds including minor elements and auxins
4. Favors the development of antagonistic organisms that serve to combat certain plant diseases

He went on to state that the fourth item was debatable.

Allison (1973) listed the important contributions of SOM as:

1. Provides the major natural source of inorganic nutrients and microbial energy
2. Acts as an ion-exchange material and a chelating agent to hold water and nutrients in available form
3. Promotes soil aggregation and root development
4. Improves water infiltration and water use efficiency

Composition of Soil Organic Matter

Schnitzer and Khan (1972) described SOM as a mixture of plant and animal residues in various stages of decomposition, substances synthesized microbiologically and/ or chemically from the breakdown of products, and the bodies of live and dead microorganisms and small animals and their decomposing remains. They also state that to simplify this complex system, SOM is usually subdivided into (1) nonhumic and (2) humic substances. The bulk of the SOM is composed of humic substances.

Campbell (1978) stated there are at least three and probably four components of SOM. In order of ease of decomposition they are fresh residues, biomass material, microbial metabolite and cell wall constituents adsorbed to clay colloids, and the old very stable humus.

Stevenson (1982) listed the following terms used to describe SOM fractions: organic residues, SOM, humus, soil biomass, humic substances, nonhumic substances, humin, humic acid, and fulvic acid.

Campbell (1978) proposed that SOM can be arbitrarily divided into relatively stable and active components. The former he more closely associated with physical stability of the soil, while the active fraction is associated with the ability of soil to cycle and supply available nutrients such as N. Stevenson (1956) showed evidence that long-term cultivation on the Morrow Plots had significantly increased the proportion of the basic AA–N nitrogen in the soil.

Depth Distribution

Much emphasis has been placed on the role of organic matter in surface soil, but less effort has been directed toward an understanding of the role of depth distribution of organic matter in soils. In addition, use of varying designations of surface soil and depths of study complicate interpreting the literature. Early work referred to "plow depth" (Epstein and Kohne, 1957). However, with the coming of conservation tillage systems, especially no-till, the term plow depth has little meaning. Hammer et al. (1995) and Conway-Nelson (1991) and others have renewed efforts to characterize organic matter with depth in various soils. Subsurface organic matter is important because its location affects nutrient and water fluxes and attenuation of chemicals within the soil profile. Roots and microbial activity are also important in the subsurface horizons. Therefore, long-term management of landscapes for improved quality of the visible surface soil may not be optimally successful without consideration of impact upon soil organic carbon in the entire solum.

Recently Recognized Functions

Rasmussen and Collins (1991) noted that SOM has recently received increased attention because of its potential to sequester carbon and its strong influence on the persistence and degradation of pesticides and organic waste in soils.

In a discussion of the importance of SOM, Gregorich et al. (1994) state: "It is the primary source of, and a temporary sink for, plant nutrients in agroecosystems and is important in maintaining soil tilth, aiding the infiltration of air and water, promoting water retention, reducing erosion and controlling the efficacy and fate of applied pesticides."

Methods of Measurement

Quantitative methods used to determine SOM contents are variable. Therefore, it is recommended that organic carbon, which can be more accurately and precisely measured by a variety of procedures, be used as a measure of SOM (Nelson and Sommers, 1982). Organic carbon can then be converted to SOM content by use of an appropriate factor, usually by multiplying the organic carbon content by 1.724. Another major problem is that bulk density values have not been measured at many sites where SOM studies have been conducted. Therefore, knowing the total amount of SOM present is impossible. When bulk density data are available, the soil organic carbon (SOC) can be converted to a weight per unit area basis, for example kilograms per meter. Because there is disagreement concerning use of the conversion factor 1.724 for all horizons, the data comparisons should be made using SOC rather than SOM for soil profile comparisons.

The objective of our review is to evaluate the impact of crop and soil management on SOM content and distribution over time. Major variables are rotations, tillage, and fertility. Confounding factors are time and residue management.

LONG-TERM STUDIES

Treatments and Rotations

Jenny (1941) emphasized that man, through cultivation and other cultural treatments, acts as an independent variable or soil-forming factor. These activities result in changes in soil properties that can greatly affect the soil system. The most obvious changes related to cultivation are significant decreases in SOM contents and deterioration of soil structure.

Mitchell et al. (1991) presented an overview of long-term agronomic research. They emphasize that when the oldest research plots were established, land and labor were plentiful and the plots were large and unreplicated. Agricultural statistics did not exist. Early experiments concentrated on crop rotations and N-providing legumes. Over time, these basic experiments have led to other studies, and both have resulted in the development of widely accepted principles of soil and crop management. However, they go on to emphasize that many records of the long-term plots are only published in state agricultural experiment station bulletins and circulars.

Brown (1989) published brief descriptions of long-term field research sites. Information compiled included the site name and location, date of initiation, thrust, and a contact person for each site. The information was compiled from responses associated with planning for the Sanborn Field Centennial, held on June 27, 1989, and the Long-Term Research Symposium, sponsored by the American Society of Agronomy in 1989. Steiner and Herdt (1993) compiled a global directory of long-term agronomic experiments. The location, initiation date, purpose, soil and climatic information, experimental design, and management practices are described for these experiments.

Morrow Plots

The Morrow Plots, now considered the oldest continuous study in the United States, are at the University of Illinois. They were established in 1876 to study the effects of crop rotations and fertilization on yields. Crop sequences, in single replication, were continuous corn, corn–oats rotation, corn–oats–clover rotation, with and without lime, manure, and rock phosphate (Stauffer et al., 1940). Table 6.1 shows that continuous corn plot with no fertility treatment decreased in SOM content by 45.6% in 55 years as compared with the adjacent sod. Changes for other cropping systems are also listed in Table 6.1. Stauffer et al. (1940) concluded that neither the cropping system nor the soil treatment had much effect on the SOC content below 23 cm.

Guernsey et al. (1969) concluded that continuous corn without fertilization had lowered yields, increased bulk density, reduced porosity, lowered SOM content, and lowered aggregate stability in the plow layer of Flanagan silt loam (fine, montomorillonitic, mesic Aquic Argiudoll), the major soil on the Morrow Plots.

Using the data reported by Odell et al. (1982), we compared the SOC differences between continuous corn, no treatment (subplot 3NC), and corn–oats–clover, LNPK 1955 to the present (subplot 5NB) for the upper 38 cm. The weighted average SOC content expressed in percentage by weight was 1.33% for the continuous corn plot and 1.92% for the rotation, a difference of 30.8% between the two plots. Using bulk density data, the former had an average of 4.8 kg/m² and the latter had 6.7 kg/m², a difference of 28%.

TABLE 6.1
Effect of Rotations and Treatments on Organic Carbon Content in Morrow Plots, 1876–1940, University of Illinois

Rotation	Treatment[a]	% organic C	% organic matter	% change
Corn	None	1.74	2.99	−45.6
	MLP	2.09	3.59	−34.7
Corn–oats	None	2.14	3.68	−33.1
	MLP	2.44	4.20	−23.6
Corn–oats–clover	None	2.28	3.92	−28.7
	MLP	3.35	5.76	+4.0
Sod	None	3.20	5.50	0.0

[a] MLP = manure–lime–phosphorus.

After Stauffer et al., 1940.

Odell et al. (1984) summarized data from the Morrow Plots for the period 1904–73. Soil from plots used for continuous corn with various treatments had an average SOC content of 1.86% (3.19% SOM). Compared with the continuous sod plot, this figure represents a 41.9% loss of SOC. The plots that were used for corn–oats from 1904 through 1966 and for corn–soybeans from 1967 to the present had an average SOC content of 2.26% (3.89% SOM). Plots used for a corn–oats–clover cropping system with various treatments had a SOC content of 2.58% (4.44% SOM). The two extremes represent a 28.2% decrease in SOM. One of their conclusions was that the LNPK treatment applied from 1955 to 1973 on previously untreated subplots significantly increased the content of SOC as compared with the continuous corn and corn–oats–clover cropping systems.

More recent data from Dr. Ted Peck (personal communication, 1996) show that on plot 3NA–continuous corn with no treatment, SOC loss has been approximately 63% (Figure 6.1), changing from 3.78% to 1.41% from 1876 to 1993. The most rapid decline in SOC occurred in the first 28 years (3.78% to 2.1%). This represents a 44% decrease. The SOC continued to decrease but at a slower rate until the 1950s (2.1% to 1.4%), a 33% decrease over approximately 50 years. Since then, the SOC appears to have reached a quasi-equilibrium.

Darmody and Norton (1993) studied the structural degradation of the soil at the Morrow Plots. Soil fabric and structural properties were used as indicators. They reported that fertilization and lime had little effect on aggregate properties or soil fabric, but that crop rotation made a considerable difference on both. Aggregate size and stability decreased in the order grass sod, corn–oats–clover, corn–soybean, and continuous corn. SOC decreased similarly, and the average values reported for all fertilization treatments for the 0- to 15-cm horizons were 4.56, 2.55, 1.95, and 1.65, respectively. The extreme values represent a 63.8% decrease in SOC content due to rotation effect.

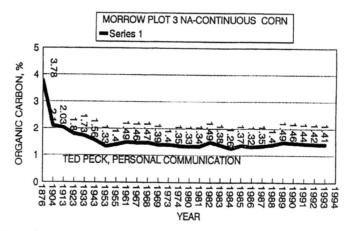

FIGURE 6.1 Morrow Plots 3 NA–continuous corn. (Organic carbon data from Ted Peck, University of Illinois, personal communication.)

Sanborn Field and Other Missouri Sites

Sanborn Field at the University of Missouri–Columbia was established in 1888 and is the oldest agricultural experiment field west of the Mississippi River (Gantzer et al., 1991). It was established for study of long-term crop rotations and management practices.

Buyanovsky et al. (1996) summarized many effects of various treatments for Sanborn Field. The greater number of times within a crop rotation the surface soil was tilled, the less was the amount of SOM remaining after 100 years (Figure 6.2). It required almost 50 years for the rotation with the most frequent tillage to reach a near stable level of SOM.

In continuous cropping without additions of inorganic fertilizers, manure at 6 tons/acre/year was just able to maintain SOM with the grain crops corn and wheat (Figure 6.3). With no treatment, continuous culture of grain crops continued to cause lowered SOM contents after 100 years. With continuous timothy, the manured treatment resulted in an increase of SOM, but SOM remained constant for more than 100 years with no treatment and with a very low level of production.

Continuous culture of corn and wheat under high fertility with total removal of aboveground growth caused a marked decline in SOM from 1914 through 1950 (Figure 6.4). When residues were returned starting in 1950, the SOM has continued to increase into the 1990s. Residue management obviously has a marked impact on SOM over time.

Gantzer et al. (1991) analyzed the topsoil remaining after 100 years of continuous cropping of plots on Sanborn Field. Soil properties within the plots were not determined at the initiation of the plots. The cropping sequences used were continuous corn, continuous timothy, and a 6-year rotation of corn–oats–wheat–clover–2 years timothy. Sites for these comparisons were near a summit divide with slope gradients ranging from 0.5 to 3.0%. The topsoil remaining was significantly less for the

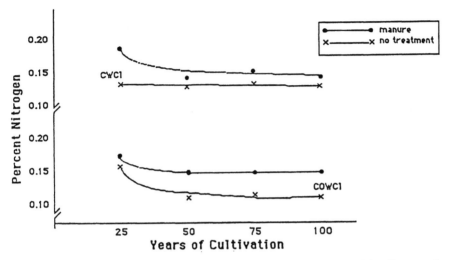

FIGURE 6.2   Changes in soil nitrogen under rotation (Sanborn Field). (After Buyanovsky et al., 1996.)

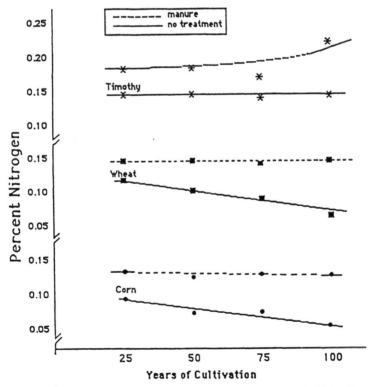

FIGURE 6.3   Changes in soil nitrogen with cultivation (Sanborn Field). (After Buyanovsky et al., 1996.)

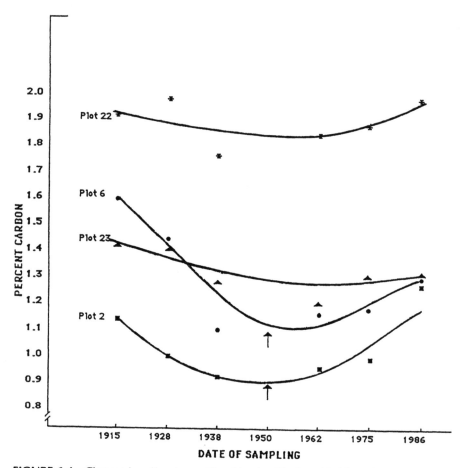

FIGURE 6.4 Changes is soil carbon with cultivation (Sanborn Field). (After Buyanovsky et al., 1996.)

continuous corn as compared with the 6-year rotation or timothy plots. The continuous corn plots had 44% as much topsoil and the rotation plots had 70% as much topsoil as the timothy plots. Clay in the plow layer of the continuous corn plots was significantly higher than in the other plots perhaps due to incorporation of the clay subsoil into the plow layer.

Observations on Taberville Prairie (Table 6.2), an unplowed prairie in southwestern Missouri, made during the work reported by Conway-Nelson (1991) showed that ped surfaces to the full depth of sampling were richly covered with a film of SOM. Such films are lost in our normal soil processing, yet these films must contribute significantly to nutrient and moisture movement in the profile. These peds form the boundaries of pores through which roots grow; thus roots benefit from the organic films while alive and replenish them upon death and subsequent decay. Missouri

TABLE 6.2
Distribution of Organic Carbon in Soils from Three Landscape Positions on Taberville Prairie (St. Clair County, Missouri)

Position	Horizon	Depth (cm)	Organic[a] carbon (%)	Bulk density (g/cm^3)	Organic carbon (kg/m^2)
Summit	A_1	7–7	2.2	$(1.3)^b$	1.5
	A_2	7–18	1.2	1.5	1.98
	A_3	18–26	1.0	1.5	1.65
	AE	26–33	0.8	1.5	0.57
	EB	33–41	0.6	1.5	0.72
	Bt_1	41–48	0.3	1.5	0.31
	Bt_2	48–67	0.3	1.5	0.85
	$2 Bt_3$	67–80	0.5	1.8	1.17
	$2 Bt_4$	80–104	0.6	1.7	2.45
	$2 BC_1$	104–122	0.2	1.8	0.65
	Total	—	—	—	11.86
Backslope	A_1	0–9	2.6	$(1.3)^b$	3.04
	A_2	9–25	1.5	1.4	3.36
	A_3	25–41	1.0	1.4	2.24
	AB	41–57	0.7	1.5	1.05
	BA	51–67	0.6	1.6	1.54
	B_{11}	67–79	0.6	1.6	1.15
	B_{12}	79–109	0.5	1.7	2.55
	$2 Btg_1$	109–135	0.3	$(1.8)^b$	1.40
	Total	—	—	—	15.52
Bottom	A_1	0–9	3.1	$(1.3)^b$	3.63
	A_2	9–22	2.0	1.51	3.90
	AB	22–34	0.9	1.5	1.62
	B_{11}	34–56	0.7	1.5	2.31
	B_{12}	56–75	0.7	1.5	2.00
	$2 B_{13}$	75–82	0.9	$(1.6)^b$	2.45
	Total	—	—	—	15.91

[a] Leco carbon analyzer.

[b] Estimated.

Modified from Conway-Nelson, 1991.

work has shown that the fraction of total carbon in sampled profiles (120–200 cm) in the surface 20 cm ranged from 22 to 87% depending upon past management, especially over species and total depth of sampling (Hammer et al. 1995; Conway-Nelson, 1991; Scrivner and Cooper, 1985). Most of these reports show that more than 50% of the SOC measured is in the portion of the solum below a depth of 20 cm. SOM contents for selected soil profiles for Sanborn Field are given in Table 6.3. Note that at 1.4 m, the SOM content averages 0.45% for these sites.

TABLE 6.3
Organic Matter[a] Content of Selected Soil Profiles from Sanborn Field (Boone County, Missouri), 1983

Depth plot increment (cm)	% organic matter					
	6	17	18	22	23	31
0–10	1.9	0.9	2.5	4.0	2.4	2.5
10–20	2.0	1.2	2.4	2.8	1.6	2.1
20–30	1.4	1.0	1.2	1.8	1.4	1.4
30–40	1.0	0.8	1.6	1.3	1.0	0.9
40–50	0.8	0.8	0.7	1.2	1.0	0.8
50–60	0.6	0.7	0.6	1.0	0.8	0.7
60–70	0.6	0.6	0.6	0.9	0.6	0.5
70–80	0.6	0.5	0.6	0.6	0.6	0.7
80–90	0.5	0.6	0.6	0.4	0.5	0.5
90–100	0.4	0.6	0.6	0.5	0.4	0.5
100–110	0.5	0.6	0.5	0.4	0.4	0.4
110–120	0.4	0.5	0.6	0.4	0.4	0.5
120–130	0.4	0.4	0.5	0.4	0.4	0.5
130–140	0.4	0.5	0.4	0.5	0.4	0.5

[a] Modified Walkley Black.

OTHER LONG-TERM STUDIES

The effects of cropping systems and management for experiments conducted at the Fort Hays Branch Experiment Station in west-central Kansas for 1916–58 were reported by Hobbs and Brown (1965). Using the 1916 levels for comparisons, SOC losses of the 0- to 18-cm horizon averaged 36.2%, ranging from a low of 25% in continuous small grain to a high of 55% in alternate row crop and fallow. Plots in continuous row crop lost 47% of the original SOC content. They concluded that the cropping system affected both nitrogen and SOC losses for the 0- to 18-cm horizon. Small grains grown continuously or alternating with fallow caused the smallest losses and row crops the greatest. Cropping systems with both small grains and row crops caused intermediate losses.

Long-term cropping system studies near Wooster, Ohio (Dick et al., 1986) show that continuous corn plots had the largest decline in SOM and nitrogen. Continuous wheat, continuous oats, a 5-year rotation of corn–oats–wheat–clover–timothy, and a 3-year rotation of corn–wheat–clover were the decreasing order of other cropping systems. They concluded that meadow crops added SOM and nitrogen, but the other nonmeadow crops depleted the SOM and nitrogen.

Peevy et al. (1940) reported the effects of rotation and manure treatments for 20 years on rotation experiments at Iowa State College. Shrader (1953) identified the soil types on each of these experiments and summarized the changes in total nitrogen. Using the above two data sources, we calculated the changes in SOM for the no-treatment plots of a modified 5-year rotation where the first 20 years were a 4-year rotation of corn–oats–clover–wheat followed by alfalfa. The data reported are for the first 20 years. Clarion (fine-loamy, mixed, mesic Typic Hapludoll) had a decrease of 10.4% SOM, 2 Nicollet soils (fine-loamy, mixed, mesic Aquic Hapludolls) decreased by an average of about 10%, Canisteo (fine-loamy, mixed, mesic Typic Haplaquoll, calcareous) by 11.6%, and a Webster (fine-loamy, mixed, mesic Typic Haplaquoll) by 26.9%.

Dodge and Jones (1948) reported changes in SOC contents for Geary silt loam (fine-silty, mixed, mesic Udic Argiustoll) near Manhattan, Kansas, for the 35-year period 1911–46. The land had been under cultivation since 1864 and before establishment of the plots in 1919 had been used for corn and occasionally oats and wheat. The experiments beginning in 1911 were set up with a 16-year rotation, a 3-year rotation, and continuous wheat. For the plots that received no fertility treatment, SOM in the 0- to 18-cm horizons of the soils in the 3-year rotation decreased 18.7%, the continuous wheat plots decreased 14.7%, and the 16-year rotation decreased 11.6%. Young et al. (1960) reported SOM change for Fargo clay (fine, montmorillonitic, frigid Typic Epiaquert). From 1913 to 1953, a 40-year period, the SOM content of the 0- to 18-cm horizon decreased by more than 22%, from 9.64 to 7.49%, when all samples were considered.

Mitchell et al. (1991) discussed long-term agronomic research in the United States. They cited 25 experiments monitored for more than 25 years, and 12 of these have been used for more than 50 years. In summarizing the data from the Magruder Plots in Oklahoma, they note that SOM for these plots declined rapidly during the first 35 years, maintained a lower rate of decline for the next 52 years, and then stabilized.

SOC and total nitrogen concentration in virgin grassland soils of the Great Plains put into cultivation declined most rapidly during the first 10 years after cultivation began (Hill, 1954; Hobbs and Brown, 1965). They continued to decline for several decades but at a gradually diminishing rate and eventually tended toward an apparent equilibrium amount. Others have reported that this equilibrium level varied with the cropping sequence (Haas et al., 1957; Ridley and Hedlin, 1968) and with type and amount of crop residues (Overson, 1966; Black, 1973; Larson et al., 1972; Rasmussen et al., 1980). Unger (1968) and Bauer and Black (1981) reported that the equilibrium level was affected by tillage practice and that greater erosion control was one reason for the observed differences. Bauer and Black (1981) suggested from their data that SOC and total N had reached a rather stable equilibrium level after the first 40 years of cultivation. The equilibrium was improved by use of stubble mulching as compared with conventi ·:al tillage management.

Larson et al. (1972) added different types and amounts of residues ranging from 0 to 16 tons to a Marshall silty clay loam (fine-silty, mixed, mesic Typic Hapludoll). The soil was cropped to corn for 11 consecutive years. Large amounts of N fertilizer were added. For the 16 tons/ha/year, average increases over the check plot for C, N, S, and P were 47, 37, 45, and 14%, respectively, for the 0- to 15-cm depth. Corn stalk residue needed to prevent loss of organic C was estimated to be 6 tons/ha/year.

Yonker et al. (1988) studied the pattern of SOC accumulation in a semiarid short grass steppe in Colorado. They concluded that there was some variation in SOC due to slope position, but differences within a toposequence need not be striking. Toeslopes usually had higher concentrations than the corresponding summits. Contrary to normal patterns, SOC concentration did not decrease at the shoulder or increase systematically downslope in their study. This suggested to the authors that the role of water as the agent of differentiation is minimized in the present-day environment. They concluded that wind was the overall dominant process in determining soil distribution in their study. Schimel et al. (1986) reported that approximately 60% of the root mass resided in the surface 10 cm in the shortgrass steppe, but surface horizons did not contain that proportionate share of soil organic C. In the Younker study, organic C in the A horizons represented 25% or less of the total organic C in the solum.

Melsted (1954) concluded that soils cropped to continuous corn with an ample supply of nitrogen decreased in SOM content under clean cultivation. The same conclusion was reached by Moldenhauer et al. (1967) on a Marshall silty clay loam soil (fine-silty, mixed, mesic Typic Hapludoll) in Iowa. When the soil had been depleted of SOM by continuous corn without fertilization, high rates of nitrogen for 12 years resulted in an increase in SOM.

Darusman et al. (1991) determined soil properties after 20 years of nitrogen fertilization with different nitrogen sources. They concluded that the primary effect was increased soil acidification and associated changes in nutrient availability and increased concentrations of nitrate and ammonium nitrogen.

PAIRED VIRGIN AND CULTIVATED SOILS

Whiteside and Smith (1941) cite several references that show that the total N and SOM in the surface horizon of tilled soils decreased relative to adjacent untilled forest or grassland areas. They reported a 23% decline in SOC (39.6% SOM) of the A1 horizon when a virgin Flanagan silt loam was compared with one cultivated intermittently from the late 1850s to 1935. These authors also concluded that those cropping systems had a great influence on the amount and direction of change in N and organic C. Greatest changes were associated with continuous row crops, followed by cereal crops, then legume and sod crops.

Anderson and Browning (1949) compared the properties of six cultivated and six virgin paired Iowa soil profiles. They concluded that the cultivated soils were less well aerated, had considerably fewer stable aggregates, and had lost almost one-third of their original nitrogen in 100 years or less of cultivation and cropping. They concluded that the changes associated with cultivation adversely affected those properties associated with good crop production. Calculation of SOM lost from data reported by Anderson (1949) shows that, for the 0- to 15-cm depth, when virgin and cultivated soils are compared, the loss of SOM by soil was Edina (fine, montmorillonitic, mesic Vertic Argialboll) = 30.2%, Grundy (fine, montmorillonitic, mesic Aquertic Argiudoll) = 36.5%, Marshall (fine-silty, mixed, mesic Typic Hapludoll)

= 29.2%, Ida (fine-silty, mixed [calcareous] mesic Typic Udorthent) = 66.3%, Carrington (fine-loamy, mixed, mesic Typic Hapludoll) = 8%, and Webster (fine-loamy, mixed, mesic Typic Haplaquoll) = 23%. The relatively small loss for the Carrington soil was attributed to a shorter time of cultivation than for the other soils. The relatively large loss for the Ida soil may have been due to accelerated erosion in addition to oxidation of SOC.

In a study of forest soils in Georgia, Giddens (1957) compared 29 locations of soils still in forest with adjacent areas cultivated for 25 years or more. Average values reported were 3.29% SOM for the uncultivated soils and 1.43% SOM for the cultivated. This is almost a 57% decrease in SOM content.

Campbell and Souster (1982) discussed the loss of SOM from cropping for some Saskatchewan soils. They observed that the active SOM fraction had incurred significantly higher losses than stable, more highly humified fractions.

Tiessen et al. (1982) studied the change in SOM in three prairie soils of different textures. Cryoborolls and Cryorthents were sampled and the C, N, and P contents of paired cultivated and uncultivated sites were determined. Reductions of about 35% in the C concentration were reported for the clay and silt loams soils after 60–70 years of cultivation. During a similar period of cultivation, a sandy loam soil had a reduction of about 46% C. Prolonged cultivation for 90 years did not result in a decrease in the rates of loss of C, N, and P on the silt loam soil. The conversion of concentration data to area-based data resulted in a decrease in the differences between cultivated and uncultivated soils. These differences were caused by an increase in the bulk density under cultivation and by an increase in the standard deviations of the data due to variability of horizon depths in cultivated fields. The authors state that their data are in conflict with those of Martel and Paul (1974) which showed a leveling off of losses of SOM after 60–70 years of cultivation. Tiessen et al. (1982) attributed the continued decline to erosive losses associated with lowered SOM contents of the soil and to an extension of the zone of depletion in lower soil horizons. They conclude that sampling the entire solum is necessary if changes are to be described accurately.

Blank and Fosberg (1989) evaluated the effects of cultivation on soil properties for six paired profiles of the Williams soil (fine-loamy, mixed, Typic Argiboroll) from north-central South Dakota. They evaluated changes in all horizons of the profiles. SOC content decreased by 26% and bulk density increased by 18% in the Ap horizons. The Ap horizons also contained 38% more very fine sand and 10% less silt. The A and Btk horizons of the virgin soils had greater wet aggregate stability, and the A horizons of the virgins soils retained 30% more water at 0 MPa tension. They also reported other physical and chemical differences among the paired profiles, which in part may be related to their finding that cultivation allows excess water to move through the soil.

Meints and Peterson (1977) reported decreases in SOC content ranging from 30.6 to 55.5% in a study that showed the effects of cultivation when paired virgin and cultivated Ustoll soil profiles were compared. The effects of cultivation could be detected to depths of more than 127 cm.

EROSION EFFECTS

A study was initiated in 1931 on Marshall soils (fine-silty, mixed, mesic Typic Hapludoll) having 9% slope near Clarinda, Iowa to determine the effects of different cropping systems on soil and water loss. Van Bavel and Schaller (1950) concluded after 19 cropping years that:

1. Changes in SOM contents were dependent upon cropping treatment and upon the amount of soil erosion.
2. A corn–oats–meadow rotation on the Marshall soil resulted in a small SOM decrease, which was probably not significant.
3. Eleven years of continuous alfalfa raised the SOM content, whereas 11 years of bluegrass did not. The changes in SOM content over time under different cropping systems are given in Table 6.4.

De Jong and Kachanoski (1988) discussed the importance of erosion in the carbon balance of prairie soil in Canada. They observed that those changes in SOC in the 0- to 0.15-m layers could be explained largely by erosion and deposition. The coarse to medium-textured soils they studied were largely on upper and middle slopes (summit and backslope) and had been cultivated since before the mid-1940s.

The literature available for documentation of the effects of erosion from long-term cropping from controlled experiments is limited. However, the effect of accelerated erosion on soils is recognized by soil scientists by using erosion classes and phases. Erosion classes are based on the percentage of the A horizon lost as compared with a comparable uneroded soil. Eroded phases are shown on soil maps of the National

TABLE 6.4
Effect of Cropping Systems on SOM Content, 1931–49, Clarinda, Iowa

Cropping system	Years	Organic matter (%)				
		1931	1937	1942	1949	% change
Continuous corn	1931–42					
Corn–oats–meadow	1943–49	3.35	2.93	2.94	2.95	−11.9
Corn–oats–meadow	1931–49	3.38	3.35	3.45	3.23	−4.4
Alfalfa	1931–42					
Continuous corn	1943–49	3.39	3.78	3.74	3.10	−8.6
Bluegrass	1931–42					
Continuous corn	1943–49	3.39	3.35	3.36	3.17	−6.5
Corn–oats–meadow	1931–49	3.28	2.87	2.90	2.33	−29.0
Desurfaced 12 in.						
Continuous corn	1931–42					
Corn–oats–meadow	1943–49	3.39	2.19	1.92	2.13	−37.2

After Van Bavel and Schaller, 1950.

Cooperative Soil Survey and are identified based on the properties of the soil that remains, although soil lost is estimated and noted (Soil Survey Staff, 1993). Olson et al. (1994) reviewed methods of identifying eroded phases of a soil and quantifying the soil lost due to erosion.

Several factors result in a decrease in surface soil thickness that can confound soil loss attributed to erosion. Cultivation of a soil with minimum erosion hazard results in decreased SOM due to the effect the disturbance associated with cropping has on the soil. In addition, most cultivated soils have higher bulk densities in the surface horizon after years of cultivation. Undisturbed soil formed under grass may have surface horizons with bulk densities of 1.1 mg/m^3 and after years under cultivation the same site may have a bulk density of 1.4–1.5 mg/m^3, thus decreasing the volume by approximately one-third. Mixing of surface and subsurface material in tillage will dilute the SOC content where surface horizons are naturally thin. Guernsey et al. (1969) reported that even on the nonerosive Morrow Plots, the A horizon became thinner and less porous, and its structure was poorer on the continuous corn plots with no fertilization as compared with the fertilized rotation plots.

FALLOW EFFECTS

Fallow is the practice of keeping land free of crops and weeds for storing water in the soil. This practice allows farmers in more arid regions to increase their yields by cropping their land every second year rather than every year. The land may be tilled to prevent weed growth, but the tillage should be only as often and as deep as necessary to kill weeds (Troeh and Thompson, 1993). In a review of cropping systems in the Great Plains, Bartholomew (1957) reported that row crops and fallow induced the largest declines in nitrogen and SOM. Small grains showed a smaller decline than row crops. Smallest declines in both SOM and nitrogen occurred when grass or alfalfa was included in the cropping systems. Manure was effective in minimizing declines of nitrogen. In some cases, soil nitrogen was maintained equal to virgin conditions when moderate applications of manure were made to small grain. Hill (Bartholomew, 1957) of the Lethbridge Experiment Station in Canada reported that the decline in SOM was closely related to the incidence of clean fallow. Wheat with alternate fallow caused the largest loss of SOM among the cropping systems studied. The smallest decline in SOM occurred under continuous wheat. The decomposition of SOM each year in a particular soil was closely related to the total amount of SOM in the soil. Bartholomew also reported that Jenny (1941) showed a decline of about 20% from the virgin condition in the first 20 years of cultivation. SOM declined about another 10% in the second 20 years and about 7% in the third 20 years. Jenny stated that at that rate of decline, an equilibrium level of SOM of about 50–60% of that present in the virgin soil would finally be reached. He estimated the change would be about 80% complete in 60 years.

Janzen (1987) reported that the SOC and nitrogen content of the soil decreased with increased frequency of fallow in the rotation at both the 0- to 15-cm depths and the 15- to 30-cm depths. The data reported were based on 33 years of cropping history. Inclusion of perennial forage in a spring wheat rotation did not increase the

SOM content above that maintained by continuous wheat under the semiarid conditions of the experiment. Distribution of organic nitrogen and carbon among labile and stable pools was strongly affected by crop rotation and frequency of fallowing in the rotation. Levels of mineralizable nitrogen and carbon associated with the continuous wheat were approximately twice those in the fallow–wheat rotation.

Campbell et al. (1991) suggest that all fallow-containing monoculture wheat systems and the unfertilized continuous wheat have continued to deplete soil organic N; unfertilized green manure–wheat–wheat has generally maintained it, and the 6-year cereal–forage and fertilized continuous wheat systems have increased the total soil N concentrations.

TILLAGE EFFECTS

Changes in tillage practices have resulted in increased interest in the relationship between types of tillage systems and soil properties, especially SOC and nitrogen. Rasmussen and Collins (1991) reviewed the long-term impacts of tillage, fertilizer, and crop residue on SOM in temperate semiarid regions. They summarized data showing that conservation tillage increased organic C and N in the top 5–15 cm of soil as compared with conventional methods of tillage. These increases average from 1 to 2% per year for both C and N in the upper 15 cm of soil. However, they cite data from Doran (1980) showing that below the upper few centimeters, the C and N have been either equal or less than that in conventional tillage.

In a review paper of stubble mulch farming, McCalla and Army (1961) cited works that suggested higher SOM contents in soils with stubble mulching as compared with plowing. They concluded, however, that the higher SOM content of the stubble mulched soils may be a less rapid decline in SOM rather than a buildup in many instances.

Rice et al. (1986) reported on a long-term tillage system study on a Maury silt loam (fine, mixed, mesic Typic Paleudalf). When the pasture sod was broken at the beginning of the continuous corn cropping, there was a rapid decline in organic nitrogen. The loss of N from the surface horizon was almost threefold greater in the conventional tillage system as compared with the no-till system. After 5–10 years of cropping, SOM concentration apparently approached a new steady-state level.

Griffith et al. (1988) reported the effects of long-term tillage and rotation effects on high and low SOM soils. They concluded that no-till had a relative advantage over the other tillage systems and attributed the advantage to improved physical conditions for low organic matter soils.

Havlin et al. (1990) compared the effects of no-till and conventional tillage on three sites in eastern Kansas. No-till treatments had greater SOC and nitrogen contents. The increases were directly related to the quantity of the residue produced and left on the soil surface. Fertilizer nitrogen increased SOC and nitrogen only slightly. They concluded that production systems that include rotations with high residue-producing crops and reduced tillage that maintains surface residue cover result in greater SOC and nitrogen.

Dormaar and Lindwall (1989) reported significant differences in SOM content of A horizons of dark brown chernozems placed in a wheat–fallow rotation in southern Alberta. After 19 years of conservation tillage, the SOC contents under three different tillage systems (no-till, herbicide plus blade cultivation, and blade cultivation) were 1.7, 1.63, and 1.37%, respectively. SOC was highest in the no-till, but the light-fraction carbon as a percentage of the total carbon was significantly higher for the cultivated treatments. They also reported data for the effect of 9 years of cultivation for three rotations (continuous wheat, winter wheat–barley–fallow, and wheat–fallow) for no-till and cultivated treatments. There were significant differences in SOC concentrations between the tillage treatments but not among the rotations. Average SOC concentration for the no-till sites was 1.72%. For the cultivated sites, average SOC concentration was 1.55%.

RATES OF DECLINE

Figure 6.1 shows the rate of change in SOC content for the continuous corn experiment on the Morrow Plots at the University of Illinois. The initial point in 1876 is based on the interpretation by Odell et al. (1982) that the SOC content of the grass sod border was close to the initial value for the continuous corn plots. Note that the rate of decline is greatest in the first 28 years and then the decrease is less rapid with increasing time.

Figure 6.5 is a plot of the Van Bavel and Schaller (1950) SOM data and shows the rate of decline for two different rotations (corn–oats–meadow and continuous corn). Because these plots were on 9% slope gradients, some loss was due to erosion. The history of these plots is not available, but uncultivated Marshall soils should have SOM contents approaching 4.5–5%. Thus, the period of anticipated rapid decline is not shown by this data set. We conclude that the plots had been cropped before the establishment of this study.

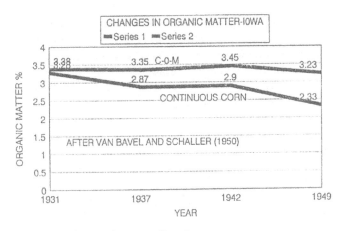

FIGURE 6.5   Changes in organic matter (Iowa).

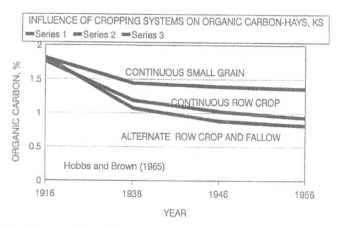

FIGURE 6.6   Influence of cropping systems on organic carbon (Hays, Kansas).

Figure 6.6 shows data from a range of cropping systems in western Kansas reported by Hobbs and Brown (1965). The rate of decline is greatest in the 1916–38 period and then the rate of change decreases. The average loss of SOC reported for all plots for 1946–56 was 2.1%.

Prairie soils in Canada declined in SOM more than twice as fast in the first 9 years as during the next 10–30 years. Declines of SOM in soils on the Morrow Plots were only about 60% as fast during the second 20 years of cultivation as compared with the first 20 years. Bartholomew (1957) concluded that not all experiments show these same trends. Occasionally, losses have been as great in the later years as the early years. It appears that, overall, annual changes have become smaller with increasing time of cultivation and SOM has tended toward levels where the annual inputs were equal to the amounts lost through mineralization.

Rasmussen and Collins (1991) prepared a summary paper on the changes in organic matter in soils of the temperate semiarid regions resulting from different management practices. By the 1900s, SOM losses were evident and the factors causing the greatest declines were identified. Frequent fallowing, intensive tillage, and removal of crop residues were cited as being responsible for accelerated loss of SOM. Manure and application of N reduced the rate of SOM loss. The effects of erosion could not be defined. In the period from 1950 to 1980, some long-term studies showed that SOM was no longer declining in some cropping systems. Annual cropping replaced fallow in some areas and this reduced SOM loss. The level of carbon returned to the soil increased due to greater fertilizer use and improved cereal varieties that increased yields of both grain and straw. Stubble mulch and no-till systems were shown to conserve up to 2% more SOM per year in the surface soil than plowing. Many long-term experiments were ended between 1950 and 1980. The 1980s showed increasing signs of environmental degradation. There was increased concern for the effects of nitrogen and pesticides in agriculture. Carbon input was shown to be critical for long-term maintenance of SOM and soil fertility. Rasmussen and Collins predicted that probably the most important area to understand in the future was the role of erosion in carbon storage and nitrogen use in soil and concluded by stating that erosion losses could negate most of the present positive strategies to increase SOM levels.

The decline in N and SOM content when the land is cultivated cannot be attributed entirely to a reduction in the quantity of plant residues available for humus synthesis (Stevenson, 1982). Improved aeration because of cultivation may cause increased microbial activity and loss of SOM. Alternate wetting and drying because of cultivation can also be an important factor in decreasing SOM content. One major effect of cultivation may be the exposure of SOM previously not accessible to microbial attack. SOM is usually lost when soils are first placed under cultivation and a new equilibrium level is reached which is characteristic of the cultural practices and kind of soil. Increased plant yields because of improved varieties, widespread use of fertilizers, and adoption of better management practices should affect equilibrium levels of SOM through return of larger quantities of plant residues. Among the important tentative conclusions Stevenson made concerning the use of ^{14}C-labeled substrates were the following:

1. Addition of fresh organic residues to the soil may result in a small priming action on the native SOM.
2. Plant residues decay rather rapidly in soil and are more or less completely transformed, even the lignin fraction.
3. Except possibly for lignins, freshly incorporated carbon first enters microbial tissue (soil biomass), the "labile fraction" of SOM, and finally complex humic polymers during advanced stages of humification.

McGill et al. (1981) reported that SOM concentration of A horizons of Chernozemic soils had decreased by 41–49% after about 80 years of cultivation.

Schimel et al. (1985) concluded that the concentration of SOC and nitrogen decreased markedly in three toposequences they studied that were derived from sandstone, siltstone, and shale. Sites were selected at five different landscape positions for each toposequence. The weighted average percentage losses of SOC to a depth of 100 cm for all landscape positions after 44 years of cultivation were 34, 46, and 35 for the sandstone-, siltstone-, and shale-derived soils.

Campbell et al. (1990) summarized long-term crop rotation studies on the Canadian prairies. They reported that most soils included in their studies exhibited a decline in concentrations of SOM following initial cultivation. However, they further state that this trend has been halted or even reversed with appropriate crop and fertilizer strategies.

MODELS

Bartholomew and Kirkham (1960) developed and evaluated a differential equation for describing the rate of change of SOM or soil nitrogen. The equation, according to them, could be used to quite accurately describe many field conditions. Estimates showed that new SOM equilibriums are established in most soils in less than 100 years following cultivation of new land.

Lucas et al. (1977) showed that for Michigan conditions, cropping practices directly affect soil carbon levels and activity. They developed a model of SOM dynamics that takes account of inputs from crop residues and manures, decomposition, and losses from erosion. In their study, in time SOC levels approach a steady-state level for a respective yield level.

The Century SOM model (Parton et al., 1983a) predicts long-term effects of climate, soil texture, and management on SOM levels and turnover rates. The Century model was developed based on data from grassland soils of North America but can simulate long-term changes in Swedish soils (Parton et al., 1983b).

Burke et al. (1989) used data from 500 rangeland and 300 cultivated soils in the U.S. Central Plains grassland to develop regional models. They concluded that SOC increased with precipitation and clay content and decreased with increased temperature, as suggested by Jenny (1941). SOC losses due to cultivation increase with precipitation and relative organic SOC losses were lowest in clay soils. Based on their analyses, it was concluded that potential SOM losses were highest in the northeastern section of their study area and decreased generally from west to east. They found the maximum predicted SOC at a mean annual precipitation of about 80 cm regardless of temperature. Minimum predicted SOC was reached at a mean annual temperature of 18°C regardless of precipitation.

Franzmeier et al. (1985) presented SOC content in the North Central Region (NCR) of the United States based on map units given on a 1:2,500,000 map of the region. The NCR comprises 1.3% of the land area of the world and contains between 0.8 and 3.5% of the total C reserves in the world.

RELATIONSHIP OF SOIL ORGANIC MATTER TO SOIL QUALITY

The concept of soil quality was introduced in the early 1990s as one component of sustainable agriculture. Recently, the National Research Council published a book on soil and water quality (National Research Council, 1993) and linked soil quality to water quality. They suggested that enhancement of soil quality should be the first step toward increasing water quality. The definition of soil quality is evolving and may continue to evolve. Most definitions consider soil quality as:

> The capacity of a specific kind of soil to function, within natural or managed ecosystem boundaries, to sustain plant and animal productivity, maintain or enhance water and air quality, and support human health and habitation (Karlen, et al., 1997).

Most measurements of soil quality are tied to soil function (Larson and Pierce, 1994; Doran and Parkin, 1994). Functions of soils include (Larson and Pierce, 1994; Doran and Parkin, 1994; Karlen et al., 1997):

1. Sustaining biological activity, diversity, and productivity
2. Regulating and partitioning water and solute flow

3. Filtering, buffering, degrading, immobilizing, and detoxifying organic and inorganic materials, including industrial and municipal by-products and atmospheric deposition
4. Storing and cycling nutrients and other elements within the earth's biosphere
5. Providing support of socioeconomic structures and protection for archeological treasures associated with human habitation

Minimum data sets for measuring or monitoring the capacity of the soil to function include SOM as a primary indicator (Larson and Pierce, 1994; Doran and Parkin, 1994; Acton and Gregorich, 1995; Gregorich, 1996). Larson and Pierce (1994) emphasized that SOM is the single most important indicator of soil quality and that SOC in the upper few centimeters of the soil is the most important. Romig et al. (1995) developed a soil health scorecard based on farmer input. Farmers in their study ranked SOM first in a list of soil health indicators. Yield was tenth in the ranking. Farmers commented that it took a long time to build organic levels in soils and recognized that SOM levels are related to kind of soil.

The reason SOM is considered so important is that it impacts all five of the soil functions and is an important part of the physical, chemical, and biological processes in soils. Schnitzer (1991) listed the following effects of SOM on soil quality for these three processes:

1. Physical effects are soil aggregation, erosion, drainage, tilth, aeration, water-holding capacity, bulk density, evaporation, and permeability.
2. Chemical effects include exchange capacity; metal complexing; buffering capacity; supply and availability of N, P, S, and micronutrients; and adsorption of pesticides and other added chemicals.
3. Biological effects are activities of bacteria, fungi, actinomycetes, earthworms, roots, and other microorganisms.

Broll (1993) suggests that SOM is most important in urban green areas as a buffer of contaminants. He stresses that effects of management on soil biological activity and humification could be improvement of quantity and quality of SOM as it affects the buffering capacity of the soil.

The impact of SOM on water-holding capacity of the soil has been a matter of debate in the literature. Hudson (1994) reviewed the literature on the effect of SOM on available water-holding capacity (AWC) and concluded that in all soil texture groups, as SOM content increased from 0.5 to 3%, AWC of the soil more than doubled. He stated that SOM is important in AWC because it makes up a significant part of the soil volume; 1–6% SOM by weight was equivalent to 5–25% by volume.

Many studies show that various fractions of SOM are important to understanding its quality and soil quality. Larson and Pierce (1994) and Gregorich (1996) suggest that the minimum data set for soil quality indicators includes, in addition to total SOC, a measure of the readily biodegradable or labile fraction. Wood et al. (1990)

used total soil carbon content, C:N ratios, mineralizable C and N, and C and N turnover as indicators of soil quality. They found that the active fractions of soil SOM in the west central Great Plains soils are very sensitive to change in management and are increased by higher levels of management (no-till) and increased fertilizer inputs.

Janssen (1984) concluded that differences in quality of SOM after 25 years of fertilizing practices were related to the proportion of "young SOM." His "young SOM" is the repeated addition of crop residues and organic manure. Tisdall and Oades (1982) suggested that correlations between total SOM and aggregate stability are not always strong because only part of the SOM fraction is involved with aggregate stability. They reviewed the effects of SOM on water-stable aggregates in soils and classified organic binding agents into transient (polysaccharides), temporary (roots and hyphae), and persistent (resistant aromatic components associated with metal cations and strongly sorbed polymers). In a review of soil structure and aggregate stability, Lynch and Bragg (1985) suggest that organic materials (such as products of decomposition of plant, animal, and microbial remains; microorganisms; and products of microbial synthesis) are the main agents of stabilization of aggregates.

SOM is the "food" that supports the living part of the soil ecosystem. Gregorich et al. (1994) stress that the multifaceted role of SOM must be taken into consideration in the assessment of soil quality. They subdivided SOM into a set of attributes that include total SOC and nitrogen, light fraction and macroorganic (particulate) matter, mineralizable carbon and nitrogen, microbial biomass, and soil carbohydrates and enzymes. The relationship of SOM attributes to soil quality is summarized in Table 6.5. For the most part, the SOM attributes are very sensitive to cropping and management systems and provide a relatively rapid measure of the impact of these systems on soil quality. Not all of the attributes need to be included in a minimum data set for the assessment of soil quality, as many are related to the same functions in soil. Choice of the SOM attribute depends on the land use and critical soil functions for that use and also the time frames for monitoring changes in the SOM attribute.

SUMMARY

SOM has long been recognized as an important factor in soil processes and crop growth. The level of SOM at any one time is dependent on a complex set of interactions of physical, chemical, and biological processes. The primary use of soils is for growing plants. The various systems that we use in crop production influence SOM levels. It has been shown that rotations, fertilization practices, residue management, and tillage systems all can be important variables in determining SOM levels.

The declines in SOM when virgin soils are converted to cultivated soils have been shown with many long-term experiments. Most of the data shows an initial rapid decline, then a period of a lower rate of decline, followed by a quasi-equilibrium under the same use and management. However, if one of the use or management

TABLE 6.5
SOM Attributes as Related to Soil Quality

SOM attribute	Attribute properties	Related soil processes	Related soil function
SOM and nitrogen	Includes a range of humified (more resistant) high-molecular-weight, complex polymeric compounds intimately associated with inorganic soil components; chemically well defined (low-molecular-weight aliphatic and aromatic acids, carbohydrates, amino acids, polypeptides, proteins, polysaccharides, and waxes that have relatively rapid turnover rates) and readily decomposable material (plant litter, roots, and dead and living organisms)	Nutrient cycling, aggregate formation and stability, carbon/nitrogen mass balance, chemical attenuation	Biomass production regulating and partitioning water, filtering and buffering, storing and cycling nutrients
Light fraction and macroorganic SOM	Mainly plant residues and residues from animals; represents from 0.1 to 4% of total SOM but contains up to 15 times more C and 10 times more N that the whole soil; a readily decomposable substrate for soil microorganisms and short-term reservoir for plant nutrients; sometimes referred to as the particulate SOM (Cambardella and Elliot, 1992)	Soil respiration, microbial and enzyme activity; very sensitive to cultivation and management practices as an early indicator of changes in soil quality	Storing and cycling nutrients, sustaining biomass production
Mineralizable carbon and nitrogen	Represents the mineralizable SOM during a growing season and comprises about 25% of the SOM; studies have shown this fraction to be associated with macroaggregates	Interface between autotrophic organisms that synthesize complex compounds from inorganic constituents and heterotrophic organisms that decompose organic compounds; sensitive to cropping system and management; the ratio of mineralizable C and N to the total is used as a measure of quality	Storing and supplying nitrogen, biomass production

TABLE 6.5
SOM Attributes as Related to Soil Quality (continued)

SOM attribute	Attribute properties	Related soil processes	Related soil function
Microbial biomass	Represents the soil microfauna part of SOM; accounts for about 1–3% of soil organic C and 2–6% of soil organic N; is sensitive to toxicities in soil	Nitrogen mineralization (can be over 50% of plant needs); very quickly responds to soil management and perturbations; ratios of microbial biomass to total organic C may be useful as an indicator of changes in SOM	Nutrient cycling
Carbohydrates	Represent from 5 to 20% of the total soil organic C; originate from plants, animals, and microorganisms; are mostly a mixture of complex polysaccharides that are decomposed to monosaccharides	Associated with the stabilization of soil aggregates; readily available energy source for microorganisms; sensitive to management practices	Regulating and partitioning water, nutrient cycling
Soil enzymes	Molecular subsystems of microorganisms; they are proteins from plants and soil organisms	Catalysts for biochemical reactions; critical indicators of SOM quality because they control nutrient release for plant and microbial growth	Nutrient cycling

Adapted from Gregorich et al., 1994.

variables is changed, the quasi-equilibrium is disturbed and the system undergoes adjustment to a new level dictated by the impact of residues.

A major problem is that many times the data available are only for the surface horizon, and those data may be limited to SOM content. Because cultivation results in an increase in bulk density over the virgin state, those data are needed by horizons within a profile to determine the change in status of SOM in the total soil. Emphasis on the changes taking place in only the surface horizon may be deceiving. Another factor that can be important on some sites is the loss of SOM due to wind and/or water erosion. Many research plots are on stable landscape positions and are not subject to water erosion. Much of the data in the literature is for the plow layer, and future studies should emphasize sampling to greater depth in soil profiles. Models developed to predict SOM dynamics seem to have limited areas of application.

The definition of soil quality is still evolving. SOM, especially that in the surface horizon, is an important component of the current concept of soil quality. In addition to the concentration data in surface horizons, other parameters that characterize SOM need to be developed and evaluated. The summation of soil properties present in soil profiles and their distribution on landscapes are also important components of soil quality. Soil scientists have a challenging task in attempting to integrate all these factors and their relative contribution to soil quality.

REFERENCES

Acton, D.F. and L.J. Gregorich. 1995. Understanding soil health. In D.F. Acton and L.J. Gregorich (eds.). *The Health of Our Soils: Toward Sustainable Agriculture in Canada.* Centre for Land and Biological Resources Research, Research Branch, Agriculture and Agri-Food Canada, Ottawa, pp. 5–10.

Allison, F.E. 1973. *Soil Organic Matter and Its Role in Crop Production.* Elsevier Scientific, New York.

Anderson, M.A. 1949. Some Physical and Chemical Properties of Six Virgin and Six Cultivated Iowa Soils. Ph.D. dissertation. Iowa State College, Ames.

Anderson, M.A. and G.M. Browning. 1949. Some physical and chemical properties of six virgin and six cultivated Iowa soils. *Soil Sci. Soc. Am. Proc.* 14:370–374.

Bartholomew, W. 1957. Maintaining organic matter. In The 1957 Yearbook of Agriculture — Soil. USDA, Washington, D.C., pp. 245–252.

Bartholomew, W. and D. Kirkham. 1960. Mathematical descriptions and interpretations of culture induced soil nitrogen changes. In *Transactions of 7th International Congress of Soil Science,* Vol. 2. International Congress of Soil Science, International Society of Soil Science, Madison, WI, pp. 471–477.

Bauer, A. and A.L. Black. 1981. Soil carbon, nitrogen, and bulk density comparisons in two cropland tillage systems after 25 years and in virgin grassland. *Soil Sci. Soc. Am. J.* 45:1166–1170.

Black, A.L. 1973. Soil property changes associated with crop residue management in a wheat–fallow rotation. *Soil Sci. Soc. Am. Proc.* 37:943–946.

Blank, R.R. and M.A. Fosberg. 1989. Cultivated and adjacent virgin soils in north central South Dakota. I. Chemical and physical comparisons. *Soil Sci. Soc. Am. J.* 53: 1484–1490.

Broll, G. 1993. The role of soil organic matter in urban ecosystems. In H J.P. Eijsackers and T. Hamers (eds.). *Integrated Soil and Sediment Research: A Basis for Proper Protection.* Kluwer Academic, Norwell, MA, pp. 85–86

Brown, J.R. 1989. A Listing of Long-Term Field Research Sites. Department of Agronomy Miscellaneous Publication No. 89–04. Missouri Agricultural Experiment Station, Columbia.

Burke, I.C., C.M. Yonker, W.J. Parton, C.V. Cole, K. Flach, and D.S. Schimel. 1989. Texture, climate, and cultivation effects on soil organic matter content in U.S. grassland soils. *Soil Sci. Soc. Am. J.* 53:800–805.

Buyanovsky, G.A., J.R. Brown, and G.H. Wagner. 1996. Sanborn Field: effect of 100 years of cropping on soil parameters influencing productivity. In E.A. Paul, K. Paustian, E.T. Elliott, and C.V. Cole (eds.). *Soil Organic Matter in Temperate Agroecosystems: Long-Term Experiments in North America.* CRC Press, Boca Raton, FL, chap. 15.

Cambardella, C.A. and E.T. Elliot. 1992. Particulate soil organic-matter changes across a grassland cultivation sequence. *Soil Sci. Soc. Am. J.* 56:777–783.

Campbell, C.A. 1978. Soil organic carbon, nitrogen and fertility. In N. Schnitzer and S.U. Kahn (eds.). *Soil Organic Matter.* Developments in Soil Science 8. Elsevier Scientific, Amsterdam, pp. 173–271.

Campbell, C.A. and W. Souster. 1982. Loss of organic matter and potentially mineralizable nitrogen from Saskatchewan soils due to cropping. *Can. J. Soil Sci.* 62:651–656.

Campbell, C.A., R.P. Zentner, H.H. Janzen, and K.E. Bowren. 1990. Crop Rotation Studies on the Canadian Prairies. Agricultural Canada Publication 1841/E. Research Branch, Agricultural Canada, Ottawa.

Campbell, C.A., M. Schnitzer, G.P. Lafond, R.P. Zentner, and J.E. Knipfel. 1991. Thirty-year crop rotations and management practices effects on soil and amino nitrogen. *Soil Sci. Soc. Am. J.* 55:739–745.

Conway-Nelson, C. 1991. Soil Variability Associated with Landscape Position in a Southwestern Missouri Prairie. Unpublished M.S. thesis. University of Missouri-Columbia.

Darmody, R.G. and L.D. Norton. 1993. Structural degradation of a prairie soil from long-term management. In A.J. Ringrose-Voase and G.S. Humphreys (eds.). *Proceedings of the IX International Working Meeting on Soil Micromorphology.* Elsevier Science, Amsterdam.

Darusman, L., R. Stone, D.A. Whitney, K.A. Janssen, and J. Long. 1991. Soil properties after twenty years of fertilization with different nitrogen sources. *Soil Sci. Soc. Am. J.* 55:1097–1100.

De Jong, E. and R.G. Kachanoski. 1988. The importance of erosion in the carbon balance of prairie soils. *Can. J. Soil Sci.* 68:111–119.

Dick, W.A., D.M. Van Doren, G.B. Triplett, and J.E. Henry. 1986. Influence of long-term tillage and rotation combinations on crop yields and selected soil parameters. II. Results obtained for a Typic Fragiudalf soil. In Res. Bull. Ohio Agric. Res. Dev. Center, Wooster, OH, p. 34.

Dodge, D.A. and H.E. Jones. 1948. The effect of long-time fertility treatments on the nitrogen and carbon content of a prairie soil. *J. Am. Soc. Agron.* 40:778–785.

Doran, J.W. 1980. Soil microbial and biochemical changes associated with reduced tillage. *Soil Sci. Soc. Am. J.* 4:765–771.

Doran, J.W. and T.B. Parkin. 1994. Defining and assessing soil quality. In J.W. Doran, D.C. Coleman, D.F. Bezdicek, and B.A. Stewart (eds.). *Defining Soil Quality for a Sus-*

tainable Environment. SSSA Special Publication No. 35. Soil Science Society of America, Madison, WI, pp. 3–21.

Dormaar, J.F. and C.W. Lindwall. 1989. Chemical differences in dark brown chernozemic Ap horizons under various conservation tillage system. *Can. J. Soil Sci.* 69:481–488.

Epstein, E. and H. Kohne. 1957. Soil aeration as affected by organic matter application. *Soil Sci. Soc. Am. Proc.* 21:585–588.

Franzmeier, D., G. Lemme, and R. Miles. 1985. Organic carbon in soils of north central United States. *Soil Sci. Soc. Am. J.* 49:702–708.

Gantzer, C.J., S.H. Anderson, A.L. Thompson, and J.R. Brown. 1991. Estimating soil erosion after 100 years of cropping on Sanborn Field. *J. Soil Water Conserv.* 45:641–644.

Giddens, J. 1957. Rate of loss of carbon from Georgia soils. *Soil Sci. Soc. Am. Proc.* 21: 513–515.

Gregorich, E.G. 1996. Soil quality: a Canadian perspective. In Proc. Soil Quality Indicators Workshop, sponsored by the New Zealand Ministry of Agriculture and Fisheries and the Lincoln Soil Quality Research Centre. Lincoln University, Christchurch, New Zealand, February 8–9.

Gregorich, E.G., M.R. Carter, D.A. Angers, C.M. Moreal, and B.H. Ellert. 1994. Towards a minimum data set to assess soil organic matter quality in agricultural soils. *Can. J. Soil Sci.* 74:367–385.

Griffith, D.R., E.J. Kladivko, J.V. Mannering, T.D. West, and S.D. Parsons. 1988. Long-term tillage and rotation effects on corn growth and yield on high and low organic matter, poorly drained soils. *Agron. J.* 80:599–605.

Guernsey, C.W., J.B. Fehrenbacher, B.W. Ray, and L.B. Miller. 1969. Corn yields, root volumes, and soil changes on the Morrow Plots. *J. Soil Water Conserv.* 24:101–104.

Haas, H.J., C.E. Evans, and E.F. Miles. 1957. Nitrogen and Carbon Changes in Great Plains Soils as Influenced by Cropping and Soil Treatments. USDA Tech. Bull. No. 1164. U.S. Government Printing Office, Washington, D.C.

Hammer, R.D., G.S. Henderson, R. Udawatta, and D.K. Brandt. 1995. Soil organic matter in the Missouri forest–prairie ecotone. In W.W. McFee and J.M. Kelly (eds.). Proc. 8th North American Forest Soils Conference. Gainesville, FL, May 10–13.

Havlin, J.L., D.E. Kissel, L.D. Maddux, M.M. Claassen, and J.H. Long. 1990. Crop rotation and tillage effects on soil organic carbon and nitrogen. *Soil Sci. Soc. Am. J.* 54: 448–452.

Hill, K.W. 1954. Wheat yields and soil fertility on the Canadian prairies after a half century of farming. *Soil Sci. Soc. Am. Proc.* 18:182–184.

Hobbs, J.A. and P.A. Brown. 1965. Effects of Cropping and Management on Nitrogen and Organic Carbon Contents of a Western Kansas Soil. Tech. Bull. No. 144. Kansas Agric. Exp. Stn., Manhattan.

Hudson, B.D. 1994. Soil organic matter and available water capacity. *J. Soil Water Conserv.* 49:189–194.

Janssen, B.H. 1984. A simple method for calculating decomposition and accumulation of "young" soil organic matter. *Plant Soil* 76:279–304.

Janzen, H.H. 1987. Soil organic matter characteristics after long-term cropping to various spring wheat rotations. *Can. J. Soil Sci.* 67:845–856.

Jenny, H. 1941. *Factors of Soil Formation.* McGraw-Hill, New York, 281 pp.

Karlen, D.L., M.J. Mausbach, J.W. Doran, R.G. Cline, R.F. Harris, and G.E. Schuman. 1997. Soil quality: a concept, definition, and framework for evaluation. *Soil Sci. Soc. Am. J.* 61:4–10.

Larson, W.E. and F.J. Pierce. 1994. The dynamics of soil quality as a measure of sustainable management. In J.W. Doran, D.C. Coleman, D.F. Bezdicek, and B.A. Stewart (eds.). *Defining Soil Quality for a Sustainable Environment.* SSSA Special Publication No. 35. Soil Science Society of America, Madison, WI, pp. 37–51.

Larson, W.E., C.E. Clapp, W.H. Pierre, and Y.B. Morachan. 1972. Effects of increasing amounts of organic residues on continuous corn. II. Organic carbon, nitrogen, phosphorus, and sulfur. *Agron. J.* 64:204–208.

Lucas, R.E., J.B. Holtman, and L. Connor. 1977. Soil carbon dynamics and cropping practices. In W. Lockeretz (ed.). *Agriculture and Energy.* Academic Press, New York, pp. 333–351.

Lynch, J.M. and E. Bragg. 1985. Microorganisms and soil aggregate stability. *Adv. Soil Sci.* 2:133–171.

Martel, Y.A. and E.A. Paul. 1974. Effect of cultivation on the organic matter of grassland soils as determined by fractionation and radiocarbon dating. *Can. J. Soil Sci.* 54: 419–426.

McCalla, T.M. and T.J. Army. 1961. Stubble mulch farming. *Adv. Agron.* 13:125–196.

McGill, W.B., C.A. Campbell, J.F. Dormaar, E.A. Paul, and D.W. Anderson. 1981. Soil organic matter losses. In Agricultural Land. Our Disappearing Heritage — A Symposium. Proc. 18th Annual Alberta Soil Science Workshop. Alberta, Agriculture, Edmonton, Alberta, pp. 72–133.

Meints, V.W. and G.A. Peterson. 1977. The influence of cultivation on the distribution of nitrogen in soils of the Ustoll suborder. *Soil Sci.* 124(6):334–342.

Melsted, S. 1954. New concepts of management of Corn Belt soils. *Adv. Agron.* 6:121–142.

Mitchell, C.C., R.L. Westerman, J.R. Brown, and T.R. Peck. 1991. Overview of long-term agronomic research. *Agron. J.* 83:24–29.

Moldenhauer, W., W. Wischmeier, and D. Parker. 1967. The influence of crop management on runoff, erosion, and soil properties of a Marshall silty clay loam. *Soil Sci. Soc. Am. Proc.* 31:541–546.

National Research Council. 1993. *Soil and Water Quality: An Agenda for Agriculture.* National Academy Press, Washington, D.C., 516 pp.

Nelson, D.W. and L.E. Sommers. 1982. Total carbon, organic carbon, and organic matter. In A.L. Page, R.H. Miller, and D.R. Keeney (eds.). *Methods of Soil Analysis,* Part 2, 2nd ed. American Society of Agronomy, Madison, WI.

Odell, R.T., W. Walker, L. Boone, and M. Oldham. 1982. The Morrow Plots: A Century of Learning. Agric. Exp. Stn. Bull. 775. University of Illinois, Urbana.

Odell, R.T., S. Melsted, and W.M. Walker. 1984. Changes in organic carbon and nitrogen of Morrow Plot soils under different treatments, 1904–1973. *Soil Sci.* 137:160–171.

Olson, K.R., L.D. Norton, T.E. Fenton, and R. Lal. 1994. Quantification of soil loss from eroded soil phases. *J. Soil Water Conserv.* 49(6):591–596.

Overson, M.M. 1966. Conservation of soil nitrogen in a wheat summer fallow farming practice. *Agron. J.* 58:444–447.

Parton, W., D. Anderson, C. Cole, and J. Stewart. 1983a. Simulation of soil organic matter formation and mineralization in semiarid agroecosystems. In R.O. Lowrance (ed.). *Nutrient Cycling in Agricultural Ecosystems.* Special Publication 23. Agric. Exp. Stn., University of Georgia, Athens.

Parton, W.J., J. Persson, and D. Anderson. 1983b. Simulation of organic matter changes in Swedish soil. In W.K. Lauenroth (ed.). *Analysis of Ecological Systems — State of-the-Art in Ecological Modeling.* Elsevier Scientific, New York, pp. 511–516.

Peevy, W., F. Smith, and P. Brown. 1940. Effects of rotational and manurial treatments for twenty years on the organic matter, nitrogen, and phosphorus contents of Clarion and Webster soils. *J. Am. Soc. Agron.* 32:739–753.

Rasmussen, P.E. and H.P. Collins. 1991. Long-term impacts of tillage, fertilizer, and crop residue on soil organic matter in temperate semiarid regions. *Adv. Agron.* 45:93–134.

Rasmussen, P.E., R.R. Allmaras, C.R. Rohde, and N.C. Roager. 1980. Crop residue influences on soil carbon and nitrogen in a wheat–fallow system. *Soil Sci. Soc. Am. J.* 44:596–600.

Rice, C.W., M.S. Smith, and R.L. Blevins. 1986. Soil nitrogen availability after long-term continuous no-tillage and conventional tillage corn production. *Soil Sci. Soc. Am J.* 50:1206–1210.

Ridley, A.O. and R.A. Hedlin. 1968. Soil organic matter and crop yields as influenced by frequency of summer fallowing. *Can. J. Soil Sci.* 48:315–322.

Romig, D.E., M.J. Garlynd, R.F. Harris, and K. McSweeney. 1995. How farmers assess soil health and quality. *J. Soil Water Conserv.* 50:229–236.

Schimel, D.S., D.C. Coleman, and K.A. Horton. 1985. Soil organic matter dynamics in paired rangeland and cropland toposequences in North Dakota. *Geoderma* 36:201–214.

Schimel, D.S., W.J. Parton, R. Adamsen, R. Woodmansee, R. Senft, and M. Stillwell. 1986. The role of cattle in the volatile loss of nitrogen from a shortgrass steppe. *Biogeochemistry* 2:39–52.

Schnitzer, M. 1991. Soil Organic Matter and Soil Quality. Tech. Bull. Agric. Canada No. 1991-1E. Ottawa, pp. 33–49.

Schnitzer, M. and S.U. Khan. 1972. *Humic Substances in the Environment.* Marcel Dekker, New York.

Scrivner, C.L. and D.T. Cooper. 1985. Organic Carbon in Missouri Soils. Missouri Agric. Exp. Stn. Res. Bull. 1055. University of Missouri, Columbia.

Shrader, W. 1953. Soil Factors Affecting Crop Yields on Clarion-Webster Soils. Ph.D. dissertation. Iowa State College, Ames.

Soil Survey Staff. 1993. Soil Survey Manual. USDA Handbook 18. U.S. Government Printing Office, Washington, D.C.

Stauffer, R.S., R. Muckenhirn, and R.T. Odell. 1940. Organic carbon, pH, and aggregation of the soil of the Morrow Plots as affected by type of cropping and manurial addition. *J. Am. Soc. Agron.* 32:819–832.

Steiner, R.A. and R.W. Herdt. 1993. *A Global Directory of Long-Term Agronomic Experiments,* Vol, 1. The Rockefeller Foundation, New York.

Stevenson, F.J. 1956. Effects of some long-time rotations on the amino-acid composition of the soil. *Soil Sci. Soc. Am. Proc.* 20:204–208.

Stevenson, F.J. 1982. *Humus Chemistry — Genesis, Composition, Reaction.* John Wiley & Sons, New York.

Tiessen, H., J. Stewart, and J. Bettany. 1982. Cultivation effects on the amounts and concentrations of carbon, nitrogen and phosphorus in grassland soils. *Agron. J.* 74:831–835.

Tisdall, J.M. and J.M. Oades. 1982. Organic matter and water-table aggregates in soils. *J. Soil Sci.* 33:141–163.

Troeh, F.R. and L.M. Thompson. 1993. *Soils and Soil Fertility,* 5th ed. Oxford University Press, New York, 462 pp.

Unger, P.W. 1968. Soil organic matter and nitrogen changes during 24 years of dryland wheat tillage and cropping practices. *Soil Sci. Soc. Am. Proc.* 32:426–429.

Van Bavel, C. and F. Schaller. 1950. Soil aggregation, organic matter, and yields in a long-time experiment as affected by crop management. *Soil Sci. Soc. Am. Proc.* 15:399–408.

Waksman, S.A. 1942. The microbiologist and soil organic matter. *Soil Sci. Soc. Am. Proc.* 7:16–21.

Whiteside, E.P. and R.S. Smith. 1941. Soil changes associated with tillage and cropping in humid areas of the United States. *Agron. J.* 33:765–777.

Wood, C.W., G.A. Peterson, and D.G. Westfall. 1990. Greater crop management intensity increases soil quality. *Better Crops with Plant Food* 74(3):20–22.

Yonker, C.M., D. Schimel, E. Paroussis, and R.D. Heil. 1988. Patterns of organic carbon accumulation in a semiarid shortgrass steppe, Colorado. *Soil Sci. Soc. Am. J.* 52: 478–483.

Young, R.A., J.C. Zubriski, and E.B. Normum. 1960. Influence of long-time fertility management practices on chemical and physical properties of a Fargo clay. *Soil Sci. Soc. Am. J.* 24:124–128.

7 Use of Winter Cover Crops to Conserve Soil and Water Quality in the San Luis Valley of South Central Colorado

J.A. Delgado, R.T. Sparks, R.F. Follett,
J.L. Sharkoff, and R.R. Riggenbach

INTRODUCTION

Potential wind erosion in the San Luis Valley (SLV) of south central Colorado is greater in the spring, especially after harvesting crops that leave little crop residue in the fall. At four sites, potential soil erosion was reduced with the use of winter cover grains (WCGs) such as wheat (*Triticum aestivum* L.) and rye (*Secale cereale* L.). Maximum soil conservation occurs with early planting of WCG, which increases dry matter production and N uptake in the fall, before growth is drastically reduced by the winter. With less effectiveness, late planting will also decrease erosion. Ungrazed early-planted winter cover rye reduced the nitrate (NO_3^-–N) available for leaching in the top 1.5 m (5 ft), from 206 to 31 kg NO_3^-–N ha^{-1} (184 to 28 lb acre^{-1}) and returned 7.5 mg ha^{-1} (3.4 t acre^{-1}) of dry plant residue into the soil. The WCG responded to spatial variability in the field, recovering higher amounts of soil NO_3^-–N from areas that contain higher residual soil NO_3^-–N. These WCGs can be used as a management tool to decrease soil erosion, return organic carbon to the soil, and scavenge soil N. They can conserve soil and water quality even after

---

1-57444-100-0/99/$0.00+$.50
© 1999 by Soil and Water Conservation Society

livestock grazing during the winter or early spring. Since these WCGs are killed in the spring and incorporated into the surface soil, improvement in the synchronization of and accountability for the N release from mineralization of WCG residue as a source of N for the following crop is also needed to ensure that WCGs are being used efficiently.

In the SLV of south central Colorado, NO_3^-–N leaching and wind erosion have been identified as events that can affect soil and water quality (USDA-SCS, 1973; Eddy-Miller, 1993). The SLV, with an average elevation of 2348 m (7700 ft), is a high-altitude intermountain desert valley (Edelmann and Buckles, 1984). With an average precipitation of 168 mm (6.6 in.) in the Alamosa County (USDA-SCS, 1973), irrigation is used extensively in the SLV. Domestic well water NO_3^-–N concentrations have been reported as high as 37 mg NO_3^-–N L^{-1} (37 ppm) in the central part of the SLV (Austin, 1993). Irrigation and National Water Quality Assessment well NO_3^-–N concentrations have been reported as high as 70 mg NO_3^-–N L^{-1} (70 ppm) near Center, Colorado (Eddy- Miller, 1993; Stogner, 1996). These NO_3^--N concentrations are higher than the allowable limit of 10 mg of NO_3^-–N L^{-1} (10 ppm) for human consumption (U.S. EPA, 1989). Although the SLV has a variety of soils, most of them are a coarse sandy texture over a coarse-textured substratum (USDA-SCS, 1973). The combination of N fertilizer use, shallow groundwater supplies, sandy soils, and irrigated agricultural systems contributes to this elevated NO_3^-–N in groundwater (Eddy-Miller, 1993; Stogner, 1996). Since the amount of crop residue after a potato (*Solanum tuberosum* L.) or vegetable harvest is low, potential wind erosion of this coarse-textured soil can be higher during the spring, when winds are stronger (USDA-SCS, 1973). Our objective was to assess the potential of WCG in the SLV to (1) conserve water quality by scavenging residual soil NO_3^-–N that may be available for leaching, (2) conserve soil quality by reducing wind erosion and by returning organic carbon and N to the surface soil, (3) conserve soil and water quality under different times of planting, and (4) conserve water and soil quality under grazed WCG.

Use of Winter Cover Crops as an N Management Tool

Winter cover crops (WCCs) can be useful tools in protecting soil and water quality (Lal et al., 1991). To some extent, there is flexibility in how to use WCCs as an N management tool. Winter cover legumes (WCLs) fix atmospheric N_2 that is added into the system, while WCGs are reported to scavenge NO_3^-–N more efficiently from the soil system than WCLs (Jones, 1942; Chapman et al., 1949; Walker et al., 1956; Jones et al., 1977; Nielsen and Jensen, 1985; Groffman et al., 1987; Meisinger et al., 1991; Shipley et al., 1992). In soils with relatively high levels of residual soil NO_3^-–N, WCGs were observed to take up to four times the amount of residual soil NO_3^-–N that legumes took up (Shipley et al., 1992). Although WCGs have been reported as more efficient scavengers of soil NO_3^-–N, in some studies WCLs have been successful scavengers of NO_3^-–N (Schertz and Miller, 1972; Morris et al., 1986), since applications of N fertilizer or high residual soil NO_3^-–N levels decrease nodulation and N_2 fixation (Viets and Crawford, 1950; Allos and Bartholomew,

1955; Schertz and Miller, 1972; Streeter, 1985). The WCL alfalfa (*Medicago sativa* L.) in its second year reduced NO_3^--N levels that were accumulated in a 1-m (3.3-ft) soil profile under continuous corn (Schertz and Miller, 1972). Morris et al. (1986) reported that WCGs scavenge N fertilizer more efficiently from the soil system early in the spring; however, later in the spring, the WCL recovered the same or a higher amount of N fertilizer than WCG. Apparently, rate and stage of growth with respect to local climate as well as residual soil NO_3^--N need to be considered when using WCC as an N management tool.

When the WCCs are killed, different factors, such as crop C/N ratios, amount of inorganic soil N, soil type, weather, and management of the crop residue, affect the intensity of the mineralization of N from the WCC residue to the following crop. Pink et al. (1945, 1948) reported greater N immobilization with crops such as WCG that in general have C/N ratios greater than 35. Frequently in the initial mineralization of the WCG, soil N is immobilized (Mitchell and Teel, 1977; Frye et al., 1985; Holderbaum et al., 1990; Doran and Smith, 1991; Decker et al., 1994). Since WCLs have on average C/N ratios lower than 20 and as low as 8, they have a greater N fertilizer equivalency (Doran and Smith, 1991).

These differences in N mineralization/immobilization between the WCG and WCL are a function of the C/N ratios and are more pronounced when compared under relatively low residual soil inorganic N. When WCLs are grown under low soil inorganic N, they can fix atmospheric N_2, while the N taken up by the WCG under low soil N levels will be diluted by photosynthesis into higher C/N ratios. Under relatively high residual soil NO_3^--N conditions, the C/N ratio of the WCG is significantly reduced. It has been reported that higher concentrations of NO_3^--N at the root surface of WCGs (wheat and rye) increase the rate of NO_3^--N uptake and the net total N uptake, which can reduce the C/N ratios (Baker and Tucker, 1971; Hojjati et al., 1972; Mugwira et al., 1980; Evanlyo, 1991). This lower C/N ratio increases N mineralization from the WCG and the availability of N to the next crop (Wagger, 1989a; Evanlyo, 1991; Shipley et al., 1992).

Huntington et al. (1985) found that when there is poor synchronization between the mineralization of N from the WCC and N uptake by the following crops, problems can occur. It is possible that poor synchronization may increase the residual soil NO_3^--N of the following crop at the end of the growing season, increasing the possibility of leaching (Meisinger et al., 1991; Torbert and Reeves, 1991). To some extent, there is flexibility as to how to manage the WCC residue. An early kill with herbicides could help the synchronization with a faster release of N from the WCC, especially for WCG if it is not allowed to dilute the N taken up (higher C/N ratios) with an early spring growth (Huntington et al., 1985; Wagger, 1989a; Evanlyo, 1991). One disadvantage of the early kill is the exposure of soil to potential soil erosion (Evanlyo, 1991). Incorporating the WCC residue after kill could accelerate the release of N (Wilson and Hargrove, 1986). Sometimes there is a synergistic response in yields by the following crop that cannot be explained by the release of mineralized N, especially in soils with low soil organic matter and available water content (Huntington et al., 1985; Decker et al., 1994). Some of these responses may vary from year to year due to environmental conditions (Wagger, 1989a,b).

Use of Winter Cover Grains as NO_3^--N Scavenger Crops to Conserve Leachable Nutrients

The WCGs contribute significantly to the conservation of water quality by recovering NO_3^--N left below the root zone of previous crops that otherwise could be susceptible to leaching (Holderbaum et al., 1990; Meisinger et al., 1991; Brinsfield and Staver, 1991; Decker et al., 1994; McCracken et al., 1995). The date of WCG planting is important in the conservation of leachable nutrients. If the leaching events occur during the fall, then rapid establishment and growth of the WCG are needed in the fall and not in the spring, when it will be too late to recover the NO_3^--N (Meisinger et al., 1991). Fowler (1982) found in the northern part of the American Great Plains that the fall planting date affects the rate of growth of wheat and rye. He reported that delaying planting between August 1 and October 15 decreased shoot dry weight of both crops. He found that during the fall, winter rye accumulated dry matter at twice the rate of wheat in the agricultural area of Saskatchewan.

Use of Winter Cover Crops to Decrease Soil Erosion

Soil quality is conserved with WCC due to the reduction in potential soil erosion by wind and water, since unsheltered soil surfaces will be more susceptible to erosion (Woodruff and Siddoway, 1965; Frye et al., 1985; Holderbaum et al., 1990; Bilbro, 1991; Langdale et al., 1991; Decker et al., 1994). Bilbro (1991) reported that winter rye planted in rows 12.7 cm (5 in.) apart reduced potential wind erosion below 11.2 mg ha^{-1} yr^{-1} (5 t acre^{-1}) in a sandy loam in the Great Plains. The use of WCC decreased rainfall soil erosion from conventional tillage cotton in Holly Springs, Mississippi, from 74 to 20 mg ha^{-1} yr^{-1} (33 to 9 t acre^{-1} yr^{-1}) and no-till losses to below 11 mg ha^{-1} yr^{-1} (5 t acre^{-1} yr^{-1}) (Mutchler and McDowell, 1990).

Use of Winter Cover Grains as a Source of Grazing

Winter wheat is also used for grazing in the south coastal plains and in other countries (Redmon et al., 1995). Cattle grazed on winter wheat gain weight during the winter; however there is a risk of death due to bloating (Clay, 1973a,b; Johnson, 1973; Bartley et al., 1975; Horn et al., 1977; Mader et al., 1983). Some SLV farmers are beginning to use WCG as a source of winter or early spring pasture for livestock before the fields are cropped again.

MATERIALS AND METHODS

Study 1: Effect of Early Planted Winter Cover Rye on Residual Soil NO_3^--N

Early-planted WCGs were defined as plantings completed before the end of the second week of September, while late-planted WCGs were defined as plantings done after the beginning of the third week of September. Study 1 was established early in the fall of 1994 at about 10 km (6 mi) north of Center, Colorado. Four plots 20.9 m² (225 ft²) each were established in the northern part of a center pivot irrigation sprinkler on a San Luis fine-loamy over sandy or sandy skeletal, mixed Aquic Natrargid (Table 7.1). Four additional plots were also established in the southern half of the same circle, but in a Kerber-like coarse-loamy, mixed, frigid Aquic Natrargid. Like polypedons for both soils were identified using field soil survey techniques (Sharkoff et al., 1995). Two transponders were placed permanently in both plots so that soils within these plots could be sampled during 1994 and 1995. Plots were located under the same sprinkler span to minimize any possible spatial variability due to sprinkler nozzles span to span. Both sets of plots were irrigated with the same rate and frequency. Planting, harvesting, cultivation, N fertilizer application, and other agricultural practices were similar for the entire circle.

Soil samples were collected in each plot during the fall of 1994 after lettuce (*Lactuca sativa* L.) harvest, during the spring of 1995 after winter cover rye kill, and during the fall of 1995 after potato harvest. At each time, soils were sampled in 0.3-m (1-ft) intervals to 1.5 m (5 ft). Plant samples for rye and potatoes were collected by harvesting 0.4 m² (4 ft²) in each plot. Aboveground rye was collected before spring kill in 1995, and potato vines, roots, and tubers were collected in the fall of 1995.

Study 2: Effect of Time of Planting on Biomass Production and N Uptake of a Grazed Winter Wheat

Study 2 was established on another center pivot located 0.8 km (0.5 mi) west of Study 1. The study area was located on a Kerber-like soil (Table 7.1) that was planted to lettuce during the spring of 1995. After an early harvest of lettuce in the north half of the circle, on August 23, 1996, winter wheat was planted at the rate of 56 kg ha⁻¹ (50 lb acre⁻¹). The south part of the circle was harvested later and winter wheat was planted on September 26, 1995 at the rate of 56 kg ha–¹ (50 lb acre⁻¹). WCCs were irrigated for cover crop establishment. Winter cover wheat was grazed by horses from late November to the middle of February 1996. Aboveground and belowground plant biomass was collected on December 20, 1995; February 15, 1996; and March 11, 1996. At each sampling date, four random 0.4-m² (4-ft²) samples were harvested in both halves of the circle.

Study 3: Biomass Production and N Uptake on a Late-Planted Winter Cover Rye

Study 3 was established late in the fall of 1995 at a center pivot located about 23 km (14 mi) southeast of Center, Colorado, and about 13 km (8 mi) from Studies 1 and 2. Four plots 20.9 m² (225 ft²) each were established on a Gunbarrel mixed, frigid Typic Psammaquent (Table 7.1), after potato harvest. Aboveground and belowground rye biomass was collected on December 20, 1995; February 15, 1996; and March 11, 1996. At each sampling date, samples for rye were collected by harvesting 0.4 m² (4 ft²) in each plot.

TABLE 7.1
Description of Sites

Study	Location (near)	Irrigation method[a]	Period	Soil type	Soil texture	Crop rotation	Time of planting
1	Center, CO	CP (south half)	1994–95	Kerber-like	Loamy sand	Lettuce–winter rye–potato	Early
	Center, CO	CP (north half)	1994–95	San Luis	Sandy loam	Lettuce–winter rye–potato	Early
2	Center, CO	CP (south half)	1995–96	Kerber-like	Loamy sand	Lettuce–winter wheat[b]–potato	Late
	Center, CO	CP (north half)	1995–96	Kerber-like	Loamy sand	Lettuce–winter wheat[b]–potato	Early
3	Center, CO	CP (whole circle)	1995–96	Gunbarrel	Loamy sand	Potato–winter rye–potato	Late
4	Blanca, CO	Furrow	1995–96	Shawa	Loam	Spinach–winter rye[c]	Early

[a] CP = center pivot sprinkler system.
[b] Grazed by livestock (horses).
[c] Grazed by livestock (steers and nonpregnant heifers).

Study 4: Biomass Production and N Uptake on an Early-Planted Winter Cover Rye

Study 4 was established about 3 km (2 mi) south of Blanca, Colorado, on an early-planted winter rye field, after spinach (*Spinacia oleracea* L.) harvesting. Four plots 20.9 m² (225 ft²) each were established at a surface-irrigated field in a Shawa fine-loamy, mixed, frigid Pachic Haploboroll to monitor the growth and N uptake of winter cover rye. Aboveground and belowground plant biomass was collected on December 20, 1995; February 15, 1996; and March 11, 1996. Since the winter cover rye was to be grazed, a livestock exclosure was built. At each sampling date, samples for rye were collected inside the exclosure by harvesting 0.4 m² (4 ft²) in each plot.

Effect of Winter Cover Crops on Potential Wind Erosion

Estimates of potential soil loss by wind erosion were made based on the wind erosion equation for each study (Woodruff and Siddoway, 1965; Skidmore et al., 1970). We estimated potential soil erosion using the methods described by Bondy et al. (1980) and USDA-SCS (1988). We determined the base erodibility index (I) of 300 mg ha^{-1} (134 t acre^{-1}) for Kerber and Gunbarrel soils, 193 mg ha^{-1} (86 t acre^{-1}) for San Luis, and 125 mg ha^{-1} (56 t acre^{-1}) for Shawa. The climatic factor (C) used was 80. The unsheltered length (L) used was 305 m (1000 ft) for surface-irrigated fields and 609 m (2000 ft) for center-pivot-irrigated fields. The erosive wind energy, equivalent flat small-grain residue, and ridge roughness values were estimated for each 2-week period from March 15 through June 15. June 15 to March 15 was considered a noncritical period. The amount of WCG biomass and percentage of area covered were taken into account for each study. The average annual erosive wind energy for the Alamosa, Colorado, area was used. The effects of different management practices such as overwinter ridges, time of bedding, and time of planting were taken into account. Since no adjustments were made for irrigation and soil crusting, these potential soil erosion calculations should overestimate the magnitude of the potential soil erosion. For the objective of a comparison between the effect of WCG and fallow, the estimates should be able to separate the main effect of the two practices. Potential wind erosion was calculated for an annual period and for the most susceptible period from March 15 to May 1.

Procedures for Collection and Measurements of Soil Inorganic Ammonium (NH_4^+–N) and NO_3^-–N and Plant Biomass and Its C/N Content

Soils collected from each 0.3-m (1-ft) depth increment were air-dried and sieved through a 2-mm (0.08-in.) sieve. The percentage weight of the coarse fragments (%RFW) was calculated with Equation 7.1. Equation 7.2 was used to calculated the percent coarse fragments by volume (%CFV), where BD was soil bulk density estimated from texture as described by the USDA-SCS National Agronomy Manual (1988). Sieved samples were extracted with 2 N KCl. The NO_3^-–N and NH_4^+–N contents of the extracts were determined colorimetrically by automated flow injection analysis. The %CFV was used to adjust the NO_3^-–N and NH_4^+–N to reflect actual field values.

$$\%RFW = \frac{\text{Wt. coarse fragments} > 2\ \text{mm}}{\text{Wt. whole sample}} \times 100 \qquad (7.1)$$

$$\%CFV = \left[\left(\frac{\%RFW}{2.65} \right) + \left(\frac{100 - \%RFW}{BD} + \frac{\%RFW}{2.65} \right) \right] \times 100 \qquad (7.2)$$

Plant samples were dried at 55°C (131°F), ground, and analyzed for C and N content by automated combustion using a Carlo Erba automated C/N analyzer. Statistical analyses were performed using the SAS analysis of variance GLM procedure (SAS Institute, 1988). Mean separation was conducted with the SAS least significant difference mean procedure. To compare the effect of WCCs on potential soil erosion, a paired t-test procedure was used.

RESULTS AND DISCUSSION

Use of Winter Cover Crops to Scavenge Residual Soil NO_3^-–N (Study 1)

In the fall of 1994, after the lettuce crop harvest, the residual 206 kg NO_3^-–N ha^{-1} (184 lb acre^{-1}) retained in the top 1.5 m (5 ft) of the finer textured San Luis sandy loam was higher than the 123 kg NO_3^-–N ha^{-1} (110 lb acre^{-1}) in the Kerber loamy sand ($P < 0.05$; Figure 7.1a). By spring 1995, the total NO_3^-–N for the top 1.5 m (5 ft) decreased significantly to about 30 kg NO_3^-–N ha^{-1} (27 lb acre^{-1}) ($P < 0.001$; Figure 7.1b). However, with the next cropping season after potato harvest, residual soil NO_3^-–N increased again in both soils to 204 and 172 kg NO_3^-–N ha^{-1} (182 and 154 lb acre^{-1}) for the sandy loam and loamy sand, respectively ($P < 0.05$; Figure 7.1c). Although a higher residual soil NO_3^-–N was found in the sandy loam (204 kg N ha^{-1} [182 lb acre^{-1}]) when compared to the loamy sand (172 kg N ha^{-1} [154 lb acre^{-1}]), differences were not significant. There was higher variability for the fall 1995 residual soil NO_3^-–N measurements (coefficient of variation = 57) when compared to the measurements taken during fall 1994 and spring 1995 (coefficient of variation <10).

Dry matter production and N uptake by aboveground rye biomass in the sandy loam were higher than in the loamy sand ($P < 0.05$; Table 7.2). In Studies 3 and 4 at spring kill, approximately 25% of the total N uptake and about 30% of the total dry biomass were in the belowground compartment. Assuming that 25% of the total N in rye was belowground, total N uptake for the rye plants grown in the sandy loam was estimated to be 179 kg N ha^{-1} (160 lb acre^{-1}). This is consistent with the reduction in residual soil NO_3^-–N of 176 kg N ha^{-1} (157 lb acre^{-1}) that was observed from fall 1994 to spring 1995 (Figure 7.1a and b and Table 7.2). We estimated 91

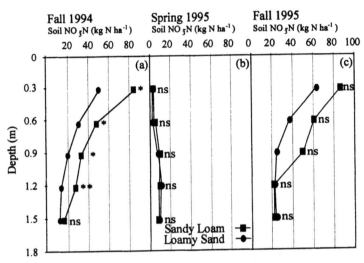

FIGURE 7.1   Study 1: Residual soil NO_3^--N in a sandy loam and loamy sand under the same irrigation, N fertilization, and agricultural management practices after lettuce harvest (a), winter rye kill (b), and potato harvest (c). Within a time period and depth, ** means that residual soil NO_3^--N in a sandy loam and loamy sand is significantly different at least significant difference 0.05 level, * means different at least significant difference 0.10 level, and ns means no significant difference.

kg N ha^{-1} (81 lb acre^{-1}) as the total N uptake by rye grown in the loamy sand. This is consistent with the reduction in residual soil NO_3^--N of 93 kg N ha^{-1} (83 lb acre^{-1}) observed for the loamy sand (Figure 7.1a and b and Table 7.2). The NH_4^+-N content for the top 0.3 m (1 ft) was <10 kg NH_4^+-N ha^{-1} (9 lb acre^{-1}) for fall 1994, spring 1995, and fall 1995.

The sandy loam had 12.2 %CFV within the 0- to 0.9-m (0- to 3-ft) soil depth, while the loamy sand had 15.9 %CFV ($P < 0.05$). The spatial variability in residual soil NO_3^--N was negatively correlated with %CFV ($P < 0.05$). This negative correlation is probably a combined effect of %CFV and soil texture. An r of -0.71 suggests that under the same management practices, such as N fertilization and irrigation, a sandy loam will retain a higher amount of residual soil NO_3^--N following harvest than will a coarser loamy sand (Figure 7.1). This assumes that losses of N from the coarser loamy sand were higher, probably due to higher NO_3^--N leaching (not measured). These data suggest that management practices such as precision agriculture that match irrigation and N application to spatial variability of coarse fragments and texture can be useful tools to reduce NO_3^--N losses for a lettuce–WCG–potato cropping sequence. Redulla et al. (1995) reported that in precision farming, variable application of N fertilizer under a center-pivot-irrigated cornfield increased N fertilizer use efficiency. Independent of management practices, rye as a WCG has the potential to conserve water quality under this spatial variability due to coarse fragments and soil texture, by recovering higher amounts of NO_3^--N in the areas with the higher residual soil NO_3^--N (Figure 7.1a and b).

TABLE 7.2

Effect of Spatial Variability Due to Soil Type on Dry Matter, N Uptake, and C/N Ratios of Winter Cover Rye and Potato Grown Under the Same Management Practices and Same Irrigation Center Pivot Sprinkler System

Center pivot[a]	Soil	Crop	Parts	Dry matter (mg ha^{-1})	N uptake (kg N ha^{-1})	C/N ratio
South half	Loamy sand	Rye	Shoots	3.2[b]	68[b]	17.7[b]
North half	Sandy loam	Rye	Shoots	5.2[b]	134[b]	14.7[b]
LSD 0.05				1.2	19	2.4
South half	Loamy sand	Potato	Stem and leaves	1.8[b]	58[b]	10.4[c]
North half	Sandy loam	Potato	Stem and leaves	1.4[b]	44[b]	10.9[c]
LSD 0.05				0.3	12	1.4
South half	Loamy sand	Potato	Roots	0.3[c]	4.3[c]	19.2[c]
North half	Sandy loam	Potato	Roots	0.2[c]	3.1[c]	20.9[c]
LSD 0.05				0.1	1.4	1.6
South half	Loamy sand	Potato	Tubers	9.1[c]	131[c]	29.2[b]
North half	Sandy loam	Potato	Tubers	7.4[c]	125[c]	22.3[b]
LSD 0.05				1.9	35	3.5

[a] LSD = least significant difference at 0.05 level.

[b] Significant differences between crop parts grown on the loamy sand and sandy loam.

[c] No significant differences between crop parts grown on the loamy sand and sandy loam.

We found spatial variability in winter rye N uptake and dry biomass production due to coarse fragments and soil texture. This higher biomass production and N uptake by rye grown in the sandy loam was a response to the higher residual soil NO_3^--N found in fall 1994 ($P < 0.05$) (Figure 7.1a and b and Table 7.2). Since rye was irrigated at planting, available water would have been expected to be higher at establishment in the surface of the sandy loam than at the coarser loamy sand (not measured). This may have been another factor that contributed to a greater dry matter accumulation.

The rye grown in the sandy loam had a lower C/N ratio (14.7) when compared to the rye grown in the loamy sand (17.7) ($P < 0.05$) (Table 7.2). Apparently the C/N ratio of the rye grown in the loamy sand was not high enough to reduce N fertilizer availability during the potato season, since tuber dry matter production and N uptake were not significantly different between the sandy loam and loamy sand soils ($P < 0.05$) (Table 7.2 and Figure 7.1). This high residual soil NO_3^--N during fall 1995 agrees with findings by other researchers who found that under high residual soil NO_3^--N and C/N ratios lower than 20, there is a potential for release of N from the WCC (Doran and Smith, 1991; Evanlyo, 1991; Shipley et al., 1992).

At harvest, pits were dug on Study 4, where rye rooting depth was measured at 0.9–1.2 m (3–4 ft). It is not clear if the rye roots were as deep as 1.5 m (5 ft) in Study 1; however, in the soil profile, NO_3^--N was significantly decreased at this depth ($P < 0.05$). If rye roots were not as deep, another possible explanation may be the possibility of upward water movement by capillarity and the absorption of NO_3^--N by the deeper WCG roots. Independent of these theories, the early-planted WCG showed a potential to scavenge NO_3^--N from greater depths than shallower root crops such as lettuce, spinach, potato, and carrot (*Daucus carota*).

Use of Date of Planting as a Management ·Tool (Studies 2 to 4)

In Study 2, early planting for winter wheat increased total dry biomass production by a factor of nine and N uptake by a factor of eight ($P < 0.05$; Figure 7.2). Assuming that most of the N was available as residual soil NO_3^--N, the wheat recovered about 147 kg NO_3^--N ha^{-1} (131 lb acre^{-1}).

In Studies 3 and 4, early planting for winter rye increased total dry biomass production and N uptake by a factor of eight ($P < 0.05$) (Figure 7.3). This higher biomass production and N uptake was also a response to N, since on December 20, 1996 there were 24 and 404 kg NO_3^--N ha^{-1} (21 and 361 lb acre^{-1}) on the top 0.3 m (1 ft) of the late- and early-planted rye, respectively. Early winter rye total dry biomass production was 8 mg ha^{-1} (3.6 t acre^{-1}) and N uptake was 300 kg N ha^{-1} (268 lb acre^{-1}) (Figure 7.3). At spring kill, the C/N ratios of the early-planted rye were as low as 8.5 and 9.3 in the aboveground and underground biomass, respectively. With these low C/N ratios, it is expected that a significant amount of N would be released during the growing season of the following crop (Doran and Smith, 1991; Evanlyo, 1991). Nine percent of the N in the early-planted rye was in the NO_3^--N form, about 27 kg NO_3^--N ha^{-1} (24 lb acre^{-1}).

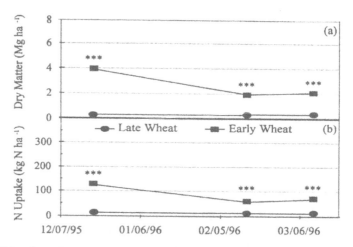

FIGURE 7.2 Study 2: total dry biomass production (a) and N uptake (b) of late- and early-planted winter cover wheat grown on a loamy sand and grazed from late November to the middle of February (Study 2). Within a time period, *** means that late and early wheat are significantly different at least significant difference 0.001 level.

Use of Winter Cover Crops to Reduce Potential Soil Erosion in the San Luis Valley (Studies 1 to 4)

The annual estimated potential soil erosion for all sites was significantly lower when WCGs were used than when all sites were fallow ($P < 0.05$) (Figure 7.4a). The

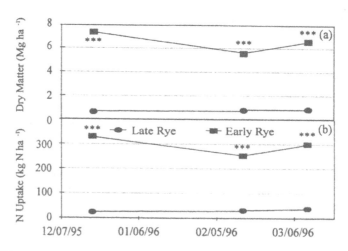

FIGURE 7.3 Total dry biomass production (a) and N uptake (b) of late- (Study 3) and early- (Study 4) planted winter cover rye. The early-planted rye was grown in a loam soil and the late was grown on a loamy sand. Within a time period, *** means that late and early rye are significantly different at least significant difference 0.001 level.

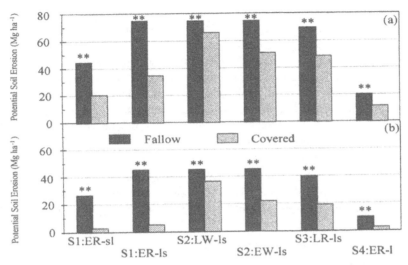

FIGURE 7.4   Potential wind soil erosion in an annual (a) and most susceptible period, from March 15 to May 1 (b). Wind soil erosion was calculated for fallow and WCC-covered fields. Sites: Study 1, early-planted rye grown on a sandy loam (S1:ER-sl) and early rye on a loamy sand (S1:ER-ls); Study 2, late wheat on a loamy sand (S2:LW-ls) and early wheat on a loamy sand (S2:EW-ls); Study 3, late rye on a sandy loam (S3:LR-sl); and Study 4, early rye on a loam (S4:ER-l). Within a site, ** means that fallow and WCC-covered fields are significantly different at least significant difference 0.05 level.

significant gains in reducing potential soil erosion were during the most susceptible period, from March 15 to May 1 (Figure 7.4b). A paired comparison between early and late planting found that early planting more effectively reduced the potential soil erosion than late planting ($P < 0.05$). Our results agree with other researchers who found that WCCs conserve soil quality by reducing potential wind erosion (Bilbro, 1991; Langdale et al., 1991).

Studies 2 and 4: Use of Winter Cover Crops for Grazing in the San Luis Valley (Preliminary Observations)

Recently, some SLV farmers have been using WCG as a source of winter and early spring grazing. However, there is a concern about grazing WCG under high soil NO_3–N levels. Apparently, winter wheat and winter rye can accumulate forage NO_3–N levels that are potentially toxic to animals (Tucker et al., 1961). The Colorado State University Agricultural Extension Service (Stanton, 1994) states that, on a dry weight basis, NO_3–N concentrations of 1150–2300 mg kg^{-1} (1150–2300 ppm) are potentially toxic and >2300 mg kg^{-1} (2300 ppm) is dangerous to livestock, when used as the only source of feed. In soils with high residual soil NO_3–N, we measured in winter rye NO_3–N levels as high as 3500 mg kg^{-1} (3500 ppm) on a dry weight basis in the aboveground plant tissue (Study 4). The farmer who grazed livestock on

Study 4 did observe bloating in some of his steers but not death, because the livestock were removed. During March 1996, the livestock grazed the rye with supplemental hay that was brought into the field.

In Study 2, from December 19, 1995 to February 15, 1996, 2 mg ha^{-1} (0.9 t acre^{-1}) of total dry biomass and 79 kg N ha^{-1} (71 lb acre^{-1}) were removed by grazing ($P < 0.05$) (Figure 7.2). Photographs of the field taken on March 11, 1996 show that even after grazing and crop dieback during the winter, wheat shoots were green and apparently growing (photographs not shown). The late-planted wheat was lightly grazed and also had green shoots on March 11, 1996 (photographs not shown). It is assumed that an unknown amount of the 79 kg N ha^{-1} (71 lb acre^{-1}) grazed by the animals was recycled as urea–N and returned to the soil (Stillwell, 1983). In these soils with pH greater than 7.7, this urea could be susceptible to losses; however, losses during the winter are expected to be lower than during the summer (Schimel et al., 1986).

CONCLUSIONS

Our results agree with other researchers' findings that WCCs can conserve water quality (Holderbaum et al., 1990; Meisinger et al., 1991; Brinsfield and Staver, 1991; Decker et al., 1994; McCracken et al., 1995). These WCGs are useful tools in the conservation of water quality when significant spatial variability, such as soil texture, %CFV, and residual soil NO_3^-–N, is observed in the field. They also return plant biomass back into the surface soil and reduce potential wind erosion. The benefits of these WCGs are significantly greater when they are planted early after harvesting of vegetables; however, benefits to the conservation of the soil and water quality were still observed when planted late.

Since farmers are using WCG for grazing, more research is needed on the effect of WCGs in livestock production, especially as to how good they are as a source of energy during winter and early spring. This was beyond the scope of this study and needs further attention by animal scientists. Alternatively, WCG can be grown for grain and not grazed. We currently do not have data from the SLV concerning production of a wheat grain following grazing; however, Study 2 shows that there is a potential for such a practice. Long-term data are needed, to evaluate how cold and relatively warmer winters will affect the responses of the livestock and the WCG.

Our results agree with other researchers who found that WCCs conserve soil quality by reducing soil erosion (Frye et al., 1985; Holderbaum et al., 1990; Bilbro, 1991; Langdale et al., 1991; Decker et al., 1994). Reduction in potential soil erosion was maximized with early-planted WCG; however, benefits in reduction of potential soil erosion were also obtained with late planting of WCG.

If farmers are to use WCG under high residual soil NO_3^-–N conditions, they will need to account for the N that is released from the WCG crop residue during the growing season of the following crop. Spatial variability in WCG dry biomass, C/N ratio, and N content should be taken into consideration to account for spatial variability in the amount of N that is returned into the surface horizon, consequently creating a potential spatial variability in the amount of N that will be released by N mineralization from the incorporated WCG residue. If the release of the N from the WCG has poor synchronization with the N uptake by the following crop, accumulation of potential leachable residual soil NO_3^-–N could occur (Meisinger et al., 1991). We did not measure how much of the residual soil NO_3^-–N in the fall of 1995 came from the N fertilizer and how much came from the mineralization of the 179 and 91 kg N ha^{-1} (160 and 81 lb acre^{-1}) that was incorporated into the surface of the sandy loam and loamy sand, respectively. Our data suggest that of the 204 and 172 kg NO_3^-–N ha^{-1} (182 and 154 lb acre^{-1}) measured in fall 1995, for the sandy loam and loamy sand, respectively, an unknown amount came from the WCG N mineralization (Figure 7.1). If a significant amount of this residual soil NO_3^-–N is from the WCG, there will need to be better synchronization between N release from the WCG and N uptake by potatoes.

SUMMARY

We found that WCGs such as rye and wheat conserve soil quality in the SLV of south central Colorado by reducing potential soil erosion. Winter cover rye showed a potential to scavenge NO_3^-–N from greater depths than shallower root crops such as lettuce, spinach, potatoes, and carrots. Maximum soil and water conservation will occur with early planting of WCGs, which increases dry matter production and N uptake in the fall before growth is drastically reduced by the winter. If farmers are to use these crops to protect soil and water quality, improvement in the synchronization and accountability for the N release from mineralization of WCC residue as a source of N for the following crop is also needed.

ACKNOWLEDGMENTS

The authors thank Ms. Anita Kear, Mr. Dave Wright, Dr. Bill O'Deen, Mr. Robert Lober, and Mr. Kevin Lee for their capable assistance during collection and analysis of soil and plant samples; Ms. Linda Warsh, Ms. Norma VanNostrand, and Mr. Larry Kawanabe for coordinating study activities with local producers; Mr. Donald "Smokey" Barker for his coordination of activities of USDA/NRCS and USDA/NRCS/SLVWQDP personnel in this project; and the San Luis Valley Water Quality Demonstration Project (SLVWQDP) for technical assistance installing water quality conservation practices.

REFERENCES

Allos, H.F. and W.V. Bartholomew. 1955. Effect of available nitrogen on symbiotic fixation. *Soil Sci. Soc. Am. Proc.* 19:182–184.

Austin, B. 1993. Report to the Commissioner of Agriculture, Colorado Department of Agriculture, Groundwater Monitoring Activities San Luis Valley Unconfined Aquifer. Colorado Department of Public Health and Environment, Denver.

Baker, J.M. and B.B. Tucker. 1971. Effects of rates of N and P on the accumulation of NO_3^--N in wheat, oats, rye and barley on different sampling dates. *Agron. J.* 63: 204–207.

Bartley, E.E., G.W. Barr, and R. Mickelsen. 1975. Bloat in cattle. XVII. Wheat pasture bloat and its prevention with poloxalene. *J. Anim. Sci.* 41:752.

Bilbro, J.D. 1991. Cover crops for wind erosion control in semiarid regions. In W.L. Hargrove (ed.). *Cover Crops for Clean Water.* Soil and Water Conservation Society, Ankeny, IA, pp. 36–38.

Bondy, E., L. Lyles, and W. Hayes. 1980. Computing soil erosion by periods using wind erosion distribution. *J. Soil Water Conserv.* 35:173–176.

Brinsfield, R.B. and K.W. Staver. 1991. Use of cereal cover crops for reducing groundwater nitrate contamination in the Chesapeake Bay region. In W.L. Hargrove (ed.). *Cover Crops for Clean Water.* Soil and Water Conservation Society, Ankeny, IA, pp. 79–82.

Chapman, H.D., G.F. Liebig, and D.S. Rayner. 1949. A lysimiter investigation of nitrogen gains and losses under various systems of covercropping and fertilization and a discussion of error sources. *Hilgardia* 19(3):57–95.

Clay, B.R. 1973a. Stoker Cattle Losses on Small Grain Pasture. Oklahoma Agric. Exp. Stn. Res. Rep. P 690. p. 8.

Clay, B.R. 1973b. Stoker cattle losses on small grain pasture. *Okla. Vet.* 25:15.

Decker, A.M., A.J. Clark, J.J. Meisinger, F.R. Mulford, and M.S. McIntosh. 1994. Legume cover crop contributions to no-tillage corn production. *Agron. J.* 86:126–135.

Doran, J.W. and M.S. Smith. 1991. Role of cover crops in nitrogen cycling. In W.L. Hargrove (ed.). *Cover Crops for Clean Water.* Soil and Water Conservation Society, Ankeny, IA, pp. 85–90.

Eddy-Miller, C.A. 1993. Evaluation of Shallow Ground Water Wells as a Method of Monitoring Nitrate Leaching in the San Luis Valley. M.S. thesis. Colorado State University, Fort Collins.

Edelmann, P. and D.R. Buckles. 1984. Quality of Ground Water in Agricultural Areas of the San Luis Valley, South Central Colorado. Water-Resources Investigations Report 83-4281. USGS.

Evanylo, G.K. 1991. Rye nitrogen cycling for corn and potato production. In W.L. Hargrove (ed.). *Cover Crops for Clean Water.* Soil and Water Conservation Society, Ankeny, IA, pp. 101–103.

Fowler, D.B. 1982. Date of seeding, fall growth and winter survival of winter wheat and rye. *Agron. J.* 74:1060–1063.

Frye, W.W., W.G. Smith, and R.J. Williams. 1985. Economics of winter cover crops as a source of nitrogen for no-till corn. *J. Soil Water Conserv.* 40:246–248.

Groffman, P.M., P.E. Hendrix, and D.A. Crossley, Jr. 1987. Nitrogen dynamics in conventional and no-tillage agroecosystems with inorganic fertilizer or legume nitrogen inputs. *Plant Soil* 97:315–332.

Hojjati, S.M., T.H. Taylor, and W.C. Templeton, Jr. 1972. Nitrate accumulation in rye, tall fescue, and bermudagrass as affected by nitrogen fertilization. *Agron. J.* 64:624–627.

Holderbaum, J.F., A.M. Decker, J.J. Meisinger, F.R. Mulford, and L.R. Vough. 1990. Fall-seeded legume cover crops for no-tillage corn in the humid east. *Agron J.* 82:117–124.

Horn, G.W., B.R. Clay, and L.I. Croy. 1977. Wheat Pasture Bloat of Stokers. Oklahoma Agric. Exp. Stn. Res. Rep. MP01. p. 26.

Huntington, T.G., J.H. Grove, and W.W. Frye. 1985. Release and recovery of nitrogen from winter annual cover crops in no-till corn production. *Commun. Soil Sci. Plant Anal.* 16:193–211.

Johnson, R.R. 1973. Cutting Cattle Losses on Wheat Pasture. Oklahoma Agric. Exp. Stn. Res. Rep. P690. p. 1.

Jones, R.J. 1942. Nitrogen losses from Alabama soils in lysimeters as influenced by various systems of green manure crop management. *J. Am. Soc. Agron.* 34:574–585.

Jones, M.B., C.C. Delwiche, and W.A. Williams. 1977. Uptake and losses of ^{15}N applied to annual grass and clover in lysimeters. *Agron. J.* 69:1019–1023.

Lal, R., E. Regnier, D.J. Eckert, W.M. Edwards, and R. Hammond. 1991. Expectations of cover crops for sustainable agriculture. In W.L. Hargrove (ed.). *Cover Crops for Clean Water.* Soil and Water Conservation Society, Ankeny, IA, pp. 1–11.

Langdale, G.W., R.L. Blevins, D.L. Karlen, D.K. McCool, M.A. Nearing, E.L. Skidmore, A.W. Thomas, D.D. Tyler, and J.R. Williams. 1991. Cover crop effects on soil erosion by wind and water. In W.L. Hargrove (ed.). *Cover Crops for Clean Water.* Soil and Water Conservation Society, Ankeny, IA, pp. 15–21.

Mader, T.L., G.W. Horn, W.A. Phillips, and R.W. McNew. 1983. Low quality roughage for steers grazing wheat pasture. Effect on weight gains and bloat. *J. Anim. Sci.* 56: 1021–1028.

McCracken, D.V., J.E. Box, Jr., W.L. Hargrove, M.L. Cabrera, J.W. Johnson, P.L. Raymer, A.D. Johnson, and G.W. Harbers. 1995. Tillage and cover crops effects on nitrate leaching in the southern Piedmont. In Proc USDA-ASAE Clean Water–Clean Environment — 21st Century, Vol. II, ASAE, St Joseph, MO.

Meisinger, J.J., W.L. Hargrove, R.L. Mikkelson, J.R. Williams, and J.W. Benson. 1991. Effects of cover crops on groundwater quality. In W.L. Hargrove (ed.). *Cover Crops for Clean Water.* Soil and Water Conservation Society, Ankeny, IA, pp. 57–68.

Mitchell, W.H. and M.R. Teel. 1977. Winter annual cover crops for no tillage corn production. *Agron J.* 69:569–573.

Morris, D.R., R.W. Weaver, G.R. Smith, and F.M. Rouquette. 1986. Competition for nitrogen-15-depleted ammonium nitrate between arrowleaf clover and annual ryegrass sown into bermudagrass sod. *Agron. J.* 78:1023–1030.

Mugwira, L.M., S.M. Elgawhary, and A.E. Allen. 1980. Nitrate uptake effectiveness of different cultivars of triticale, wheat, and rye. *Agron. J.* 72:585–588.

Mutchler, C.K. and L.L. McDowell. 1990. Soil loss from cotton with winter cover crops. *Trans. ASAE* 33:432–436.

Nielsen, N.E. and H.E. Jensen. 1985. Soil mineral nitrogen as affected by undersown catch crops. In Assessment of Nitrogen Fertilizer Requirement. Proc. NW-European Study Group for the Assessment of Nitrogen Fertilizer Requirement. Netherlands Fert. Inst., Haren, The Netherlands.

Pink, L.A., F.E. Allison, and U.L. Gaddy. 1945. Greenhouse experiments on the effect of green manures upon N recovery and soil carbon content. *Soil Sci. Soc. Am. Proc.* 10:230–239.

Pink, L.A., F.E. Allison, and U.L. Gaddy. 1948. Greenhouse experiments on the effect of green manures crops of varying carbon–nitrogen ratios upon nitrogen availability and soil organic matter content. *Agron J.* 40:237–248.

Redmon, L.A., G.W. Horn, E.G. Krenzer, Jr., and D.J. Bernardo. 1995. A review of livestock grazing and wheat grazing yield: boom or bust? *Agron. J.* 87:137–147.

Redmon, L.A., E.G. Krenzer, Jr., D.J. Bernardo, and G.W. Horn. 1996. Effect of wheat morphological stage at grazing termination on economic return. *Agron J.* 88:94–97.

Redulla, C.A., J.L. Havlin, G.J. Kluitenberg, N. Zhang, and M.D. Schrock. 1995. Variable nitrogen management for improving groundwater quality in south central Kansas. *Agron. Abstr.* 87:320.

SAS (Statistical Analysis System) Institute. 1988. *SAS/STAT Users Guide,* Version 6.03, 3rd ed. SAS Institute, Cary, NC.

Schertz, D.L. and D.A. Miller. 1972. Nitrate–N accumulation in the soil profile under alfalfa. *Agron. J.* 64:660–664.

Schimel, D.S., W.J. Parton, F.J. Adamsen, R.G. Woodmansee, R.F. Senft, and M.A. Stillwell. 1986. The role of cattle in the volatile loss of nitrogen from a shortgrass steppe. *Biogeochemistry* 2:39–52.

Sharkoff, J.L., R.J. Ristau, R.R. Riggenbach, and M.L. Hynden. 1995. San Luis Valley Water Quality Demonstration Project Annual Report. USDA/NRCS/SLVWQDP, Monte Vista, CO.

Shipley, P.R., J.J. Meisinger, and A.M. Decker. 1992. Conserving residual corn fertilizer nitrogen with winter cover crops. *Agron. J.* 84:869–876.

Skidmore, E.L., P.S. Fisher, and N.P. Woodruff. 1970. Wind erosion equation: computer solution application. *Soil Sci. Soc. Am. Proc.* 34:931–935.

Stanton, T.L. 1994. Nitrate Poisoning in Livestock. Service in Action No. 1.610. Colorado State University Cooperative Extension, Fort Collins.

Stillwell, M.A. 1983. The Effect of Bovine Urine on the Nitrogen Cycle of a Shortgrass Prairie. Ph.D. dissertation. Colorado State University, Fort Collins.

Stogner, R.W. 1996. Nitrate monitoring of the shallow unconfined aquifer in the San Luis Valley. In 13th Annual Potato/Grain Conf. Colorado State University Cooperative Extension, p. 123.

Streeter, J.G. 1985. Nitrate inhibition of legume nodule growth and activity. II. Short term studies with high nitrate supply. *Plant Physiol.* 77:325–328.

Torbert, H.A. and D.W. Reeves. 1991. Benefits of a winter legume cover crop to corn: rotation versus fixed-nitrogen effects. In W.L. Hargrove (ed.). *Cover Crops for Clean Water.* Soil and Water Conservation Society, Ankeny, IA, pp. 99–100.

Tucker, J.M., D.R. Cordy, L.J. Berry, W.A. Harvey, and T.C. Fuller. 1961. Nitrate Poisoning in Livestock. California Agric. Exp. Stn. Circ. 506.

U.S. Department of Agriculture–Soil Conservation Service. 1973. J.P. Pannell, J.M. Yenter, S.O. Woodyard, and R.E. Mayhugh. Soil Survey of Alamosa Area, Colorado. USDA-SCS, Washington, D.C.

U.S. Department of Agriculture–Soil Conservation Service. 1988. National Agronomy Manual, 2nd ed. USDA-SCS, Washington, D.C.

U.S. Environmental Protection Agency. 1989. *Federal Register* 54 FR 22062, 22 May. U.S. EPA, Washington, D.C.

Viets, F.G., Jr. and C.L. Crawford. 1950. The influence of nitrogen on the growth and nitrogen fixation of hairy vetch. *Soil Sci. Soc. Am. Proc.* 14:234–237.

Wagger, M.G. 1989a. Time of desiccation effects on plant composition and subsequent nitrogen release from several winter annual cover crops. *Agron. J.* 81:236–241.

Wagger, M.G. 1989b. Cover crop management and nitrogen rate in relation to growth and yield of no-till corn. *Agron. J.* 81:533–538.

Walker, T.W., A.F.R. Adams, and H.D. Orchiston. 1956. Fate of labeled nitrate and ammonium nitrogen when applied to grass and clover grown separately and together. *Soil Sci.* 81:339–351.

Wilson, D.O. and W.L. Hargrove. 1986. Release of nitrogen from crimson clover residue under two tillage systems. *Soil Sci. Soc. Am. J.* 50:1251–1254.

Woodruff, N.P. and F.H. Siddoway. 1965. A wind erosion equation. *Soil Sci. Soc. Am. Proc.* 29:602–608.

8 Soil Quality: Post Conservation Reserve Program Changes with Tillage and Cropping

M.J. Lindstrom, T.E. Schumacher, D.C. Reicosky, and D.L. Beck

INTRODUCTION

The Conservation Reserve Program (CRP) has successfully reduced erosion on highly erodible lands by establishment of a permanent vegetative cover and development of a stable soil structure. This study was started to determine the effects of tillage systems on water infiltration after CRP lands are returned to crop production. CRP land enrolled in the program in 1988 was converted from CRP to cropland in 1994 and returned to a winter wheat– (*Triticum aestivum* L.) fallow rotation by varying levels of tillage intensities. Water infiltration measurements were conducted on selected treatments in 1994 after a condensed fallow period of 3 months and in 1995 after a 12-month fallow period immediately after planting winter wheat. Water was applied at a rate of 63.5 mm h^{-1} for 1 hr using a sprinkling-type infiltrometer for a dry run on antecedent soil moisture and then again the following day (wet run). No differences in water infiltration characteristics were observed in 1994 between the no-tilled, intensively tilled, or bare CRP (vegetation removed). Differences were observed during the wet runs in 1995, with the tilled treatments exhibiting less infiltration (36.8 and 41.9 mm) than the no-till or the vegetated CRP (62.2 and 63.5 mm). These results show a decrease in water infiltration characteristics, a component of soil quality, within a 1-year fallow period after conversion to cropland for the tilled treatments.

1-57444-100-0/99/$0.00+$.50

The CRP was initiated in 1985 under the Food Security Act with the intention of converting up to 18 million ha of highly erodible land to permanent cover. By 1993, 14.8 million ha or 8% of the cropland in the United States was enrolled in the CRP. By 1997, 8.9 million ha will be released from their CRP contracts. Fifty-five percent of the CRP is located in the ten Great Plains states. Average erosion reduction is estimated to be 42.6 mg ha^{-1} yr^{-1} for land enrolled in the CRP (Osborn, 1993).

The 10-year period in grass is likely to have resulted in changes in soil organic carbon (Gebhart et al., 1994), soil structure (Kay, 1990), and the ability to absorb water (Bouma and Hole, 1971). The estimated half-life for net improvement in soil structural stability after establishment of grass ranged from 4.5 years for a clay loam to 7.8 years in a sandy loam (Rasiah and Kay, 1994). A review by Kay (1990) on the effects of grass on the rate of soil structural stability improvements suggests that a minimum of 10 years is needed for maximum benefits. Improvements are likely to continue more slowly after this period.

The degree of soil improvement during the CRP period is likely to be soil and site dependent. Soils placed in the CRP were generally degraded soils lower in organic matter than surrounding soils because they were placed into the CRP based on a highly erodible classification due to wind, water, or a combination of the two. Highly eroded soils tend to have reduced productivity, degraded soil structure, lower organic matter, and a less than ideal rooting environment (Frye et al., 1982; Lindstrom et al., 1992).

Ideally, the soil quality improvements derived from the CRP should be maintained as CRP lands are returned to crop production. Most of the lands in the CRP had suffered much erosion and soil structural degradation from cropping. Although these lands have improved greatly during the period in grass, the structural advantages gained during the CRP can be rapidly lost with the onset of intensive tillage. The first change that occurs with tillage is breaking the continuity of established soil pores formed by roots and soil fauna (Packer and Hamilton, 1993). Natural weathering of exposed soil aggregates proceeds rapidly, resulting in a decrease in aggregation and lower aggregate stability (Low, 1972). Soil carbon will also decline rapidly due to short-term tillage-induced fluxes of carbon dioxide and increased microbial activity (Reicosky and Lindstrom, 1993). The net result will be a degrading soil which may rapidly come to equilibrium at the pre-CRP soil condition.

The objective of this study was to determine the effects of various soil management practices on soil structure, organic carbon, and water infiltration after CRP lands are returned to crop production. This chapter focuses primarily on water infiltration characteristics related to methods of conversion of CRP land to cropland. Water infiltration is a key indicator of soil quality as described by Karlen and Stott (1994) in their conceptual soil quality model.

STUDY DESCRIPTION

A 15-ha site was established in 1994 in Lyman County approximately 60 km southeast of Pierre, South Dakota (lat. 45° 05' N, long. 100° 10' W). The site was planted to CRP in 1988. Plant species established were intermediate wheatgrass

(*Agropyron intermedium* Host.) and alfalfa (*Medicago sativa* L.). The soil is a Promise clay (very fine, montmorillonitic, mesic Udic Haplustert). As a vertisol, the soil is subject to extensive cracking when dry and surface sealing when wet. The slopes at the site are between 2–4%. The soil is classified as highly erodible based on wind erosion estimates. However, water erosion is a problem caused by severe thunderstorms which are common during the growing season. Average annual precipitation is 450 mm.

The study site was set up in a randomized complete block design with four main blocks. Each block has six treatments as follows:

1. CRP — undisturbed CRP sod
2. CT — conventional tillage, winter wheat– (*Triticum aestivum* L.) fallow rotation
3. CP — conservation tillage to meet the conservation compliance plan (20% residue cover after planting), winter wheat–fallow rotation
4. NT1 — no-till, winter wheat–green fallow rotation
5. NT2 — no-till, winter wheat–corn– (*Zea mays* L.) green fallow rotation
6. NT3 — conventional tillage breakout of CRP followed by no-till after harvest of the first winter wheat crop, winter wheat–green fallow rotation

The green fallow was lentils (*Lens culinaris* Medik.) established in the spring of the fallow year and killed chemically during the growing season. Each crop in each rotation is present every year, resulting in 12 plots per block. Surface residue cover of 20% is the accepted value to meet conservation compliance for a fall-planted crop.

The grass–alfalfa CRP mix was harvested for hay May 30, 1994 on all plots except the CRP treatments. Tillage of the CT, CP, and NT3 plots to be planted to winter wheat in 1994 was accomplished by disking (two passes) with a heavy tandem disk in mid-June followed by a chisel plow equipped with 45-cm sweep blades as required for weed control. Three passes were required during the 1994 fallow season. Tillage during the fallow period for CT, CP, and NT3 treatments was identical during the initial conversion to cropland. The CRP vegetation on the no-till plots (NT1 and NT2) to be planted to winter wheat in 1994 was killed by application of herbicides on July 1. This was followed with an additional herbicide application on September 6. Weed growth on the tilled fallow treatments was also chemically controlled on this date. Winter wheat was planted in the first year of the rotation on September 15, 1994.

Conversion to cropland for the 1995 fallow season plots was initiated in September 1994. The CRP vegetation on the no-till fallow (NT1 and NT2) was killed by application of herbicides on September 6. The tilled plots (CT, CP, and NT3) were disked twice at this same time with a heavy tandem disk followed by a chisel plow equipped with 45-cm sweep blades approximately 1 month later. The CT treatment was chisel plowed three times during the 1995 season to control weeds. Weed growth on the no-till plots (NT1 and NT2) was again chemically controlled on May 25, planted to lentils on June 9, and then the lentils were killed on August 9. The

CP and NT3 plots were chisel plowed once in May, planted to lentils on June 9, and then sprayed to kill the lentils on August 9. All tilled 1995 fallow plots were again chisel plowed on September 10 and planted to winter wheat on September 20.

Water infiltration measurements in 1994 and 1995 were conducted with a sprinkling-type infiltrometer (Onstad et al., 1981) immediately after planting winter wheat. Water application rate was 63.5 mm h^{-1} for 1 hr. Water infiltration test plot size was 91.4 cm by 152.4 cm established by driving a steel frame into the soil around three sides of the test area. A steel cutoff wall was driven into the soil on the downslope side with the upper edge even with the soil surface. A collection trough was attached to the cutoff wall where the runoff water is collected, passed over a tipping bucket, and then transferred to a tank via a vacuum system (Onstad et al., 1981). Volumes of the tipping buckets were 80 cm^3.

Two infiltration measurements were made on each plot: the first measurement was on antecedent soil moisture (dry run), and the second measurement was made the following day (wet run). Infiltration measurements in 1994 were made on the CRP plots with and without vegetation, no-till (NT1), and tilled (CT) planted plots. In 1995, infiltration measurements were made on the CRP plots with vegetation and on the planted NT1, NT3, and CT plots.

Time domain reflectometery (TDR) (Topp et al., 1982) was used to measure changes in soil water content during the 1995 sprinkling infiltrometer measurements. Duplicate TDR probes were inserted horizontally into the soil profile at depths of 10, 20, 30, and 40 cm. A pit was dug along one edge to the infiltration plots for insertion of the TDR probes. The pit was covered with a plastic shield to protect against water running along the face of the pit. Readings were collected at 5-min intervals starting at the time of water application for the dry run and continued until the next day after the wet run.

Tension infiltrometer measurements (Ankeny et al., 1991), calibrated to tensions of 30, 60, and 120 mm H$_2$O, were made in conjunction with the infiltration measurements in 1995. The tension infiltration readings were taken at four sites in the near vicinity of the infiltration test site within 24 hr after the wet runs. Surface residue coverage was measured by the line transect method (Sloneker and Moldenhauer, 1977) immediately after planting winter wheat both years.

RESULTS AND DISCUSSION

Surface residue cover for all plots is shown in Table 8.1. In 1994, the major effect was related to tillage and the length of time since conversion to cropland. The planted plots were in CRP sod until late May of 1994, whereas the fallow plots were in CRP sod until just prior to surface residue cover measurements. The tilled treatments showed a decrease in surface residue cover compared to the no-till treatments, but the residual effect of the CRP vegetation was still evident by the relatively high residue cover (31%) in the tilled 1994 planted treatments. In 1995, surface residue cover was high for all plots after winter wheat harvest. Tillage after harvest on the CT and CP plots only reduced surface residue cover by about 15%. The green fallowed no-till plots (NT1 and NT2) maintained a high surface residue

TABLE 8.1
Percent Surface Residue Cover After Planting Winter Wheat in 1994 and 1995

Treatment	Tillage	Residue cover (%)
1994		
Tilled (CT, CP, NT3), $n = 12$	Planted	31
Tilled (CT, CP, NT3), $n = 12$	Fallow	66
No-till (NT1, NT3), $n = 8$	Planted	90
No-till (NT1, NT3,) $n = 9$	Fallow	100
CRP, $n = 4$		100
1995		
Tilled (CT, CP, NT3), $n = 12$	Planted	16
Tilled (CT, CP), $n = 8$	Fallow	80
No-till (NT1, NT2), $n = 8$	Planted	60
No-till (NT1, NT2, NT3), $n = 12$	Fallow	93
CRP, $n = 4$		100
Least significant difference (0.05)		10

cover (approximately 60%), but were lower than the 1994 values. This decrease reflects the increased length of the fallow period (12 months in 1995 versus 3 months in 1994). All tilled plots (CT, CP, and NT3) showed a large reduction in surface cover and were below the conservation compliance value of 20% after planting, even with the green fallow.

Water infiltration results for 1994 and 1995 are shown in Table 8.2. No runoff was observed for the vegetated CRP plots either year. The CRP plots in which surface vegetation had been removed in late May and again prior to the test exhibited runoff (Table 8.2), suggesting the importance of surface cover in protecting soil surface

TABLE 8.2
Water Runoff Expressed as Percent Runoff of Applied 63.5 mm for 1994 and 1995

	1994		1995	
Treatment	Dry	Wet	Dry	Wet
CRP	0	0	0	0
CRP–(bare)	26	54	—	—
NT1	28	47	1	2
NT3	—	—	4	34
CT	23	53	6	42
LSD (0.05)[a]	NS	NS	NS	13.5

[a] Least significant difference (0.05) for comparison of treatments producing runoff (excludes CRP with 0% runoff).

TABLE 8.3
Tension Infiltration Rates Measured on the 1995 Test Plots
($n = 16$)

Treatment	Infiltration rate ($\mu m\ s^{-1}$)		
	Tension 30 mm	Tension 60 mm	Tension 120 mm
CRP	14	6	2
NT1	15	6	2
NT3	15	8	4
Chisel	19	10	5
Least significant difference (0.05)	NS	NS	NS

aggregate and reducing surface seal formation. In 1994, except for the runs on the vegetated CRP plots, no differences in water infiltration were observed.

Differences in runoff were observed during the 1995 runs. The tilled treatments (CT and NT3) exhibited more runoff (less infiltration) than the no-till treatment (NT1) during the wet run. The tilled plots (CT and NT3) had significantly less surface residue than the no-till (NT1) and therefore would be more susceptible to aggregate breakdown and surface sealing. Tillage that caused disruption of the soil pore network could also be an important factor. The increased length of fallow also is a factor. Whatever the cause for the observed differences in infiltration, the reduction in tilled plots versus no-till treatments indicates a change in soil physical properties 1 year after conversion of CRP to cropland.

Tension infiltration rates (Table 8.3) and calculated hydraulic conductivity (Table 8.4) measured in 1995 showed no differences between treatments, indicating similar pore size distribution in the effective water transmission range. Since the tension infiltration measurement eliminates surface seal formation, the lack of differences between treatments suggests surface seal formation was the probable cause for infiltration differences.

TABLE 8.4
Unsaturated Hydraulic Conductivity Calculated from Tension Infiltration
Water on the 1995 Test Plots ($n = 16$)

Treatment	Hydraulic conductivity ($\mu m\ s^{-1}$)		
	Tension 30 mm	Tension 60 mm	Tension 120 mm
CRP	8	2	1
NT1	9	3	1
NT3	8	4	2
Chisel	13	4	2
Least significant difference (0.05)	NS	NS	NS

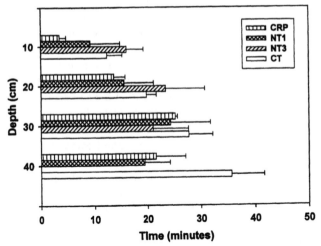

FIGURE 8.1 Dry run: wetting front arrival times during and after application of 63.5 mm h^{-1} of water. Zero time represents the beginning of water application for a 1-hr period. Error line represents the standard error of the mean (average $n = 5$). Date not available for the NT3 development at 40 cm.

The wetting front arrival times at the TDR probes by depths are shown in Figures 8.1 and 8.2, respectively, for the dry and wet runs in 1995. In general, the wetting front advancement was faster in both the dry and wet runs for the CRP and NT1 plots than for the tilled treatments (CT and NT3) ($P \leq 0.09$ for the tillage main

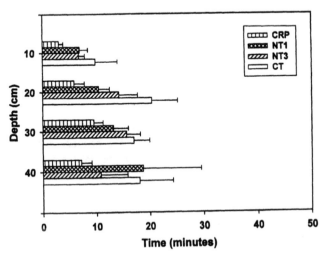

FIGURE 8.2 Wet run: wetting front arrival times during and after application of 63.5 mm h^{-1} of water. Zero time represents the beginning of water application for a 1-hr period. Error line represents the standard error of the mean (average $n = 6$).

effect). Time for wetting front appearance generally increased with depth of measurement ($P \leq 0.01$ for the depth main effect). However, some discrepancies in wetting front movement with depth were observed which may be attributed to bypass flow (Tsuyoshi, 1993). These discrepancies were not consistently observed for any one treatment ($P \leq 0.33$ for the depth by tillage interaction). Large cracks observed to a depth of 60 cm were present within the soil profile at the start of the dry run, which provided channels for saturated water flow.

Similar tension infiltration rates suggest little change in soil internal structure since this soil had been converted to cropland. The treatment differences in water infiltration during the wet run in 1995 do show reduced infiltration for the tilled treatments and suggest the effect of surface residue cover absorbing raindrop impact, preventing surface aggregate breakdown and surface seal formation. The faster wetting front advancement with CRP and no-till treatment suggests soil structural differences.

The CRP plots exhibited the ability to absorb and transmit the applied water in both the dry and wet runs (a total of 127 mm), indicating a high soil quality in reference to water infiltration. As the CRP condition was modified by removing the surface vegetation in 1994, the ability to absorb and transmit water declined, showing the importance of surface cover in maintaining surface structure. Additional changes in the soil's ability to absorb and transmit water were observed with the tilled plots compared to the CRP and no-till plots and were related to the increase in the length of fallow period. These changes may reflect surface seal formation or tillage-induced change in soil properties. In any case, a decline in infiltration and redistribution (an important component of soil quality) was observed within 1 year after tillage of CRP land. Further work will be required to determine the magnitude and rate of these changes with time. The soil surface on the no-till and CRP treatments stabilizes water infiltration relative to the conventional tillage treatment. The greater surface residue observed on the no-till and CRP treatments helps to maintain soil quality by providing organic carbon that is likely to promote microbial activity at the soil surface, enabling improved surface stability (Karlen et al., 1994).

SUMMARY AND CONCLUSIONS

The similar treatment infiltration rates in 1994 suggest little change in soil properties during the condensed (3-month) fallow period. The primary difference was in the CRP treatment with and without surface vegetation. Water runoff from the CRP treatment without vegetation emphasizes the importance of surface residues to prevent surface sealing and to maintain the soil's ability to adsorb water.

Increased water runoff was observed for the wet run in the 1995 tilled treatments compared to the vegetated CRP and the no-till treatment. No differences were observed in tension infiltration rates or calculated conductivity between treatments. These results suggest the importance of surface residue cover in absorbing raindrop impacts to maintain aggregate stability and reduce surface sealing. The wetting front advanced faster at all depths for both dry and wet runs on CRP and no-till treatments than on the tilled treatments, indicating a higher quality soil condition in regard to water transmission and redistribution. Reasons for the faster wetting front advance rate have not been clearly identified. The results show a decrease in water infiltration within a 1-year fallow period after conversion of CRP to cropland for the tilled treatments and indicate a decline in soil quality with tillage.

REFERENCES

Ankeny, M.D., M. Ahmed, T.C. Kaspar, and R. Horton. 1991. Simple field method for determining unsaturated hydraulic conductivity. *Soil Sci. Soc. Am. J.* 55:467–470.

Bouma, J. and F.D. Hole. 1971. Soil structure and hydraulic conductivity of adjacent virgin and cultivated pedons at two sites: a Typic Argiudoll (silt loam) and a Typic Eutochrept (clay). *Soil Sci. Soc. Am. J.* 35:316–319.

Frye, W.W., S.A. Ebelhar, L.W. Murdock, and R.L. Blevins. 1982. Soil erosion effects on properties and productivity of two Kentucky soils. *Soil Sci. Soc. Am. J.* 46:1051–1055.

Gebhart, D.L., H.B. Johnson, H.S. Mayeux, and H.W. Polley. 1994. The CRP increases soil organic carbon. *J. Soil Water Conserv.* 49:488–492.

Karlen, D.L. and D.E. Stott. 1994. A framework for evaluating physical and chemical indicators of soil quality. In J.W. Doran, D.C. Coleman, D.F. Bezdicek, and B.A. Stewart (eds.). *Defining Soil Quality for a Sustainable Environment.* SSSA Special Publication No. 35. Soil Science Society of America, Madison, WI, pp. 53–72.

Karlen, D.L., N.C. Wollenhaupt, D.C. Erbach, E.C. Berry, J.B. Swan, N.S. Eash, and J.L. Jordahl. 1994. Long-term tillage effects on soil quality. *Soil Tillage Res.* 32:313–327.

Kay, B.D. 1990. Rates of change of soil structure under different cropping systems. *Adv. Soil Sci.* 12:1–52.

Lindstrom, M.J., T.E. Schumacher, A.J. Jones, and C. Gantzer. 1992. Productivity index model for selected soils in North Central United States. *J. Soil Water Conserv.* 47:491–494.

Low, A.J. 1972. The effect of cultivation on the structure and other physical characteristics of grassland and arable soils (1945–1970). *J. Soil Sci.* 23:363–386.

Onstad, C.A., J.R. Radke, and R.A. Young. 1981. An outdoor portable rainfall erosion laboratory. In *Erosion and Sediment Transport Measurement.* Publication No. 133. International Association Hydrologic Science, Paris, pp. 415–422.

Osborn, T.C. 1993. The Conservation Reserve Program: status, future, and policy options. *J. Soil Water Conserv.* 48:271–279.

Packer, I.J. and G.J. Hamilton. 1993. Soil physical and chemical changes due to tillage and their implications for erosion and productivity. *Soil Tillage Res.* 27:327–339.

Rasiah, V. and B.D. Kay. 1994. Characterizing the changes in aggregate stability subsequent to the introduction of forages. *Soil Sci. Soc. Am. J.* 58:935–942.

Reicosky, D.C. and M.J. Lindstrom. 1993. Fall tillage method — effect on short-term carbon dioxide flux from soil. *Agron. J.* 85:1237–1243.

Sloneker, L.L. and W.C. Moldenhauer, 1977. Measuring the amounts of crop residue remaining after tillage. *J. Soil Water Conserv.* 32:231–236.

Topp, G.C., J.L. Davis, and A.P. Annan. 1982. Electromagnetic determination of soil water content using TDR. I. Applications to wetting fronts and steep gradients. *Soil Sci. Soc. Am. J.* 46:672–678.

Tsuyoshi, M. 1993. *Water Flow in Soils.* Marcel Dekker, New York, pp. 93–120.

9 Soil Quality and Environmental Impacts of Dryland Residue Management Systems

O.R. Jones, L.M. Southwick, S.J. Smith, and V.L. Hauser

INTRODUCTION

No-tillage (NT) management in a dryland winter wheat– (*Triticum aestivum* L.) sorghum– (*Sorghum bicolor* [L.] Moench) fallow grain production system in the southern Great Plains resulted in reduced infiltration and increased runoff due to soil crusting when compared to stubblemulch (SM) tillage on Pullman clay loam (fine, mixed, thermic Torrertic Paleustoll). However, surface residues on NT resulted in an improvement in overall water conservation because of reduced evaporation. An average of 10–18% more plant-available soil water was stored with NT than with SM during fallow. Soil quality factors affected most by tillage management included dry aggregate and water-stable aggregate size, soil bulk density, surface residue cover, random roughness, and distribution of organic carbon (OC). NT-managed soil had larger dry aggregate size, higher bulk density, greater residue cover, and OC was concentrated in the surface 2 cm (0.75 in.) in comparison to SM. The environmental impacts of using NT were minimal. NT reduced total sediment loss by 54% in comparison to SM, although runoff volumes were greatest from NT. Nutrient concentrations and losses in runoff were extremely low from both tillage systems, with loss <5 kg/ha (4.5 lb/acre) N and <1 kg/ha (0.9 lb/acre) P per year on these unfertilized watersheds. There was no evidence of atrazine accumulating in the soil or leaching below the root zone, and atrazine loss in runoff amounted to a maximum of 0.26% of total application. However, up to 1.5% of propazine applications was lost in runoff. Propazine, applied to both NT and SM sorghum when runoff probability was high, appears to have a greater potential for negatively impacting the environment under semiarid conditions than does atrazine, applied when runoff probability is low. Propazine accumulated in the soil profile but was not detected below 0.6 m (2 ft). Nitrate N leached deeper in the soil profile with NT due to improved water conservation and a wetter soil.

SM and NT residue management systems were developed for the southern Great Plains to protect the soil from wind and water erosion, but they can also affect both soil and water quality (National Research Council, 1993). More than 75% of conservation farm plans for highly erodible lands use variations of mulch-till, NT, or ridge-till residue management systems to control erosion (Conservation Tillage Information Center, 1993). SM is a version of mulch-till that uses sweeps or blades to undercut the soil surface at a depth of 10–15 cm (4–6 in.), leaving 75–80% of residues on the surface after each operation. With SM management, most residues have been incorporated and disappeared from the soil surface by planting time due to the four to five sweep tillage operations that are required in fallow systems to control weeds between crops.

In wetter regions where larger amounts of crop residues are produced, NT management has the potential to increase infiltration, reduce runoff, and improve water conservation through reduced evaporation (Allmaras et al., 1985). When crop residues from NT are not adequate to protect the soil surface from the kinetic energy of rainfall, as is often the case with dryland cropping in semiarid areas, then soil crusting can occur. Surface crusts can reduce infiltration, increase surface runoff, and reduce water conservation on drylands (Unger, 1992; Steiner, 1994; Jones et al., 1994). With increased use of agricultural chemicals with NT management, the potential exists to negatively impact surface water quality through chemical or nutrient loss in runoff and to degrade soil quality through an accumulation of pesticides or nutrients in or below the soil profile. Alternatively, NT management has the potential to improve soil quality by increasing soil OC relative to SM management and to improve surface water quality by greatly reducing erosion (Jones et al., 1996). NT management of sorghum and wheat in the southern plains reduced nutrient losses in runoff to levels similar to those from unfertilized native grass (Karlen and Sharpley, 1994).

Our objectives were (1) to compare the effects of NT and SM management systems on infiltration, runoff, erosion, and water conservation for dryland grain production and (2) to determine the impacts of adopting NT management on soil and water quality, with emphasis on herbicide and nutrient accumulation in the soil profile as compared to SM management. We consolidate and update information previously reported by Jones et al. (1994, 1995).

MATERIALS AND METHODS

The research was conducted on field-size contour-farmed graded-terraced watersheds located on Pullman clay loam (fine, mixed, thermic Torrertic Paleustoll) at

Bushland, Texas, in the semiarid southern Great Plains. Average annual precipitation is 466 mm (18.3 in.) and average 6-month (April through September) evaporation is 1270 mm (50 in.). Six 2- to 4-ha (5- to 10-acre) watersheds were cropped in a 3-year winter wheat– (*Triticum aestivum* L.) grain sorghum– (*Sorghum bicolor* [L.] Moench) fallow (WSF) sequence with NT management on three of the watersheds and SM management on the other three watersheds. Each phase of the rotation was present each year. The WSF sequence produces two crops in 3 years, with an 11-month fallow (noncropped) period preceding each crop. All watersheds were cropped in an SM-tilled WSF sequence for more than 30 years prior to establishing the NT treatment on three of the watersheds in 1981. Beginning in 1984, each of the watersheds was equipped with a 0.76- or 0.91-m (2.5- or 3-ft) H-flume, water stage recorder, and automatic pumping sampler to measure and sample storm runoff.

Weeds were controlled and seedbeds prepared on SM watersheds with a 4.6-m- (15-ft-) wide sweep machine equipped with three 1.8-m (6-ft) V-shaped blades. Agricultural chemicals applied with NT management included atrazine [6-chloro-N-ethyl-N'-(1-methylethyl-1,3,5-triazine-2,4-diamine)] at 2.7 kg/ha (2.4 lb/acre) a.i. and 2,4-D [(2,4-dichlorophenoxy)acetic acid] at 0.8 kg/ha (0.7 lb/acre) a.i. after wheat harvest, chlorsulfuron [2-chloro-N-{[(4-methoxy-6-methyl-1,3,5-triazine- 2-yl)amino]carbonyl]}benzenesulfonamide] at 0.8 kg/ha (0.7 lb/acre) a.i. during fallow after sorghum, and glyphosate [N-(phosphonomethyl)glycine] at 0.6 kg/ha (0.5 lb/acre) a..i. applied prior to crop planting and to control weed escapes during fallow. Propazine [6-chloro-N,N'-bis(1-methylethyl)1,3,5-triazine-2,4-diamine] at 1.7 kg/ha (1.5 lb/acre) a.i. was applied preemergence to both NT- and SM-managed sorghum.

Runoff-weighted storm runoff samples were composited and analyzed for sediment, nutrient, and triazine herbicide contents using procedures described by Jones et al. (1995). Infiltration rates on selected watersheds were measured with a rotating disk rainfall simulator using cistern-stored rainwater. Characterization measurements of soil quality indicators taken near rainfall simulation sites included dry aggregate and water-stable aggregate size distributions of the top 2 cm (0.75 in.) of soil, OC concentration incrementally to a 15-cm (6-in.) depth, random roughness, residue cover, soil bulk density, and gravimetric soil water content to a 0.9-m (3-ft) depth. Tillage system effects on water conservation were evaluated by comparing soil water contents taken gravimetrically to a 1.8-m (6-ft) depth at planting of sorghum and wheat, which was also the end of the fallow after wheat and fallow after sorghum phases of WSF. Additional experimental details, sampling, and laboratory procedures are contained in Hauser and Jones (1991) and Jones et al. (1994).

Data were analyzed using a randomized complete block one-way analysis of variances comparing tillage treatments. Nonreplicated data associated with runoff (volume, sediment, and nutrient content) were analyzed treating WSF phases as blocks (SAS Institute, 1990).

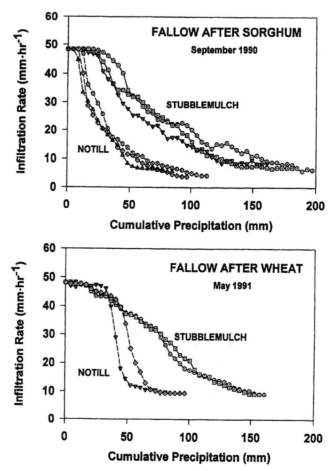

FIGURE 9.1 Tillage system effects on infiltration rate measured by rainfall simulator during dry runs on fallow after sorghum and fallow after wheat near the end of fallow. Rainfall simulator application rate was 48 mm h^{-1} at two or three sites on each watershed.

RESULTS AND DISCUSSION

Soil Quality

Soil Crusting and Infiltration

NT management reduced infiltration and increased storm runoff in comparison to SM management as a result of soil crusting (Figure 9.1). Cumulative infiltration on SM sites was 62% greater than on NT sites at the end of 1 hr of water application with a rainfall simulator and 90% greater at the end of 2 hr for fallow after sorghum. Infiltration differences due to tillage method were less for fallow after wheat because of greater surface residue cover than fallow after sorghum, but were substantially reduced with NT.

TABLE 9.1
Means for Soil Quality Parameters Determined Before or After Applying Simulated Rainfall to Pullman Clay Loam During Fallow Periods[a]

Condition	Time of measurement	Tillage treatment		ANOV	
		SM	NT	N	Prob > F
Dry aggregate size, MWD[b] (mm)					
Fallow/sorghum	Before dry run	14.76	6.57	4	0.001
Fallow/wheat	Before dry run	7.75	2.56	4	0.001
Water-stable aggregate size, MWD (mm)					
Fallow/sorghum	Before dry run	1.24	2.02	4	0.002
Fallow/wheat	Before dry run	0.76	0.66	4	0.29
Organic carbon concentration, 0–0.15 m (g kg^{-1})					
Fallow/sorghum	Before dry run	9.5	9.0	4	0.167
Fallow/wheat	Before dry run	8.7	8.2	4	0.226
Soil water content, 0–0.10 m (m^3 m^{-3})					
Fallow/sorghum	Before dry run	0.11	0.17	4	0.13
Fallow/wheat	Before dry run	0.10	0.12	4	0.22
Soil water content, 0–0.91 m (m^3 m^{-3})					
Fallow/sorghum	Before dry run	0.25	0.27	4	0.048
Fallow/wheat	Before dry run	0.27	0.30	4	0.020
Random roughness (mm)					
Fallow/sorghum	Before dry run	15.2	6.3	4	0.001
	After wet run	8.7	6.3	2	0.037
Fallow/wheat	Before dry run	16.7	9.9	4	0.013
	After wet run	12.1	8.8	1	c
Soil bulk density, 0–0.10 m (mg m^{-3})					
Fallow/sorghum	Before dry run	0.92	1.21	4	0.005
Fallow/wheat	Before dry run	0.83	1.04	4	0.033
Surface residue cover (%)					
Fallow/sorghum	Before dry run	25.0	56.7	4	0.002
Fallow/wheat	Before dry run	73.1	86.1	4	0.002

[a] Measurements were taken in September 1990, near the end of fallow after sorghum, and in May 1991, near the end of fallow after wheat.

[b] Mean weight diameter.

[c] For the wet run, rainfall simulator was operated only at one site on each tillage treatment.

Surface random roughness and bulk density of the 0- to 10-cm (0- to 4-in.) depth were soil quality factors most affected by tillage and appear responsible for much of the difference in infiltration between SM and NT treatments (Table 9.1). After large amounts of rainfall or a high-intensity storm, the soil surface of both SM and NT treatments becomes smooth and crusted. However, when tillage is performed on SM to control weeds, the soil surface is loosened, soil crusts are disrupted, and porosity, surface roughness, and depressional storage capacity are increased. Thus, infiltration rate and cumulative infiltration are increased and runoff is reduced with SM. Since sweep-tillage was performed four to five times during fallow periods on SM treatments, higher average infiltration rates were maintained on SM than on NT where the soil surface remained smooth and consolidated after rainstorms.

Yields of dryland crops were excellent through the experimental period. Surface coverage with residues exceeded 50% on NT (Table 9.1); however, surface smoothing and soil crusting still occurred because the Pullman clay loam Ap soil horizon has a high silt content (17% sand, 53% silt, and 30% clay) and is very susceptible to crusting (Unger and Pringle, 1981). Also, possibly in response to a crusted surface and the dry climate, earthworm activity is low; thus macropore influence on infiltration is reduced, at least in comparison to macropore effects reported with NT in more humid areas (Edwards et al., 1988).

Organic Carbon

Average OC concentration to a 15-cm (6-in.) depth was not affected by tillage system (Table 9.1); however, the distribution of OC within the sampling profile was affected. With NT, OC concentration averaged 2–3 g kg^{-1} (0.2–0.3%) higher near the surface (0–2 cm [0–0.8 in.]) than in the rest of the sampling profile. These results agree with Potter et al. (1997), who also found that after 10 years of NT management, OC concentrations were greater near the surface. However, total OC sequestration in the 0- to 20-cm (0- to 0.8-in.) profile was not significantly different between SM and NT management in dryland cropping systems that included fallow. They did find, however, that NT management with annual cropping of wheat or sorghum did increase carbon sequestration relative to SM.

Water Conservation

Average total plant-available water from a 0- to 1.8-m (6-ft) soil depth was 33 mm (1.3 in.) greater with NT than with SM at sorghum planting (end of fallow after wheat) and 18 mm (0.7 in.) greater with NT at wheat planting, which is also the end of fallow after sorghum (Figure 9.2). Although more runoff occurred from NT than from SM watersheds, 10–18% more soil water was stored with NT, indicating greatly reduced evaporation with NT in comparison to SM management. The difference in evaporation between NT and SM can probably be attributed to a combination of the following: (1) reduced porosity and increased diffusion resistance to vapor flow due to consolidation and crusting of the NT surface (Jalota and Prihar, 1990), (2) reduced wind speed and vapor movement due to more standing surface residues with NT (Smika and Unger, 1986), (3) increased albedo and reduced temperature at the soil surface with NT (Smika, 1983; Steiner, 1994), and (4) increased evaporation with SM due to tillage (Smika and Unger, 1986).

Triazines in the Soil Profile

Soil profile triazine contents from an April 1993 incremental sampling of paired watersheds are shown in Table 9.2. Both watersheds received a 1.34-kg/ha (1.2-lb/

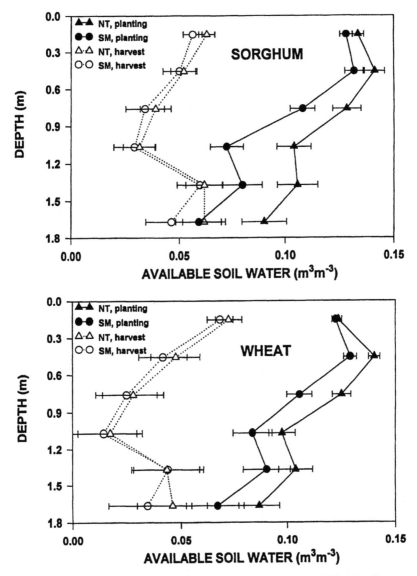

FIGURE 9.2 Tillage effects on 1986–93 average plant-available (0.03–1.5 mPa potential) soil water content at crop planting (end of fallow) and harvest (beginning of fallow). Error bars = standard error of the mean.

acre) application of propazine at sorghum planting and the NT watershed received a 2.24- to 3.36-kg/ha (2.0- to 3.0-lb/acre) application of atrazine after wheat harvest every 3 years from 1982 to 1993. The last herbicide applications before sampling were propazine on June 17, 1992 (286 days before sampling) to both watersheds and atrazine on the NT watershed on July 1, 1991 (638 days before sampling). Propazine

TABLE 9.2
**Concentration and Distribution of Triazines in the Soil Profiles of
Selected Watersheds Incrementally Sampled to a 3-m depth, April 1993**

Soil depth increment (m)	Propazine conc[a] (μg kg^{-1})		Propazine amount (g ha^{-1})		Atrazine conc[b] (μg kg^{-1})	
	SM	NT	SM	NT	SM	NT
0–0.15	29	46	66	103	—	ND[c]
0.15–0.30	9	8	21	19	—	ND
0.30–0.61	ND[b]	6	0	28	—	ND
0.61–0.91	ND	ND	0	0	—	ND
0.91–3.00	ND	ND	0	0	—	ND

[a] Propazine was last applied 286 days before sampling with 447 mm of precipitation occurring during the time interval.

[b] Atrazine was last applied (NT treatment only) 638 days before sampling with 1033 mm of intervening precipitation.

[c] No triazine detected in sample.

moved in the soil, but all propazine was retained within the top 0.6 m (2 ft) of soil. The profile accumulation was 87 g/ha (1.2 oz/acre) on SM and 150 g/ha (2.1 oz/acre) on NT watersheds, about 6.4 and 11.1%, respectively, of the last application. Propazine, when applied at small rates every third year, accumulated to some extent in the soil profile, but did not move more than 0.6 m (2 ft) into the soil profile. In contrast, on a nearby conservation bench terrace (Jones, 1975), propazine was detected at a 3-m (10-ft) depth in April 1993 at a concentration of 38 μg/kg (ppb), with total cumulative propazine of 867 g/ha (12.4 oz/acre) in the 3-m (10-ft) profile. The conservation bench terrace received an average of 70 mm (2.7 in.) per year supplemental runoff in addition to precipitation and had been cropped annually to sorghum since 1957, with annual applications of about 1.3 kg/ha (1.2 lb/acre) propazine since 1965. Thus, propazine moved with water through the soil profile and potentially poses a threat to groundwater quality if applied annually where leaching occurs.

Atrazine was not detected in the soil profile at any depth increment sampled. This was a little surprising since atrazine was applied at a rather large rate when the soil was cracked and rainstorms could move atrazine to the depth of cracking in the soil profile. The 3-year interval between applications apparently provides ample time for degradation of atrazine in the Pullman soil.

Environmental Quality

Storm Runoff

Storm runoff measurements from H-flumes verify rainfall simulator data. Most runoff from WSF occurred during the two fallow phases that comprise 64% of the

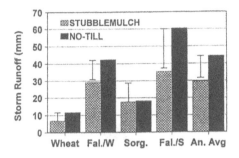

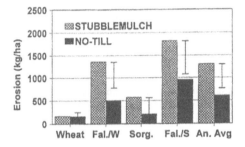

FIGURE 9.3   Tillage system effects on mean storm runoff and sediment loss (erosion) from 1984 to 1993 by WSF phase and annual average, graded terraces, Bushland, Texas. Bars indicate least significant difference ($p = 0.05$) for each pair.

3-year rotation. NT resulted in a substantial increase in runoff over SM for both the fallow after wheat and fallow after sorghum phases (Figure 9.3). The largest average increase in runoff due to NT was 24 mm (0.9 in.), which occurred on fallow after sorghum because sorghum residues provided less protection against crusting than did wheat residues. Averaged across all four phases of the WSF rotation for 10 years, runoff from SM management was 30 mm (1.2 in.) per year compared to 44 mm (1.7 in.) per year from NT management. Precipitation averaged 517 mm (20.3 in.) per year.

Erosion

The smooth consolidated surface on NT that reduced infiltration and promoted runoff was resistant to water erosion and, in combination with the presence of residues on the surface, resulted in much lower sediment concentrations in runoff compared with SM. Total sediment loss from WSF was reduced 54% with NT. However, average annual soil losses from both tillage systems were <1300 kg/ha (1160 lb/acre) (Figure 9.3), well below the soil loss tolerance (T) of 11 mg/ha/yr (5 ton/acre/yr). Obviously, both NT and SM are effective tillage systems for controlling soil loss due to water erosion when used in conjunction with contour tillage and terracing. Both management systems maintained sufficient residues on the surface to control wind erosion.

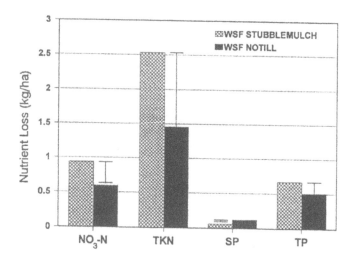

FIGURE 9.4   Average annual amount of nutrients lost in runoff, 1984–93, Bushland, Texas. TKN = total Kjeldahl N, SP = soluble P, and TP = total P. Bars indicate least significant difference ($p = 0.05$) for each pair.

Nutrients in Runoff

The total amount of nutrients lost in runoff from these unfertilized watersheds was very low (Figure 9.4), usually less than the amount of nutrients input from rainfall. The maximum average concentrations of nutrients (mg/L, ppm) observed in runoff for any WSF phase were NO_3^-–N, 3.71; total Kjeldahl N, 8.13; soluble phosphorus, 0.77; and total phosphorus, 2.70. Nitrate concentrations are well below maximum EPA guidelines. However, P concentrations were high enough to contribute to eutrophication of ponds or lakes and exceed EPA guidelines for P concentrations in water (U.S. EPA, 1973, 1976). Soluble P would be the primary form contributing to eutrophication. The finding of relatively high P concentrations in runoff from these unfertilized, terraced, and well-managed watersheds is largely due to fallow effects on P availability. The P contents we found are useful as baseline values in evaluating chemical quality of runoff water from cultivated land in semiarid areas. Potential hydrogen ion concentration (pH) ranged from 7.17 to 7.55 and maximum electrical conductivity of runoff water was 12.4 mS m^{-1} (1.2 µmhos), indicating high-quality water with low salt content.

Triazines in Runoff

Propazine was surface applied (no incorporation) to both SM and NT fields at sorghum planting when soil profile water contents were usually wet and runoff was likely if storms occurred. As expected, the initial runoff-producing storm following herbicide application produced greater herbicide concentrations, with subsequent storms having smaller concentrations. Propazine concentrations were greatest on SM for the initial sample-producing storm, but concentrations were similar on SM and NT watersheds for subsequent storms (Table 9.3). The maximum concentrations observed were 105 µg/L (ppb) in runoff from SM and 98 µg/L (ppb) from NT. The total amount of propazine lost in runoff from watersheds largely reflected the amount of runoff that occurred. Losses of 10–15 g/ha (0.1–0.2 oz/acre) were observed when large storms occurred within a month of application. Maximum loss in runoff from any application was 1.5%, which occurred with SM in 1990 and with NT in 1992.

Atrazine, surface applied at a rate nearly three times greater than propazine, had little loss (Table 9.3) because applications occurred immediately after wheat harvest, when the soil profile was dry and cracks extended 0.75 m (2.5 ft) into the soil. Rapid water infiltration rates with this soil condition resulted in few runoff-producing storms. Maximum atrazine loss observed was 5.9 g/ha (0.08 oz/acre) or 0.26% of application rate. The concentration of atrazine in runoff was large (160 µg/L, ppb) for the July 15, 1993 storm, but runoff volume was small; thus, total loss of atrazine was small.

Nitrate–Nitrogen in the Soil Profile

The soil profile contains large amounts of NO_3^-–N as a result of N mineralization in excess of plant requirements during the 30+ years the watersheds were farmed in a dryland WSF sequence with SM before establishing the current experiment in 1982 (Figure 9.5). Total NO_3^-–N contents for the 5.5-m (18-ft) profile were similar: 480 kg/ha (430 lb/acre) on NT and 548 kg/ha (490 lb/acre) on SM. However, the NO_3^-–N distribution within the profile has been significantly altered with NT. Improved water conservation and greater soil water contents with NT resulted in the leaching of over 80% of profile NO_3^-–N to depths below 1.8 m (6 ft), the maximum rooting depth of wheat or sorghum on Pullman clay loam. Maximum amounts of NO_3^-–N are accumulated at the 2.0-m (6.6-ft) depth with SM compared with a 3.4-m (11.1-ft) depth with NT. Our results agree with those of Eck and Jones (1992) in that differences in NO_3^-–N content of the root zone with dryland cropping appear to be primarily due to preferential leaching rather than to differential nitrification.

SUMMARY AND CONCLUSIONS

Environmental impacts of adopting NT management on dryland compared with the impacts of the lower chemical input SM management appear to be minimal for the southern High Plains. Storm runoff was greater from NT, which could benefit wildlife and playa (small wet-weather lakes) ecology, since 85% of land in the southern High Plains drains into playas rather than into streams. The sediment load in runoff was reduced 54% with NT, thus reducing potentially detrimental off-site effects of erosion. Nutrient concentrations and amounts in runoff were small for both NT and SM management, often less than precipitation inputs.

TABLE 9.3
Concentrations and Loss of Triazines in Runoff from Selected Watersheds and Storms

Date	Days after application	Precipitation	Runoff (mm) SM	Runoff (mm) NT	Triazine conc[a] (μg L^{-1}) SM	Triazine conc[a] (μg L^{-1}) NT	Triazine loss (g ha^{-1}) SM	Triazine loss (g ha^{-1}) NT	Triazine loss (% of applied) SM	Triazine loss (% of applied) NT
Propazine[a]										
July 2, 1990 (applied 1.10 kg ha^{-1} propazine, a.i.)										
July 16, 1990	14	5	0	0	—	—	—	—	—	—
July 20, 1990	18	51	23	8	65	47	15.1	3.9	1.37	0.35
August 16, 1990	45	29	15	6	11	19	1.6	1.2	0.14	0.10
June 17, 1992 (applied 1.34 kg ha^{-1} propazine, a.i.)										
June 21, 1992	4	19	1	4	105	73	0.5	2.8	0.04	0.21
June 22, 1992	5	25	11	14	102	98	11.0	13.3	0.82	0.99
June 27, 1992	10	19	5	4	71	78	3.3	3.2	0.25	0.24
July 10, 1992	23	31	3	3	35	35	0.9	1	0.07	0.07
June 14, 1993 (applied 1.34 kg ha^{-1} propazine, a.i.)										
June 19, 1993	5	39	8	3	—[b]	—[b]	—	—	—	—
July 15, 1993	31	73	36	31	11	47	3.9	14.4	0.29	1.07
Atrazine[c]										
June 30, 1989 (applied 3.36 kg ha^{-1} atrazine, a.i.)										
August 6, 1989	37	37	—	5	—	26	—	1.1	—	0.03
September 12, 1989	74	75	—	32	—	3	—	1.0	—	0.03
July 31, 1992 (applied 2.80 kg ha^{-1} atrazine, a.i.)										
August 26, 1992	26	47	—	10	—	4	—	0.4	—	0.01
July 12, 1993 (applied 2.24 kg ha^{-1} atrazine, a.i.)										
July 15, 1993	3	73	—	4	—	160	—	5.9	—	0.26

[a] Propazine was surface applied preemergence or sorghum.
[b] No samples obtained.
[c] Atrazine was surface applied on NT watersheds only.

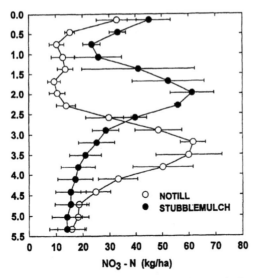

FIGURE 9.5   Soil profile NO$_3^-$–N contents of nonfertilized watersheds cropped in a dryland WSF rotation with SM or NT management. Sampled September 1993. Error bars = SD.

Increased leaching of NO$_3^-$–N to depths below the plant root zone, because of improved moisture conservation with NT management, could pose an environmental hazard when a water table is close to the surface. Fortunately, the depth to water tables range from 20 to 150 m (66 to 300 ft) in the southern High Plains and is about 75 m (246 ft) below the surface at the research site. Our data show the wetting front has not penetrated past the 5.5-m (18-ft) depth on the research watersheds; thus, with dryland management the wetting front may never reach the water table.

No evidence of atrazine accumulation in the soil or leaching below the root zone was found, and atrazine loss in runoff amounted to a maximum of only 0.26% of total atrazine applied and was usually much less. Propazine, applied for seasonal weed control in sorghum with both NT and SM management systems, appears to have a greater potential than atrazine for negatively impacting the environment because soil is wetter at time of application and greater runoff occurs. Losses up to 1.5% of the total propazine application were measured in storm runoff. Some propazine accumulation occurred in the soil, although it was undetected at depths greater than 0.6 m (2 ft) on either SM or NT watersheds.

Tillage management affected several soil quality indicators; however, the major effect was on soil crusting. With NT, surface crusts remained intact, severely reducing infiltration rate and cumulative infiltration, whereas with SM, crusts were destroyed by tillage and infiltration capacity was restored. Overall water conservation, however, was enhanced with NT management due to reduced evaporation.

NT management for grain production on dryland in the southern Great Plains has the potential to sequester carbon and improve soil quality relative to conventional SM management, albeit slowly, without adversely degrading the environment.

ACKNOWLEDGMENT

The authors gratefully acknowledge the contributions of Grant L. Johnson, Biological Technician, USDA-ARS, Bushland, Texas, in conducting this study.

REFERENCES

Allmaras, R.R., P.W. Unger, and D.W. Wilkins. 1985. Conservation tillage systems and soil productivity. In R.F. Follett and B.A. Stewart (eds.). *Soil Erosion and Crop Productions,* ASA-CSSA-SSSA, Madison, WI, pp. 387–412.

Conservation Tillage Information Center. 1993. Our HEL attitude. *Conserv. Impact* 11(10):1.

Eck, H.V. and O.R. Jones 1992. Soil nitrogen status as affected by tillage, crops, and crop sequences. *Agron. J.* 84:660–668.

Edwards, W.M., L.D. Norton, and C.E. Redmond. 1988. Characterizing macropores that affect infiltration into non-tilled soil. *Soil Sci. Soc. Am. J.* 52:483–487.

Hauser, V.L. and O.R. Jones. 1991. Runoff curve numbers for the southern High Plains. *Trans. ASAE* 34:142–148.

Jalota, S.K. and S.S. Prihar. 1990. Bare soil evaporation in relation to tillage. *Adv. Soil Sci.* 12:187–216.

Jones, O.R. 1975. Yields and water-use efficiencies of dryland winter wheat and grain sorghum production systems in the southern Great Plains. *Soil Sci. Soc. Am. J.* 39: 98–103.

Jones, O.R., V.L. Hauser, and T.W. Popham. 1994. No-tillage effects on infiltration, runoff, and water conservation on dryland. *Trans. ASAE* 37:473–479.

Jones, O.R., S.J. Smith, L.M. Smith, and A.N. Sharpley. 1995. Environmental impacts of dryland residue management systems in the southern High Plains. *J. Environ. Qual.* 24:453–460.

Jones, O.R., B.A. Stewart, and P.W. Unger. 1996. Management of dry-farmed southern Great Plains soils for sustained productivity. In E.A. Paul, K. Paustian, E.T. Elliot, and C.V. Cole (eds.). *Soil Organic Matter in Temperate Agroecosystems: Long Term Experiments in North America.* Lewis Publishers, Boca Raton, FL.

Karlen, D.L. and A.N. Sharpley. 1994. Management strategies for sustainable soil fertility. In J.L. Hatfield and D.L. Karlen (eds.). *Sustainable Agriculture Systems,* Lewis Publishers, Boca Raton, FL, pp. 47–108.

National Research Council. 1993. *Soil and Water Quality, An Agenda for Agriculture.* National Academy Press, Washington, D.C.

Potter, K.N., O.R. Jones, H.A. Torbert III, and P.W. Unger. 1997. Crop rotation and tillage effects on carbon sequestration in the semiarid southern Great Plains. *Soil Sci.* 162: 140–147.

SAS Institute. 1990. *SAS/STAT User's Guide. Version 6,* 4th ed., Vol. 2. SAS Institute, Cary, NC.

Smika, D.E. 1983. Soil water change as related to position of wheat straw mulch on the soil surface. *Soil Sci. Soc. Am. J.* 47:988–991.

Smika, D.E. and P.W. Unger. 1986. Effect of surface residues on soil water storage. *Adv. Soil Sci.* 5:111-138.

Steiner, J.L. 1994. Crop residue effects on water conservation. In P.W. Unger (ed.). *Managing Agricultural Residues.* Lewis Publishers, Boca Raton, FL, pp. 41–76.

Unger, P.W. 1992. Infiltration of simulated rainfall: tillage systems and crop residue effects. *Soil Sci. Soc. Am. J.* 56(1):283–289.

Unger, P.W. and F.B. Pringle. 1981. Pullman Soils: Distribution, Importance, Variability, and Management. Texas Agric. Exp. Stn. Bull. 1372.

U.S. Environmental Protection Agency. 1973. Water Quality Criteria 1972. USEPA Report R3-73-033. U.S. Government Printing Office, Washington, D.C.

U.S. Environmental Protection Agency. 1996. Quality Criteria for Water. U.S. Government Printing Office, Washington, D.C.

10   R for Soil Quality = Long-Term No-Till

Bobby G. Brock

INTRODUCTION

It is generally agreed that (1) any given soil can reach its best quality only through optimum levels of physical, chemical, and biological properties; (2) organic matter is the one soil component which most positively impacts the three aforementioned properties; and (3) no-till is the most feasible production technique to effect measurable, sustainable organic matter increases, especially in the top 5 cm of the profile.

It has been deemed feasible, until recently, to continue a no-till method only until it becomes necessary to incorporate certain elements or to alleviate an increased bulk density condition. This usually was necessary after only 3–4 years of no-till. Unfortunately, this conventional tillage produced a setback, or interruption, in the slow development of soil quality improvements.

Innovative farmers and research scientists, however, have stretched the imaginary limits and have shown that continuous no-till can extend, uninterrupted, upwards of 25–30 years and beyond.

Through both scientific scrutiny and on-farm application, it has been noted that significant, sustainable increases in organic matter content are achievable only in an extended absence of tillage with significant levels of residues maintained on the soil surface. Even a brief interruption of the no-till technique with conventional tillage tends to significantly increase the oxidation of organic matter, in addition to an interruption of other slowly developing improvements (Dean, personal communication, 1996).

The following is a discussion of the improvements to be gained in physical, chemical, and biological soil properties through long-term no-till.

1-57444-100-0/99/$0.00+$.50
© 1999 by Soil and Water Conservation Society

PHYSICAL

The development and maintenance of water-stable aggregates at the soil surface is, perhaps, the best indication of physical condition improvements. Such aggregate development cannot be adequately sustained through conventional cultivation, as there is not sufficient residue to offer protection against the development of surface crusting. Even when large amounts of organic material are incorporated, a naked soil surface will soon crust sufficiently to impede rainfall and/or irrigation water infiltration. This results in splash erosion, a re-sortment of surface particles, increased erosion, and a general deterioration of soil quality. Adequate surface cover is necessary before a given soil can reach its best quality.

Studies have shown that water-stable aggregate increases are achieved under long-term no-till. For example, on a Typic Hapludult of the southern piedmont, Langdale et al. (1992) reported a 50% increase in water-stable aggregates.

Even more striking were runoff losses from Hurricanes Erin and Opal in 1995, again from the P-1 watershed at Watkinsville, Georgia. After more than 20 years of no-till, these storms produced 0 and 4.26 mm from 14.96 and 12.7 cm, respectively (Crovetto, 1992).

Increases in rainfall infiltration and/or irrigation applications have been recorded by researchers and noted by farmers. Some of the most dramatic changes were documented by Langdale et al. (1992).

Average runoff from conventional and long-term no-till was 16.2 and 1.8%, respectively, while runoff losses from low-return-frequency storms were 52.6 and 21.8% (Langdale et al., 1992).

The free movement of air and water is essential for optimum plant growth. While research measurements have documented increases in bulk density under long-term no-till, air and water movements seem to remain adequate or even improve. Personal observations by this writer have noted dramatic contrasts in both infiltration and aeration between conventional and no-till fields, side by side. These differences were noted in clay loam and loamy sand soil types.

Increased organic matter content and greater infiltration combine to give greater available water. An example of this was reported by Crovetto (1992). Under conventional tillage, the top 20 cm had about 5% available water, while no-till after 7 years had in excess of 8%.

CHEMICAL

One of the more significant improvements through long-term no-till is the increase in the cation exchange capacity of the soil. This improvement has not been widely reported. However, Crovetto (1992) reported the cation exchange capacity to be 11 under conventional tillage, while it was 18 with no-till after 7 years on an Alphasol on his farm in Chile.

Increased nutrient levels were noted in a soil after 8 years of no-till in southeast Virginia (Virginia Polytechnic Institute, 1995). Greater quantities of phosphorus and potassium were found in the top 8 in. of the soil, as compared to conventional cultivation, and additionally, increased levels of calcium, magnesium, zinc, and manganese were found in the top 2 in. Percent organic matter was greater to a depth of 8 in. and was significantly greater at the 0- to 2- and 2- to 4-in. depths.

According to Griffith and Reetz (1994), conservation tillage is one practice most closely associated with a buildup and maintenance of soil organic matter levels. "As a result, a soil capacity to retain plant nutrients in the rooting zone is enhanced."

It has generally been considered that it is not feasible to achieve significant increases in organic matter in the sandy soils of the southeastern United States. However, Hunt et al. (1996) have achieved significant increase in carbon content of such soils through long-term no-till studies at the USDA-ARS experiment station near Florence, South Carolina. These results were much slower to be achieved than is common in finer textured soils. Strong differences were noted after only 7 years of continuous no-till. After 14 years, the mean carbon content at the 0- to 5-cm depth was nearly double that of conventional tillage plots. Numerical increases were evident down to the 15-cm depth.

BIOLOGICAL

This component of soil quality is rapidly catching the attention of research scientists and farmers alike. Studies and observations are beginning to establish a strong link between soil biology improvements and long-term no-till. For example, it was reported by the University of Georgia that no-till fields have 3.5 times more earthworms than conventional tillage fields (Dean, personal communication, 1996). Similar observations have been made in numerous studies in various locations.

Perhaps one of the more complete studies showing the relationship between long-term no-till and soil biology was made by Crovetto (1992). After 7 years of no-till on his farm near Concepcion, Chile, he notes significant increases in yeasts, algae, actinomycetes, both free-living and nitrogen-fixing bacteria, and several other forms of mesofauna.

Additionally, one of his more interesting findings was more than double the number of spores of vesicular arbuscular mycorrhiza fungi. These fungi serve as an extension of the crop root system by extending hyphae into the surrounding soil. These fungi do not establish well under cultivation or in a short time frame. Hence, long-term no-till is quite conducive to the development of this symbiotic relationship.

SUMMARY

This chapter is not intended to fully document the case for no-till as a requirement for a productive, healthy soil. No doubt such can be achieved under cultivation. But observation and research alike tend to show beyond any doubt that a soil cannot achieve a maximum, sustainable level of quality as long as it is naked and subjected to rapid losses of organic carbon through the stirring action of cultivation.

ACKNOWLEDGMENTS

Appreciation is expressed to Wendy Rudd, Richmond SWCD, Rockingham, NC; Jeff Raifsnyder, USDA-NRCS, Raleigh, NC; and Dorothy Moebius, USDA-NRCS, Raleigh, NC, for their assistance in the preparation of this chapter.

REFERENCES CITED

Crovetto, C.L. 1992. *Rastrojos Sobre El Suelo*. 301 pp.

Dean J.E., 1996. Conservation Agronomist, USDA-NRCS, Athens, GA, personal communication.

Griffith, W.K. and H.F. Reetz, Jr. 1994. Improved nutrient efficiency and organic matter build-up. *Better Crops* 78(2):7.

Hunt, P.G., D.L. Karlen, T.A. Matheny, and V. Quisenberry. 1996. Changes in carbon content of a Norfolk loamy sand after 14 years of conservation or conventional tillage. *J. Soil Water Conserv.* 51:255–258.

Langdale, G.W., W.C. Mills, and A.W. Thomas. 1992. Use of conservation tillage to retard erosive effects of large storms. *J. Soil Water Conserv.* 47:257–260.

Virginia Polytechnic Institute. 1995. Unpublished data. Results of chemical analyses on samples from several depths in two fields that were in no-till or conventional tillage for eight seasons.

11 Whole-Soil Knowledge and Management: A Foundation of Soil Quality

Robert McCallister and Peter Nowak

INTRODUCTION

Soil quality discussions have brought to the forefront the idea that soil is in a sense an "organism" (entity) with vital connections within its composition across space (King, 1895; Balfour, 1944; Rodale, 1945; Haberern, 1992). This soil organism has breadth and depth. Because of this three-dimensional nature, soils have differences in production capacity, susceptibility to degradation, and regenerative capabilities (Wolman, 1985).

The central idea of this chapter is that soil management by land users, and public programs to support land-users' attempts to achieve sustainable soil quality, should consider the importance of connections *within the whole soil profile* as it changes across the landscape. The link between soil erosion and its effect upon whole-soil quality is a prime reason for better understanding of the full soil profile by farmers. A Wisconsin case study is used to underscore the importance of whole-soil knowledge.

EROSION AND SOIL QUALITY

This chapter does not discuss detailed specifics about the effects of erosion upon agricultural soil quality. Other authors within this collection attend to these issues with care and expertise. As background, however, a selection of works that have examined connections between soil quality, soil erosion, and the whole soil profile is presented.

1-57444-100-0/99/$0.00+$.50

Soil erosion by humans is a problem because it occurs more rapidly than the process of soil formation. People-induced erosion plunders quickly, while pedogenesis creates methodically. With loss of fertile topsoil rich in available nutrients, the reduction of crop production capacity is well documented (Odell, 1950; Larson et al., 1983; Lal, 1987). Loss of organic matter and accompanying soil life adversely affects soil physical properties for plant growth, decreasing pore space and soil/root hair connectivity and increasing soil bulk densities (Karlen et al., 1990; Edwards, 1991). These management-caused injuries to topsoil quality can increase the inherent erodibility (K factor) of the soil (Gersmehl et al., 1989), creating a self-perpetuating downward spiral in soil quality.

Erosion may profoundly influence pedogenic processes, leading to textural changes and horizonation transformation throughout the whole soil profile (Mokma et al., 1996). Crusts formed by transported sediments and rain-pounded surfaces (Morin and Van Winkel, 1996) may block water from entering into the mineral soil as readily, changing the weathering and biochemical interactions within the horizons below (Bruce et al., 1988; Power, 1990). Erosion also reduces a soil's buffering capacity to mitigate water pollution from land-applied chemicals (National Research Council, 1993).

These erosion-caused changes in soil morphology clearly indicate that crop-related impacts of soil erosion are not confined to the surficial topsoil. As topsoil thickness decreases, the remaining soil solum becomes more important as a rooting medium. Root lengths of many common crops and forages extend well beyond topsoil depths to find vital water and nutrients within the lower soil solum (Taylor and Terrell, 1982; Aldrich et al., 1975). In eroded conditions, rooting environments based primarily within the subsoil have a higher likelihood of being a stressful environment for plant roots. Inadequate fertility, chemical imbalances, physical restrictions, and too much or too little water may be limiting factors heightened by a "subsoil-only" root zone (Winters and Simonson, 1950; National Research Council, 1993). Literally, the depth of the solum is a primary determinant of the available water capacity of the root zone and crop productivity (Pierce et al., 1983; Pierce, 1991).

Passage of time (Jenny, 1980) makes soil a four-dimensional body. Over time, soils are either developing or degrading. Impacts of time can be concentrated. This is particularly true in the interplay between human activity and, relative to geomorphological processes, its rapid influence on soil quality. Fifty-plus years of honest effort has shown us that soil erosion mitigation and soil quality improvement are obviously not matters that we can fix and then move away from. Soil quality improvement and erosion control are matters that must be engaged over long time frames. They need steadfast attention. An appreciation and understanding of the whole soil by those who manage it is an important part of this ongoing need.

Whole-soil quality in agricultural settings requires whole-soil management. Whole-soil management should begin with knowledge of the whole soil that one has to work with, not just the topsoil. What is the current status of whole-soil knowledge by farmers and the status of whole-soil concepts within government programs? What might help whole-soil management become a part of ecologically based whole-farm planning? These are critical questions to ponder if soil quality is ever to become a central part of ecologically based farm and ranch management.

WHOLE-SOIL KNOWLEDGE AND MANAGEMENT

Farmer Soil Knowledge

Farmers have a vast amount of practical knowledge about how their soils affect crop productivity. The strengths of their knowledge should be important contributions to soil education. Currently, this resource is greatly underutilized and should be much more vigorously pursued. A brief review (Table 11.1) of farmers' classification of soil indicates that published research about farmer soil knowledge has largely been conducted in less mechanized agricultural societies outside of North America.

Over the centuries, nonwestern agriculture has amassed an enormous cache of practical soil information that each successive generation taps to produce crops (Rajasekaran and Warren, 1995). Some local soil knowledge, such as in the Colca Valley of Peru (Sandor and Furbee, 1996), has developed into multilevel taxonomies that help farmers make field-level management decisions. As these societies become westernized, indigenous knowledge, including soil knowledge, is rapidly vanishing. In view of this rapid cultural transformation, Birmingham (1996) states that too few studies have investigated how local peoples classify and distinguish land and soil. Pawluk and colleagues (1992) concur that indigenous soil knowledge is still understudied, even though there have been recent improvements in the situation (they describe the formation of a new center to study agricultural indigenous knowledge in nonwestern societies at Iowa State University).

Chambers and colleagues (1989) explain that the categories and names used by nonwestern farmers to describe types of land, plants, and pests usually differ from those used by scientists. The farmer criteria for these classifications are usually functionally related to use, unlike the morphologic categorizations derived by soil scientists (Chambers et al., 1989). Pawluk and colleagues (1992) agree, adding that classifications resulting from astute observation of the local environment may lead to solutions in production problems. In her research of local soil knowledge in western Africa, Birmingham (1996) also reports that local soil classification schemes are closely related to crop productivity.

Given that the topic of soil knowledge has been formally studied in many other settings, the lack of published study in the North American setting is puzzling. Does this imply that indigenous knowledge of the soil in North America has been viewed as inconsistent with an industrializing agriculture? Or is the implication that soil knowledge has been whittled away into a remnant of its former self no longer worth studying (if the knowledge ever widely existed).

An exception to this dearth of North American study of soil knowledge is the renewed interest in soil quality/health. Recent reviews of research results suggest physical and chemical soil health indicators (Arshad and Coen, 1992), biological (Visser and Parkinson, 1992) and nutritional indicators (Karlen et al., 1992), and reference point soil ecosystems (Granatstein and Bezdicek, 1992) as standards for

TABLE 11.1
Local Systems for Recognition of Different Soil Types

Researcher and date	Location	Dominant soil features classified	Classification function
Williams and Ortiz-Solomo (1981)	Mexico	Surface soil distinctions in color, texture, consistence, moisture, workability, vegetation	Soil capabilities for crop production
Rozas (1985) in DeBoef et al. (1993)	Peru	Indicator plants, location related to altitude and frosts, climate, topography	Cropping, allocation of growing land according to social status and age
Kean (1987)	Zambia	Riverine strips and dambos (areas of retained moisture)	Microenvironments important to local farming systems
Edwards, R. (1987)	Botswana	Mothlaba, mokata, and seloko soils	Distinctions for sorghum production
Barrow (1987)	Kenya	Lowland areas, hill areas with perennial grasses, hard-time hills	Grazing land distinctions
Mosi et al. (1991)	India	Over 30 distinct soils: soil depth (because of drought-prone area), tillage attributes, changes in hardness, crusting	Stability of crop performance, crop productivity
in Tabor (1992)	Senegal, Mauritania, Mali	Kolanga: lower floodplain and flood longer, khare: floodplain higher	Differentiate location for rice and sorghum
in Tabor (1992)	Haiti	Farmer designation of soils by color, texture, landscape position	Productivity
in Tabor (1992)	Senegal	Riparian soils — walo: annually floods by Senegal, changoul: lower watersheds, salka: upper watersheds	Crop productivity ratings, timing of management activities
Zimmerer (1994)	Bolivia	Three nested levels: (1) field or pasture, (2) surface texture, (3) surface color	Crop suitability, water and soil conservation goals
Sandor and Furbee (1996)	Peru	Terrace agriculture: textural and plant-related >50 soil types, some knowledge of rooting in subhorizons, depths	Assist overall agricultural management of terraced fields under traditional methods

interpreting changes in soil health. A common thread to these recommendations is that usable soil quality indexes should fit, to some degree, the sensory abilities of the managers of the land where the soil lies. A good example of meeting this practical need is occurring in Nebraska, where on-farm tool kits are being devised

TABLE 11.2
Ten Most Important Descriptors of Soil Health from
Romig et al. (1995)

Rank	Soil health property (plow layer emphasis)	Rank	Soil health property (plow layer emphasis)
1	Organic matter	6	Tillage ease
2	Crop appearance	7	Surface soil structure
3	Erosion	8	pH of topsoil
4	Earthworms	9	Plow layer soil test
5	Drainage	10	Crop yield

From Table 1 in Romig et al., 1995.

so that farmers can measure soil respiration, bulk density, soil water content, electrical conductivity, pH, nitrate nitrogen, infiltration, and water-holding capacity (Doran, 1994).

Specific to farmer knowledge about soil health, the Wisconsin Soil Health Program collected information from a number of interested farmers about their perspective of surface soil quality/health. To build recognition of "what makes a healthy soil," Romig and colleagues (1995) collaborated with Wisconsin farmers who were cooperators with a university-sponsored integrated cropping system research project. The farmer/researcher team derived and weighted soil health criteria of the plow layer (Table 11.2).

While the topsoil health criteria from 28 motivated farmers with university ties are not representative of the whole-soil health knowledge of farmers in general, the project provided an empirical base of ideas for testing with a wider farm audience. The work resulted in a prototype for soil health scorecards to be used by farmers to measure plow layer features and crop/surface soil interactions (Romig et al., 1995). The project also exemplifies the "learning organization" method (Vorley and Keeney, 1995) of moving toward mutually beneficial research in agriculture.

Although local farmer knowledge may be voluminous and underappreciated, farmer knowledge can, however, be overromanticized as to its wisdom. In his study of Honduran farmers and their plants and pests, Bentley (1989) explains that what farmers misconceive or do not know will likely not help them. For example, some Honduran farmers believe that "weevils" spontaneously generate, growing from pinhead size to adults within the grain plants. The young "weevils" that one farmer showed Bentley were actually wasp larvae that live among and parasitize the real weevil pests (Bentley, 1989). This incorrect knowledge will certainly not help this farmer alleviate any weevil problem, and the farmer could gain from understanding the benefit of the wasp larvae that he thought were pests.

The same negative values for misconceptions apply to how soil may be understood by farmers. To illustrate, a number of midwestern farmers have the conception that groundwater flows through soil in large Mississippi River–sized confluences. While this phenomenon might occur in some very select locations, hydrogeologic explorations tell us that this is generally not the way that groundwater moves from one location to the next (Freeze and Cherry, 1979). This misconception of groundwater flow will not help these farmers keep their wells from becoming polluted. "If one of these rivers is not running under my fields," a farmer might think, "how could any of my field inputs get into the groundwater?" When these individuals watch a backhoe dig a pit on their land, they are surprised to see most of the visible water flowing into the pit coming from discreet cracks and channels rather than the proverbial "underground river from Canada" (it is an exhilarating "teachable moment").

Ask these individuals what it takes to produce a good corn crop on the same field and their knowledge is top-notch. From purely a production standpoint, who could argue it, with their high yields year after year. However, this application of production-related knowledge may not be using inputs or money efficiently or in an environmentally benign fashion. A strictly pragmatic crop production view of soil neglects aspects important to maintenance of water and soil quality over many generations. In the 1990s, farmers are being asked to acknowledge more than their soils' production capabilities and to also consider how their management will affect soil quality for the coming generations.

Taking responsibility for improving the land ecosystem in mind of the distant future, the concept of land stewardship is based on the underlying presumption that farmers know the capabilities and limitations of their soils. Farmers need a basic understanding of how their soils change with depth and breadth in order to appreciate their soils' strengths and weaknesses. Therefore, a fundamental question aimed at improving long-term soil quality must be asked: How well do farmers understand their three-dimensional soil landscape?

THE WISCONSIN CASE

A representative sample of Wisconsin corn farmers were asked about the soil within one of their cornfields. The 745 farmers who participated in the study related soil knowledge and management information for the cornfield and general socioeconomic information for the farm operation. Farmer responses were compared to data derived from USDA county soil surveys for the same cornfield (refer to McCallister [1996] for further details).

Overview of Case Study Methods

This study combined data that were site specific within both the social science and physical science realms. Data collection would not have been possible without the cooperation of USDA's National Resource Conservation Service (NRCS) and Consolidated Farm Service Agency and the Wisconsin Geological and Natural History Survey.

Figure 11.1 is a "road map" of how a diverse array of information collection methods were integrated and carried out. These disparate data sets were pooled into a unified format for statistical analysis. Within the separate data sets, information

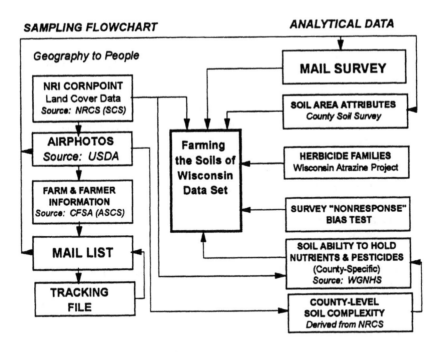

FIGURE 11.1   Information collection.

was tagged with an ID number to enable matching of same-case data across the data sets. Each case was then constructed by collecting all data tagged with the same ID number.

Sampling

Locational point sampling has evolved to combine the positive aspects of both random and systematic techniques, while minimizing the negatives (Holmes, 1967; Haggett, 1990). Berry and Baker (1968) developed a technique that benefits from the spatially unaligned aspect of random point sampling, while providing the blanket coverage of a systematic sample. The general design for this stratified systematic unaligned sampling had been used earlier by Berry (1962) for studying agricultural land within floodplains. The USDA also uses a stratified systematic unaligned framework to choose the sample points for its periodic National Resource Inventories (NRI).

National inventories of rural land have been conducted numerous times over the past 70 years by the U.S. government (Soil Conservation Service [SCS], 1989). The 1987 NRI sample frame (used for this research) was influenced by five previous geographically specific inventories in 1958, 1967, 1975, 1977, and 1982. Because of the overall sampling framework and data content, the 1987 NRI was a good fit for the project because it provided:

1. A well-designed pseudo-random geographic sample for inferential statistical work
2. Coverage of whole state with prospects of project being generalizable to state level
3. Site-specific physical data for each sample location
4. Ability to include only cornfields in project sample and data set
5. Disperse sites representative of cornfield density, neither clumped nor too far apart
6. Access to air photos of familiar scale for mail survey that matched sample sites
7. Possibility of linking to Agricultural Stabilization and Conservation Service farmer and farm data via air photos
8. Linkage to more specific soil data via air photos and county soil survey air photos

In essence, the NRI database was used to derive a representative sample that mirrors the characteristics of the land and the people managing it. It provided a sample of a certain type of place (cornfields) and access to the people (principal operators) who manage them.

This project was given clearance by the USDA to allow linkage of NRI sample point data to its geographic location under a confidential agreement. A reciprocal benefit to NRCS matches NRCS's goal that the NRI can be an integrating framework for socioeconomic and physical analyses related to soil and water conservation (SCS, 1989). Recently, Paul Johnson, chief of NRCS, called upon an expert panel to analyze the information network of NRCS. The panel recommended that partnerships be pursued so that external clients (such as this project) have ready access to NRCS data for mutually beneficial research (NRCS, 1995).

The 62 main corn-growing counties of Wisconsin were represented in the NRI-derived data set. Permission to contact operators of NRI sample points was obtained through correspondence by phone and letter to NRCS and Farm Service Agency officials. Table 11.3 shows the numbers within the sample at each step of the sampling process. In all, 67% or 745 of the 1118 surveys were usable returns to infer findings to the state level (3.5% sampling error at the 95% confidence level).

TABLE 11.3
Sample-Related Numbers: From NRI Points to Final Valid Sample

1987 NRI points	1987 NRI corn points	1987 NRI corn points connectable to entity	Nonredundant entities prior to any contact	Valid sample farms/farmers in project
6534	1795	1725	1372	1118

Note: Cases became invalid because farmer deceased, farmer redundant, farm field invalid, information unavailable.

Survey Procedure

A mail survey instrument was developed to collect information from the farmers operating the NRI-chosen cornfields. The 745 farmers who completed the survey provided socioeconomic information for their operations and soil knowledge and management data for the cornfields containing the NRI sampling point: (1) field description, (2) soil at the NRI corn point, (3) other soils in the chosen cornfield, (4) field input management and soil variation, (5) farm planning, and (6) farmer comments and request for information bulletins.

Each mail survey contained a USDA air photo (scale: 1 in. = 660 ft) that was glued to the inside of the foldout front cover of the survey. A small red dot, at the NRI sample point, was affixed to the air photo. To help farmers get their bearings, each survey air photo had a label with county, township, and section, as well as north arrow and scale. Prior to the first survey question, the farmers were advised to leave the air photo folded out for visual reference as they filled out the survey.

Survey questions were written in a closed-ended multiple choice format. While this causes certain limitations due to restrictions of possible responses, it provides benefits in usability and analysis of the data (Sheatsley, 1983).

Dillman's attention to detail in mail survey research (Dillman, 1978; Dillman et al., 1993) served as a guiding principle to best fulfill the potential of the mail survey format, while acknowledging the limitations of the format: (1) advance letter, (2) first copy of survey, (3) reminder letter, (4) second copy of survey, (5) second reminder letter, (6) third copy of survey with incentive offer, and (7) report synopsis and incentives mailed to participants.

Pretesting of a questionnaire by cohorts of the target audience generally leads to improvement of the instrument and answers the basic question, "Does it work?" (Backstrom and Hursh, 1963; Weisberg and Bowen, 1977). Six face-to-face pretests of the survey were conducted to refine content, wording, and structure. The six farmer pretesters were chosen by a project staffer to emulate the array of ages, farm enterprises, and aptitudes that would be found in the project sample of farmers. Pretest interviews were conducted on the farm or at the local town hall meeting building. These farmer pretests significantly improved survey content and overall survey palatability.

After the survey process was completed, five questions were selected from the survey and mailed to nonrespondents. This nonresponse bias test was used to determine if the project was missing a significant part of the soil knowledge story because these people did not fully participate. No statistically significant differences were found between the respondent and nonrespondent groups.

Collecting Comparative Soil Information at the NRI Sample Point

Farmer ideas about the soil at the NRI sample point were compared with soil profile descriptions for the soil mapped at that point within the county soil surveys. Generalizations within county soil survey soil mapping units (Mausbach and Wilding, 1991; McCallister, 1992) on aerial photographs are no exception to the scale-related limitations inherent in any maps (Edmonds et al., 1985). Modern soil surveys, such

TABLE 11.4
Data Collected from County USDA Soil Survey Descriptions

Overall soil	Topsoil	Subsoil	Parent material
Soil series name, yield potential, subsoil fertility	Texture of A horizon	Texture of B horizons	Texture of C horizon
Soil mapping unit including slope and erosion phase	Depth of A horizon	Subsoil permeability based upon particle size class	Depth of solum or depth to top of C horizon
Soil drainage class	A horizon pH category		
WGNHS rank and score for attenuation of agrichemicals	% organic matter content		

as the Soil Survey of Monroe County, Wisconsin (Barndt and Langton, 1981), refer to the limitations of scale in mapping and data collection in their description of "inclusions" within soil mapping units. At any given location within a polygon, such as at the 50-ft-wide NRI sample point, there is some unknown probability that the soils there have characteristics that fall outside the parameters for the soil designated in that mapping unit. With recognition of these limitations, this best available information was used to compare to farmer ideas about the soil at the NRI sample point.

Profile descriptions in county soil surveys are used by the Wisconsin Geological and Natural History Survey (WGNHS) to assess the ability of soil to attenuate (hold) agrichemicals.

Cates and Madison (1991) use seven soil characteristics in this soil assessment. The data collected for the 745 NRI sample points followed the county soil survey–based WGNHS data collection format with some additions, as shown in Table 11.4.

SPSS Data Entry, Verification, and Cleaning

Data from the mail surveys and soils information were entered into Statistical Package for Social Sciences (SPSS) using the PC version of the SPSS data entry program. Data entered into SPSS format were cleaned and verified. All detected mistranscriptions were repaired. Descriptive and inferential statistical analyses were carried out using SPSS release 4.0 on a UNIX mainframe system.

RESULTS OF THE WISCONSIN CASE

Topsoil

Knowledge of topsoil is important to farmers because topsoil characteristics greatly influence soil quality in terms of vulnerability to erosion, crop suitability, workability, water-holding capacity, soil warming, natural fertility and nutrient/pesticide efficiencies, rooting patterns, planting densities, and soil biota. Knowledge of topsoil conditions is practically useful for timely management of fields, from seedbed preparation to harvest.

- Farmers' knowledge of the soil at the sample location in their cornfields matched the SCS soil survey data for mineral topsoil textures as follows: loamy or silty textures were matched most frequently (47% match), sandy textures were matched next most frequently (40% match), and clayey textures were matched least frequently (34% match). Although no match exceeded 50%, the trend indicates that farmers may recognize the presence of loamy/silty topsoil textures better than sandy or particularly clayey topsoil textures.

- Most farmers appear to think that their topsoil is about the same depth as the reach or plowing depth of their tillage implements. Figure 11.2 indicates the following for farmers reporting a topsoil depth. When topsoil is shallow (up to 4 in. deep), farmers tend to think their topsoil is deeper, that is, 74% of these cases indicated that the topsoil is within tillage implement depth (4–12 in. deep). When topsoil is at tillage implement depth (4–12 in. deep), most farmers (83% of cases) concur. When topsoil depth is greater than 1 ft deep, 71% of these farmers think their topsoil is shallower, again up in the range of tillage implement depth. While plow-disturbed A (Ap) horizons are common, most farmers do not appear to recognize enduring A horizon depths that are either shallower or deeper than tillage implement depth. Recall that topsoil depth is a critical determinant of productivity (Odell, 1950) and soil buffering of pollutants.

- Five comparative organic matter categories were derived from Cates and Madison's (1991) figures that categorize organic matter content by soil order, erosion state, wet soils (Aquollic, aquic), organic soils

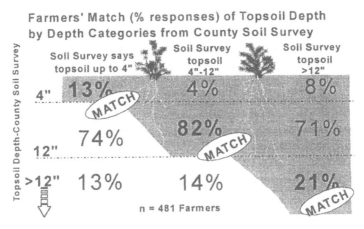

FIGURE 11.2 Farmer estimate of topsoil depth at sample location in cornfield.

(Histosols), or lithic soils (bedrock within 20-in. depth). Only 5% of the
farmers accurately matched the organic matter content category for the
location in their cornfields. In fact, over one-half (51%) of the farmers
answering this question had no idea what the organic matter content was
and checked the "not sure of organic matter content" response. Given that
Karlen and colleagues (1990), Edwards (1991), and many others have
shown that organic matter is critically important to soil quality as measured by fertility, porosity and infiltration, soil biota, soil/root hair connectivity, and bulk densities, farmers' knowledge about organic matter
needs vast improvement.

Subsoil and Soil Parent Material

Knowledge of subsoil is important to farmers because subsoil greatly influences
availability of water to crop roots particularly during peak demands in mid to late
summer, nutrient/pesticide efficiencies and leaching potential, natural fertility and
biota of the soil, and whether the soil encourages or impedes optimum crop rooting
patterns (Winters and Simonson, 1950; National Research Council, 1993). Fewer
earthworms may lead to loss of tilth/workability and less water infiltration to be
stored in the subsoil (Kladivko and Timmenga, 1990). Rooting may be affected by
physical (e.g., claypans) and chemical (e.g., unweathered carbonates) limitations or
low water-holding capacity of subsoils/parent materials (Black, 1957) that may be
brought closer to the surface by erosion. As "source" material of the soil solum, the
depth to and substance of the soil parent material affects all the soil above and is a
defining factor in the overall morphology of the whole soil profile.

- Textural knowledge of the subsoil at the location in the cornfield is less
  than textural knowledge of the topsoil. Only 30% matched three simplified texture categories of mostly sandy, mostly silty, or mostly clayey
  subsoil textures as compared to the county soil survey. Farmers appear to
  recognize clayey subsoil textures better than sandy or particularly loamy/
  silty subsoil textures.
- According to Pierce and colleagues (1983), the depth of the solum is a
  primary determinant of the available water capacity of the root zone and
  crop productivity. This key idea of solum depth, or depth to parent material at the sample location in the cornfield, was a real mystery to
  farmers. As Figure 11.3 shows, 92% of farmers did not match within
  ±25% of the solum depth given in the county soil survey. As a whole, it
  appears that farmers' understanding of the soil falls off dramatically with
  depth into the soil profile.

Seeing Surficial Change But Not Seeing Erosion

The farmers indicated that there were on average nearly three (2.9) different soils in
the cornfield (average size of field = 24 acres). This is a slight underestimate when

Farmer & County Soil Survey Correspondence:
DEPTH TO TOP OF PARENT MATERIAL

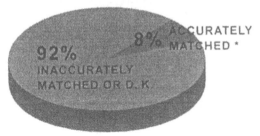

n = 742 Farmers   *farmers matched to within +/- 25% of depth in county soil survey

FIGURE 11.3 Farmer estimate of depth to parent material at sample location in cornfield.

compared to the average number of soil series (3.6) or average number of soil mapping units (3.7) in the cornfields as a whole. The question then arises how farmers recognize soil variation in their cornfields.

A Pearson's correlation between farmer "best estimate" of number of different soils and the separate components of the USDA soil mapping units (soil series, slope class, erosion phase) yielded insignificant correlations. When all three components (soil series, slope class, erosion phase) were combined for a composite number of different soil mapping units in the cornfield, the correlation coefficient was +0.26 (significance = <0.001). Yet, this is not a high level of correlation. At most, farmers overall may have a slight ability to denote soil variation coincident with variation in the "bundle of characteristics" that make a soil mapping unit. Farmers do not appear to make strong distinctions in soil variation that coincide with number of mapped erosion phases on their cornfields. Of any single soil characteristic, number of different erosion phases within the cornfield had the lowest correlation (+0.05) with farmer estimates of number of different soils.

If they don't think in soil mapping unit characteristics, what do farmers use to denote differences in soil? Farmers were asked which of a list of 17 possible clues "tell you there are different soils on the cornfield." Over 40% of all farmers responding to the question indicated that they use one of the clues listed in Figure 11.4 to recognize different soils on the cornfield. Note that these surficial and production-oriented clues are largely interpretable from the tractor seat. Although changes in surface soil color may denote differences in amount of erosion, less than 25% of respondents checked "differences in soil erosion" as a clue. This is interesting in that soil color (surface) was the most often cited clue to distinguish different soils in the cornfield. It is not inconceivable that farmers may generally be "seeing erosion" as color changes but not thinking of these color changes as denoting soil quality differences associated with erosion.

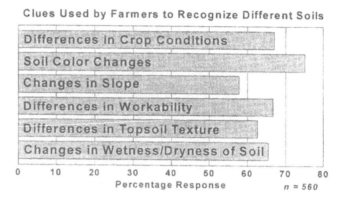

FIGURE 11.4 How Wisconsin farmers denote soil changes in crop fields.

BRIGHT SPOTS FOR WHOLE-SOIL KNOWLEDGE AND MANAGEMENT

Although subfield management is not yet common, many farmers manage separate fields differently at least partly due to their concepts of how soils vary from field to field. Farmers were asked what soil factors caused them to change application rates of herbicides, fertilizer, and manure between different cornfields. Figure 11.5 shows how frequently ten different soil factors were cited by farmers as causing them to change application rates between separate cornfields. Over half of the farmers answering this question cited changes in fertility of the crop root zone (68%), differences in topsoil texture (60%), and organic matter changes (59%). Changes in drainage condition and differences in soil ability to "hold" nutrients/herbicides were

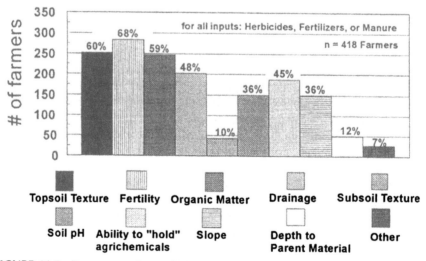

FIGURE 11.5 Farmers say these soil characteristics help to determine application rates.

cited by over 40% of responding farmers. Note that subsoil texture and depth to parent material, factors with very low knowledge scores, are very infrequently cited as causing farmers to change input rates.

Recall that farmers often use differences in topsoil texture and changes in wetness/dryness of soil as clues to distinguish different soils. It makes sense that they would cite these same most identifiable characteristics when listing soil factors that cause changes in management. However, farmers had very low knowledge scores for organic matter content and did not often cite either organic matter, or factors related to the crop root zone such as fertility, or attenuation ability as clues to distinguish different soils. Does this indicate that these are soil characteristics that farmers understand least well, but that they also claim are important to their management decisions? It may be so. Farmers may be simply repeating production-related soil characteristics that are used by private (crop consultants, agrichemical dealers) and public (extension, conservation agencies) sources. The farmers may understand that these soil characteristics are important without understanding how to identify them within their crop fields or how production decisions may in turn influence the nature of these soil features. To a certain extent, farmers may be able to identify "what *I should* know about soil," but still not understand the full soil profile very well.

NEEDS RELATED TO WHOLE-SOIL KNOWLEDGE AND MANAGEMENT

Farmers have a decent understanding of surface soil characteristics associated with crop productivity. However, they generally do not have a good understanding of the whole-soil mechanisms that influence long-term soil quality for sustainable crop productivity. Soil knowledge falls off considerably with depth into the soil profile. Farmers' overall knowledge of soil and how it varies could be much better.

Ultimately, and based on a solid scientific foundation, the research emphases pertaining to soil quality need to be translated into practical recommendations that can be applied on a farmer's field. The applicability and utility of those recommendations will largely depend on farmers' knowledge of the whole soil profile across the land they operate. As scientific knowledge of soil quality characteristics continues to emerge, it needs to be integrated with the practical types of soil knowledge that many farmers possess. Preliminary results from this research suggest the following factors be taken into consideration:

- Farmers read the soil from its visual surface clues and its workability. Use soil surface color as an entrance concept to other soil characteristics, such as organic matter content, erosion and its effects, natural fertility, subsoil differences, and drainage conditions.
- Variation in the lower root zone needs emphasis. Use the concept of soil quality/health to pull farmers' thoughts downward into the soil profile along the root channels and earthworm burrows to encourage whole-soil stewardship.

- Keep a dynamic spark in soil survey information. Some soil scientist(s) should remain local after county soil survey mapping is complete for interpretation of soil information to an array of users.
- Do not neglect the usefulness of county soil surveys. Use the county soil surveys in paper (often underutilized) and in new computerized formats.
- Publish simplified and picture-oriented *Field Guide to Soils* to accompany more encyclopedic county soil survey. The overall format could be set nationally, yet each county could plug in its own descriptive narrative and photos.
- Fully utilize the educational tools that have already been developed. For example, update Roe's (1952) comprehensive work that was aimed at farmers and expand upon Romig and colleagues' (1995) soil health scorecard.
- Public agency people need to encourage listening between conventional and low-input/organic farmers, agribusiness, consultants, and farm suppliers. Collaborative learning between this mix of people and soil scientists should occur on agency/cooperator farms using whole soil profile hands-on bull sessions.
- Farmer-oriented site-specific management technologies may help some people better understand how their soils vary at the subfield level. Ultimately, site-specific management (e.g., GPS, yield maps, VRT) will move beyond simplistic correlations of nutrient applications to yields and begin to acknowledge the importance of the distribution of soil quality features across a field. Data on spatially referenced yields across time will become the mode for teaching about the impact of different soil quality indicators on crop yields under different growing conditions. Larger cash grain operations appear to be best prepared for this technology.

WHOLE-SOIL QUALITY POLICIES AND PROGRAMS?

Much of the knowledge gained about soil quality will be organized and disseminated by government agencies. Yet, government policies and agency programs have often been primarily targeted toward a particular layer of the soil: the topsoil (e.g., soil erosion farm planning), the subsoil (e.g., salinity control programs), or the soil parent material (e.g., groundwater protection programs). If farmers are to become better equipped to manage their land with the whole soil in mind, then agency programs should consider moving away from piecemeal approaches to help farmers attain a renewed emphasis on holistic soil management.

By knowing their soils well, farmers may be more empowered to help direct soil quality programs, suggest particular conservation practices, or deflect unreasonable requirements on their management plans. For example, some farmers worry that minimum or no-tillage systems will become required practices to grow corn on any ground. As a case in point, one farmer commented, "I would like to see the bottom line on no-tilling heavy clay ground before this is pushed on us with no profit return to pay for our high taxes and expenses. Heavy clay cannot be run the same as nice drained loam ground." This farmer will have an even better argument if he can explain, from a full soil profile perspective, why no-till does not work as well on his particular soils.

From federal land retirement programs to local property tax assessments, farmers who have a fundamental knowledge of their soils have the power to help define how these programs and policies affect them. As opposed to arm-waving diatribes, reasonable arguments based upon the facts of the land are stronger justifications for administrators to take note of a farmer's unique situation and grant waivers to generic rules that might be unchallenged by those less knowledgeable.

BRIGHT SPOTS FOR WHOLE-SOIL QUALITY PROGRAMS

Planting flexibility in the 1996 Farm Bill could promote more diverse crop rotations to build soil. Also in the Farm Bill, EQUIP targets environmentally sensitive land for cost-share practices to meet soil as well as water quality objectives. Nutrient management plans (590) are labor intensive but consider surface and groundwater as well as erosion control. Formation of the NRCS soil quality institute should lead to innovations to help farmers think about the whole soil.

NEEDS RELATED TO WHOLE-SOIL QUALITY POLICIES AND PROGRAMS

Conservation programs must consider the quality of the landscape across horizontal and vertical dimensions. All soil layers are important in whole-soil quality improvement programs. Suggestions for considering soil quality information within government programs are as follows:

- Public agencies must understand current levels of soil knowledge in the general farm populace and design flexible programs accordingly.
- Farmers and nonfarmers alike need to hear more about the abundant and long-standing scientific evidence that whole-soil quality is critical for sustainable productivity and environmental protection.
- Programs should emphasize that erosion may affect the whole soil profile, beyond the physical loss of topsoil and its fertility. For example, fewer earthworms may lead to loss of tilth/workability and less water infiltration to be stored in the subsoil. Rooting may be affected by physical (e.g., claypans) and chemical (e.g., unweathered carbonates) limitations or low water-holding capacity of subsoils/parent materials brought closer to the surface by erosion.
- GIS under the direction of experienced field personnel may help target priority areas for conservation agency programs considering topsoil, subsoil, and parent material. Farmer abilities, needs, and current management strategies should be considered in any GIS-aided program development.
- Some agency programs should be aimed at the private sector to provide educational materials and to encourage discussion with customers about

management impacts upon the whole soil, not just on production-oriented output.
- Whole-soil management campaigns should be essential components of any serious attempts at programs aimed at ecologically based whole-farm planning.

CONCLUSION

This discussion has advanced that farmers' current knowledge of the whole soil must be considered in research, policies, and programs to promote long-term soil quality. The example case indicates that the soil knowledge base upon which to build whole-soil quality stewardship is in need of fundamental improvement.

Just as conservation structures need maintenance and upkeep, there is no such thing as the "permanent installation" of soil knowledge and a conservation ethic behavior among land managers such as farmers. Instead, soil knowledge and soil-related behavior need regular upkeep. The tools for upkeep and maintenance include cooperative education and policies to advance our understanding of soil as a four-dimensional entity. Soil quality/soil erosion researchers should address this practical need by collaborative learning with a wide variety of farmers under the real-world conditions of their farms.

Abundant scientific evidence exists to show why conservation of whole-soil quality is a critical concern for all humanity. Beyond collaboration with farmers, researchers must vigorously share this evidence with an aim to engage the concerns of the general public and then guide them toward beneficial actions.

REFERENCES

Aldrich, S.R., W.O. Scott, and E.R. Leng. 1975. *Modern Corn Production,* 2nd ed. A&L Publications, Champaign, IL.

Arshad, M.A. and G.M. Coen. 1992. Characterization of soil quality: physical and chemical criteria. *Am. J. Altern. Agric.* 7(1 and 2):25–32.

Backstrom, C.H. and G.D. Hursh. 1963. *Survey Research.* Northwestern University Press, Evanston, IL.

Balfour, Lady Eve. 1944. *The Living Soil.* Faber & Faber, London.

Barndt, W.D. and J.E. Langton. 1981. Soil Survey of Monroe County, Wisconsin. U.S. Department of Agriculture, Soil Conservation Service in cooperation with the Research Division of the College of Agriculture and Life Sciences, University of Wisconsin-Madison.

Barrow, E.G.C. 1987. Extension and learning examples from the Pokot and Turkana. Presented at IDS Workshop on Farmers and Agricultural Research: Complementary Methods. University of Sussex, Brighton, England, July 26–31.

Bentley, J.W. 1989. What farmers don't know can't help them: the strengths and weaknesses of indigenous technical knowledge in Honduras. *Agric. Human Values* 6(3):25–31.

Berry, B.J. 1962. Sampling, Coding, and Storing Flood Plain Data. Agriculture Handbook No. 237. U.S. Department of Agriculture. U.S. Government Printing Office, Washington, D.C.

Berry, B.J. and A.M. Baker. 1968. Geographic sampling. In B.J.L. Berry and D.F. Marble (eds.). *Spatial Analysis: A Reader in Statistical Geography*. Prentice-Hall, Englewood Cliffs, NJ, pp. 91–100.

Birmingham, D. 1996. Local Knowledge of Soils: The Case of Contrast in Cote Divoire and Its Considerations for Extension. Ph.D. dissertation. University of Wisconsin-Madison.

Black, C.A. 1957. *Soil–Plant Relationships*. John Wiley & Sons, New York.

Bruce, R.R., A.W. White, Jr., A.W. Thomas, W.M. Snyder, G.W. Langdale, and H.F. Perkins. 1988. Characterization of soil–crop yield relations over a range of erosion on a landscape. *Geoderma* 43:99–116.

Cates, K. and F. Madison. 1991. Site Evaluation: Worksheet #11 of Farm-A-Syst, Farmstead Assessment System. University of Wisconsin-Extension Bulletin G3536-11W. University of Wisconsin-Madison.

Chambers, R., A. Pacey, and L.A. Thrupp (eds.). 1989. *Farmer First*. Intermediate Technology Publications, London.

DeBoef, W., K. Amanor, K. Wellard, and A. Bebbington. 1993. *Cultivating Knowledge: Genetic Diversity, Farmer Experimentation and Crop Research*. Intermediate Technology Publications, London.

Dillman, D.A. 1978. *Mail and Telephone Surveys: The Total Design Method*. John Wiley & Sons, New York.

Dillman, D.A., M.D. Sinclair, and J.R. Clark. 1993. Effects of questionnaire length, respondent-friendly design, and a difficult question on response rates for occupant-addressed census mail surveys. *Public Opinion Q.* 57:289–304.

Doran, J. 1994. On-Farm Measurement of Soil Quality Indices — Bulk Density, Soil Water Content, Water-Filled Pore Space, EC, pH, NO_3–N, Infiltration, Water Holding Capacity, and Soil Respiration. USDA-ARS, University of Nebraska, Lincoln.

Edmonds, W.J., J.B. Campbell, and M. Lentner. 1985. Taxonomic variation within three soil mapping units in Virginia. *Soil Sci. Soc. Am. J.* 49:394–401.

Edwards, R. 1987. Farmers' knowledge: utilization of farmers' soil and land classification. Presented at IDS Workshop on Farmers and Agricultural Research: Complementary Methods. University of Sussex, Brighton, England, July 26–31.

Edwards, W.M. 1991. Soil structure: processes and management. In R. Lal, and F.J. Pierce (eds.). *Soil Management for Sustainability*. Soil and Water Conservation Society, Ankeny, IA, pp. 7–14.

Freeze, R.A. and J.A. Cherry. 1979. *Groundwater*. Prentice-Hall, Englewood Cliffs, NJ.

Gersmehl, P.J., B. Baker, and D.A. Brown. 1989. Land management effects on innate soil erodibility: a potential complication for compliance planning. *J. Soil Water Conserv.* 44:417–420.

Granatstein, D. and D.F. Bezdicek. 1992. The need for a soil quality index: local and regional perspectives. *Am. J. Altern. Agric.* 7(1 and 2):12–16.

Haberern, J. 1992. A soil health index. *J. Soil Water Conserv.* 47(1):6.

Haggett, P. 1990. *The Geographer's Art*. Basil Blackwell, Oxford, England.

Holmes, J. 1967. Problems in location sampling. *Ann. Assoc. Am. Geogr.* 57:757–780.

Jenny, H. 1980. *The Soil Resource*. Ecological Studies, Vol. 37. Springer-Verlag, New York.

Karlen, D.L., D.C. Erbach, T.C. Kaspar, T.S. Colvin, E.C. Berry, and D.R. Timmons. 1990. Soil tilth: a review of past perceptions and future needs. *Soil Sci. Soc. Am. J.* 54:153–161.

Karlen, D.L., N.S. Eash, and P.W. Unger. 1992. Soil and crop management effects on soil quality indicators. *Am. J. Altern. Agric.* 7(1 and 2):48–55.

Kean, S. 1987. Developing a partnership between farmers and scientists: the example of Zambia's Adaptive Research Planning Team. *Exp. Agric.* 24(3):289–299.

King, F.H. 1895. *The Soil: Its Nature, Relations, & Fundamental Principles of Management.* Macmillian, New York.

Kladivko, E.J. and H.J. Timmenga. 1990. Earthworms and agricultural management. In J.E. Box, Jr. and L.C. Hammond (eds.). *Rhizosphere Dynamics.* American Association for the Advancement of Science, Washington, D.C., pp. 192–216.

Lal, R. 1987. Effects of soil erosion on crop productivity. *CRC Crit. Rev. Plant Sci.* 5(4): 303–367.

Larson, W.E., F.J. Pierce, and R.H. Dowdy. 1983. The threat of soil erosion to long-term crop production. *Science* 219:458–465.

Mausbach, M.J. and L.P. Wilding (eds.). 1991. *Spatial Variabilities of Soils and Landforms.* Soil Science Society of America, Madison, WI.

McCallister, B. 1992. Is This Soil Wet? Pedologic Clues to Soil Drainage Conditions: A Review of Research with Cases from Northern Illinois (unpublished manuscript). Institute for Environmental Studies, University of Wisconsin-Madison, 105 pp.

McCallister, R. 1996. How Wisconsin Farmers Understand and Manage Their Soil Landscapes: A Site-Specific People and Place Methodological Analysis. Ph.D. dissertation. University of Wisconsin-Madison.

Mokma, D.L., T.E. Fenton, and K.R. Olson. 1996. Effect of erosion on morphology and classification of soils in the north central United States. *J. Soil Water Conserv.* 5(2):171–175.

Morin. J, and J. Van Winkel. 1996. The effect of raindrop impact and sheet erosion on infiltration rate and crust formation. *Soil Sci. Soc. Am. J.* 60:1223–1227.

Mosi, D., M. Janikiraman, and H. Eswaran. 1991. Communicating soil survey information to traditional farmers. *Soil Surv. Horizons* 32(2):31–33.

National Research Council. 1993. *Soil and Water Quality: An Agenda for Agriculture.* National Academy Press, Washington, D.C.

Natural Resource Conservation Service. 1995. Data Rich and Information Poor. A Report to the Chief of the Natural Resources Conservation Service by the Blue Ribbon Panel on Natural Resource Inventory and Performance Measurement. USDA, Washington D.C.

Odell, R.T. 1950. Measurement of productivity of soils under various environmental conditions. *Agron. J.* 42:282–292.

Pawluk, R.R., J.A. Sandor, and J.A. Tabor. 1992. The role of indigenous soil knowledge in agricultural development. *J. Soil Water Conserv.* 47(4):298–302.

Pierce, F.J. 1991. Erosion productivity impact prediction. In R. Lal and F.J. Pierce (eds.). *Soil Management for Sustainability.* Soil and Water Conservation Society, Ankeny, IA, pp. 35–52.

Pierce, F.J., W.E. Larson, R.H. Dowdy, and W.A.P. Graham. 1983. Productivity of soils: assessing long-term changes due to erosion. *J. Soil Water Conserv.* 38:39–44.

Power, J.F. 1990. Erosion effects on soil chemistry and fertility. In W.E. Larson, G.R. Foster, R.R. Allmaras, and C.M. Smith (eds.). *Proceedings of Soil Erosion and Productivity Workshop.* University of Minnesota, St. Paul, pp. 27–30

Rajasekaran, B. and D.M. Warren. 1995. Role of indigenous soil health care practices in improving soil fertility: evidence from South India. *J. Soil Water Conserv.* 50(2): 146–149.

Rodale, J.I. 1945. *Pay Dirt; Farming & Gardening with Composts.* Devin-Adair, New York.

Roe, J. (ed.). 1952. *A Better Living from Your Soil.* Meredith Publishing, Des Moines, IA.

Romig, D.E., M.J. Garlynd, R.F. Harris, and K. McSweeney. 1995. How farmers assess soil health and quality. *J. Soil Water Conserv.* 50(3):229–236.

Rozas, J.W. 1985. El Sistema Agricola Andino de la CC de Amnaru. Tesis Fac. de Antropologia. Universidad del Cusco, Peru.

Sandor, J.A. and L. Furbee. 1996. Indigenous knowledge and classification of soils in the Andes of southern Peru. *Soil Sci. Soc. Am. J.* 60:1502–1512.

Sheatsley, P.B. 1983. Questionnaire construction and item writing. In P.H. Rossi, J.D. Wright, and A.B. Anderson (eds.). *Handbook of Survey Research.* Academic Press, Orlando, FL.

Soil Conservation Service. 1989. Summary Report —1987 National Resources Inventory. Statistical Bulletin No. 790. U.S. Government Printing Office, Washington, D.C.

Tabor, J.A. 1992. Ethnopedological surveys — soil surveys that incorporate local systems of land classification. *Soil Surv. Horizons* 33(1):1–5.

Taylor, H.M. and E.E. Terrell. 1982. In J. Rechigl, Jr. (ed.). *CRC Handbook of Agricultural Productivity,* Vol. 1. CRC Press, Boca Raton, FL.

Visser, S. and D. Parkinson. 1992. Soil biological criteria as indicators of soil quality: soil microorganisms. *Am. J. Altern. Agric.* 7:33–37.

Vorley, W.T. and D.R. Keeney. 1995. *Sustainable Pest Management and the Learning Organization.* Leopold Center for Sustainable Agriculture, Iowa State University, Ames.

Weisberg, H.F. and B.D. Bowen. 1977. *An Introduction to Survey Research and Data Analysis.* W.H. Freeman, San Francisco.

Williams, B.J. and C.A. Ortiz-Solomo. 1981. Middle American folk soil taxonomy. *Ann. Assoc. Am. Geogr.* 71(3):335–358.

Winters, R. and R.W. Simonson. 1950. The subsoil. *Adv. Agron.* 3:1–92.

Wolman, M.G. 1985. Soil erosion and crop productivity: a worldwide perspective. In R.F. Follett and B.A. Stewart (eds.). *Soil Erosion and Crop Productivity.* Soil Science Society of America, Madison, WI, pp. 9–21.

Zimmerer, K.S. 1994. Local soil knowledge: answering basic questions in highland Bolivia. *J. Soil Water Conserv.* 49(1):29–34.

12 The Effects of Forest Management on Erosion and Soil Productivity*

William J. Elliot, Deborah Page-Dumroese, and Peter R. Robichaud

INTRODUCTION

In forest conditions, surface runoff and soil erosion are generally low because of the surface litter cover. Hydraulic conductivities are in excess of 15 mm/hr, and erosion rates are generally less than 0.1 mg ha^{-1}. If the litter layer is disturbed, then runoff and erosion rates can increase by several magnitudes. Disturbances can be natural, such as wildfire, or human induced, such as harvesting or prescription burning for ecosystem management, where conductivities can drop to under 5 mm/hr and erosion rates can exceed 20 mg ha^{-1}. Roads adversely impact forest soil productivity by directly reducing the productive area and by causing the greatest amount of soil erosion. Conductivities of roads have been measured to be less than 1 mm/hr, and erosion can be in excess of 100 mg ha^{-1}. Harvesting activities reduce surface cover, and compact the soil, leading to increased runoff and erosion. Erosion generally decreases productivity of forests by decreasing the available soil water for forest growth and through loss of nutrients in eroded sediment. The Water Erosion Prediction Project (WEPP) model is shown to be a useful tool in predicting the erosion impacts of different levels of vegetation removal at harvest and different levels of compaction. WEPP predicted that the nutrients lost through the organic matter in sediments are significant, but less than nutrient loss through tree removal. Work is ongoing to collect long-term site productivity data from numerous sites to aid in the analysis of forest management on soil erosion and site productivity.

For many years research has related soil erosion to productivity, with most activities focusing on agricultural or rangeland conditions. The Pierce (1991) over-

---

* This paper was written and prepared by U.S. Government employees on official time and therefore is in the public domain and not subject to copyright.

1-57444-100-0/99/$0.00+$.50
© 1999 by Soil and Water Conservation Society

view includes over 60 references to research on impacts of erosion on agricultural production. Pierce concluded that exact relationships between erosion and productivity are unclear and that to define any such relationship, considerable research is necessary over a wide range of soil and plant conditions.

Research is ongoing into the effects of management practices on forest soil productivity. Table 12.1 summarizes some of this research. Relationships between disturbance and productivity are not simple but rather are extraordinarily complex, reflecting interactions among disturbance levels, soil water-holding capacities, nutrient cycling properties, and climate. Therefore, the effect of a given disturbance is highly dependent on site-specific soil properties and microclimate and may also be influenced by year-to-year variation in climate. Table 12.1 shows that generally disturbances reduce long-term productivity, but there are cases where short-term productivity has increased following disturbances (e.g., Harvey et al., 1996; Corns, 1988). Research on the impacts of soil erosion on forest productivity is limited. This chapter provides an overview of current knowledge on the influence of forest management activities on soil erosion and related on-site impacts and the subsequent effects of those impacts on forest productivity.

Soil erosion in an undisturbed forest is extremely low, generally under 1 mg ha^{-1} yr^{-1} (0.5 ton/acre/year). Disturbances, however, can dramatically increase soil erosion to levels exceeding 100 mg ha^{-1} yr^{-1} (50 tons/acre/year). These disturbances include natural events such as wildfires and mass movements and human-induced disturbances such as road construction and timber harvesting. Soil erosion, combined with other impacts from forest disturbance, such as soil compaction, can reduce forest sustainability and soil productivity

TABLE 12.1
Typical Effects of Forest Disturbances on Productivity

Practice	Impact	Productivity response
Roads	Area removed from production	Up to 30% of forest area lost[a]
Fire	Organic matter loss Disease reduction	Long-term effects not measured; observed loss of organic matter leading to growth reduction from water and nutrient stress[b]
Compaction	Reduced water availability and increased runoff	Height reduction of 50%[c] or more Volume reduction up to 75%[d]
Tree harvest	Loss of organic matter and site disturbance	Up to 50% reduction if site is severely compacted[e]

[a] Megahan and Kidd, 1972.

[b] Harvey et al., 1979.

[c] Reisinger et al., 1988.

[d] Froehlich, 1978.

[e] Amaranthus et al., 1996.

Forest Practices

Soil erosion in forests generally follows a disturbance such as road construction, a logging operation, or fire. In undisturbed forests, erosion is most often due to epochal events associated with fire cycles, landslides, and geologic gully incision.

Ground cover by forest litter, duff, and organic material is the most important component of the forest environment for protecting the mineral soil from erosion. Forest litter provides most of the nutrients needed for sustainable forestry. Ground cover amounts can be reduced by the logging operation (harvesting and site preparation) and burning by either wildfire or prescribed fire. For example, skidder traffic on skid trails can reduce ground cover from 100 to 10–65%. Burning can reduce ground cover from 100 to 10–90% depending on the fire severity.

Roads

In most managed forest watersheds, most eroded sediment comes from roads which have no vegetative protection and tend to have low hydraulic conductivities, leading to runoff and erosion rates that are greater than in the surrounding forests (Elliot et al., 1994a). Numerous researchers, including Swift (1988) and Bilby et al. (1989), have quantified the major role of roads on sedimentation in forests. In addition to erosion, roads reduce forest productivity by the land they occupy. A kilometer (0.6 mi) of road in 1 km^2 (250 acres) of forest represents a 0.5% loss in area and removal from productivity. Forest roads can occupy up to 10% of the forest area if there is a history of intensive logging. Roads are assumed to be unproductive in forest plans, regardless of any erosion impacts.

Currently, the USDA Forest Service has a major program to close roads. Closure methods vary from locking a gate to completely removing the road prism in an effort to reduce sedimentation and related hydrologic problems. The productivity of closed or removed roads has not been directly measured, but additional mitigation measures such as ripping and replanting are frequently included in any closure scenario to encourage maximum regrowth rates (Moll, 1996).

Timber Management

Traditionally, forest management practices focused on fire suppression and clear-cut logging methods. With an increased understanding of forest ecosystems, the USDA Forest Service is applying ecosystem management principles to forest management. These principles include partial-cut management systems and increased use of prescribed fires. Such practices, however, require more frequent operations in the forest environment.

Harvesting Effects

Harvesting methods vary in degree of disturbance. On steeper slopes (generally >35% slope), helicopter, skyline, or ground cable logging systems are common. Trees may be felled and removed with full suspension of logs via a helicopter or cable system and carried to landing sites. With a ground cable system, one end of the log is suspended and the other end is slid on the ground to a landing area. On less steep slopes (generally <35% slope), wheeled or tracked forwarders or skidders remove felled trees. A forwarder loads and carries trees to a landing area in one operation. A skidder drags the logs to the landing generally on designated skid trails. Skid trails cause the most disturbance by displacing the ground cover and compacting the mineral soil. Additional disturbance is caused by skidder tires loosening the soil, especially on slopes over 20%.

Tree cutting by itself does not cause significant erosion, and timber harvest operations usually cause less erosion per unit area than roads, but the area of timber harvest is usually large relative to roads so that the total erosion from timber harvest operations may approach that from roads (Megahan, 1986). However, the decrease in the number of trees results in a decrease in evapotranspiration, which contributes to increased subsurface flow, streamflow, and channel erosion. Field research has found that timber harvesting tends to compact the soil. Compaction increases soil erosion and adversely impacts forest productivity (Yoho, 1980). Most erosion comes from skid trails on timber harvest units because of the reduced infiltration rates and disturbance to the organic layer (Robichaud et al., 1993b). Therefore, the accelerated erosion caused by timber harvesting may result in deterioration of soil physical properties, nutrient loss, and degraded stream water quality from sediment, herbicides, and plant nutrients (Douglas and Goodwin, 1980).

Nutrient Impacts

Harvesting trees removes nutrients from a generally nutrient-deficient environment (Miller et al., 1989). Table 12.2 shows the effect of tree harvest on nitrogen availability. Increasing harvest intensity from bole-only through whole-tree and complete biomass harvesting doubled nitrogen loss on the average quality site, but more than tripled loss at the poor quality site. Leaching losses are also greater on the poorer site. Researchers generally agree that harvesting only the bole will not greatly deplete nutrient reserves, but shorter rotations and whole-tree harvesting remove more nutrients than can be replaced in a rotation. Harvesting crowns is undesirable because they contain a large portion of the stand nutrient content.

Fire Effects

The most common method of site preparation in the United States is prescribed burning. Mechanical site preparation methods, however, are common in southern forests to physically destroy or remove unwanted vegetation from the site and to facilitate machine planting. Burning is conducted alone, and in combination with other treatments, to dispose of slash, reduce the risk of insects and fire hazards, prepare seedbeds, and suppress plant competition for natural and artificial regeneration. Fire has long been a natural component of forest ecosystems (Agee, 1993), and current research is finding that fire helps maintain forest health. The use of prescribed fire will increase with the current emphasis on ecosystem management.

TABLE 12.2
Comparison of Height, Diameter, and Nitrogen Pools
After Harvest Treatments of Varying Intensities from
Two Sites of Differing Site Quality, Pack Forest, Washington[a]

Harvest treatment	Height growth (m)	Diameter growth (mm)	kg/ha			
			Total N	Harvest loss	3-year leaching loss	% loss
Average site quality						
Bole only	1.7	29	2935	470	4.4	16
Whole tree	1.9	32	2827	678	0.5	24
Complete[b]	1.8	36	2719	870	0.7	32
Poor site quality						
Bole only	1.4	22	984	157	2.1	16
Whole tree	1.1	16	903	289	4.7	32
Complete[b]	1.1	15	934	486	5.5	53

[a] From Miller et al., 1989.

[b] Complete removal of all aboveground biomass.

Erosion following fires can vary from extensive to minimal, depending on the fire severity and areal extent (Robichaud and Waldrop, 1994). Fire severity refers to the effect of the fire on some component of the forest ecosystem, such as nutrient loss or amount of organic material consumed (litter and duff). Erosion from high-severity fires can cover large areas, and fires may create hydrophobic or water-repellent soil conditions. Erosion from low-severity fires may be minimal to none (Robichaud et al., 1993a; Robichaud and Waldrop, 1994).

EROSION MODELING

Since the late 1950s, soil erosion models have provided natural resource managers with tools to predict the impacts of management practices on soil erosion. Earlier models tended to focus on Midwest and Southeast agricultural conditions, where erosion was considered a severe problem associated with farming practices. Models for range and forest lands have only recently received widespread interest as managers focus on off-site sediment impacts as well as on-site erosion rates.

Sediment Yield Models

Most of the early models, which culminated in the empirically based Universal Soil Loss Equation (USLE), focused on upland soil erosion rates (Wischmeier and Smith, 1978). The USLE was developed to predict soil erosion from small, relatively homogenous plots (Mutchler et al., 1994). Forest environments tend to have much greater spatial variability in vegetation and soils (Elliot et al., 1996), making application of the USLE difficult. Dissmeyer and Foster (1985) developed a subfactor approach to predict soil erosion from forest conditions for areas where intensive operations such as tillage are carried out, and harvest areas can be considered similar to intensively managed farming systems. The erosion–productivity impact calculator (EPIC) model was developed to apply the USLE prediction technology to long-term productivity impact predictions (Williams et al., 1984). The EPIC model, however, was developed for applications to croplands only.

Forest Service specialists have developed watershed models to aid in predicting the cumulative effects of road and harvest area erosion on stream sedimentation (such as WATSED [Range, Air, Watershed and Ecology Staff Unit, 1991]). The strength of these models is in assessment of cumulative effects on stream sedimentation in a large watershed. WATSED, however, was not developed to predict site-specific effects.

More recent physically based soil erosion models, including the Chemicals Runoff and Erosion from Agricultural Management Systems (CREAMS) model (Knisel, 1980) and the Water Erosion Prediction Project (WEPP) model (Laflen et al., 1991), provide estimates of sedimentation for predicting both on-site and off-site impacts. The WEPP model, in particular, shows considerable promise to assist in predicting soil erosion and sediment yields in a forest environment (Elliot et al., 1996).

The WEPP model predicts upland erosion and off-site effects from erosion events influenced by management activities (Laflen et al., 1991). Erodibility values have been measured on forest roads and disturbed harvest areas, and validation activities with the WEPP model for forests have been encouraging (Elliot et al., 1994b). The WEPP model predicts both erosion and the textural and organic composition of the eroded sediment.

PRODUCTIVITY RESPONSE TO MANAGEMENT

A coordinated national research effort was implemented on a broad spectrum of benchmark sites across the nation to separate impacts of soil organic matter reduction from soil compaction resulting from management activities (Powers et al., 1990). These sites were relatively undisturbed prior to study installation. An extensive range of pre- and postharvest measurements are being taken. This study alters site organic matter and total soil porosity over a range of intensities encompassing a number of possible management scenarios. It creates a network of comparable experiments producing nil to severe soil disturbance and physiological stress in vegetation over a broad range of soils and climates. Establishing and monitoring this network creates a research opportunity of unusual scope and significance. Early results indicate that immediate postharvest biomass declines are most likely caused by compaction and not organic matter removal, whereas long-term productivity changes will be more dependent on organic matter losses.

Erosion Loss

A close tie clearly exists between surface organic matter and forest soil productivity (Jurgensen et al., 1996). As a rooting medium for higher plants, soils provide the essentials of water, structural support, nutrients, and soil biota. Mixing and/or short-distance displacement of topsoil and surface organic matter from a site can decrease productivity. Logging generally disturbs less than 30% of the total harvested area (Rice et al., 1972; Miller and Sirois, 1986), but the impact can be severe.

Erosion reduces forest productivity mainly by decreasing the soil water availability. This is a result of changing the water-holding capacity and thickness of the root zone (Swanson et al., 1989). Erosion removes plant-available nutrients. Fertilizer applications can partly offset these losses, but they greatly increase costs and are uncommon. Another impact of erosion on productivity is degraded soil structure. Removal of the loose, organic surface materials promotes surface sealing and crusting that decrease infiltration capacity and may increase erosion (Childs et al., 1989). Erosion also results in loss of important soil biota, such as mycorrhizal fungi, which facilitate nutrient uptake by plants (Amaranthus et al., 1989, 1996).

Surface erosion proceeds downward from the surface soil horizons. Because the highest concentrations of nutrients and biota and the maximum water-holding capacity are in the uppermost horizons, incremental removal of soil nearer the surface is more damaging than subsoil losses. Productivity may inevitably decline on most shallow forest soils as erosion causes root-restricting layers to be nearer the surface and as organic matter is washed away. Consequently, the largest declines in productivity are most likely to occur in marginal, dry environments.

Assessing how erosion affects site productivity is often difficult. Erosion rates are poor indicators of loss in productivity because most soil is redistributed within a watershed and not necessarily lost to production. Soils differ in their tolerance to erosion loss. For instance, Andisols have relatively high water-holding capacity and natural fertility. Erosion may be severe on these sites, but productivity may decline little. In contrast, Lithosols are shallow and generally less productive, so a small rate of erosion can lead to a significant decline in water-holding capacity and productivity.

Compaction Impacts

Compaction is a reduction in total porosity. Macroporosity is reduced, while microporosity increases as large pores are compacted into smaller ones. An increase in microporosity can lead to greater available water-holding capacity throughout a site, but this increase is usually at the expense of aeration and drainage (Incerti et al., 1987).

Compaction of forest soil is a serious concern for managers because of the use of heavy equipment to harvest timber and to prepare a site for planting. Usually, the more porous the soil initially, the greater the compaction depth. For example, volcanic ash soils of the western United States are highly productive in their undisturbed condition but are prone to compaction because they have a low-volume bulk density and relatively few coarse fragments (Geist and Cochran, 1991). Once sensitive sites have been disturbed through timber harvest activities and site preparation, porosity (Dickerson, 1976) and hydraulic conductivity decline (Gent et al., 1984). Compaction depth can exceed 450 mm (Page-Dumroese, 1996).

Compaction reduces productivity through reduction in root growth, height, and timber volume (Greacen and Sands, 1980; Froehlich and McNabb, 1984) and may be produced by a single pass of logging equipment across a site (Wronski, 1984). Productivity losses have been documented for whole sites (Wert and Thomas, 1981) and for individual trees (Froehlich, 1979; Helms and Hipkin, 1986). Decreases in important microbial populations have been observed in compacted soils (Amaranthus et al., 1996). In general, however, the environmental degradation observed in the field results from both compaction and disturbance or removal of surface organic horizons (Childs et al., 1989).

Soil compaction may also increase surface runoff because of reduced infiltration (Greacen and Sands, 1980). However, because of increased soil strength, compacted soils may have lower erodibility and consequently suffer less erosion for the same amount of runoff (Liew, 1974). A significant amount of erosion after harvest activities has been attributed to compaction but may be attributable to both compaction and the removal of vegetative cover (Dickerson, 1976).

PREDICTED EROSION RATES AND PRODUCTIVITY

We carried out a series of WEPP runs for a productivity study site in central Idaho to allow comparison of a range of management effects on soil erosion. We compared the predicted effects on erosion from wildfires to different levels of harvest and compaction, to better understand the interactions among natural events, human activities, soil erosion, soil productivity, and ultimately forest ecosystem sustainability.

Harvesting Impacts

For the modeling study, we modeled a slope length of 100 m (328 ft), with a steepness of 61%, typical of the site. Soil properties of the site are presented in Table 12.3. The WEPP management file described a forest in the first year, a disturbance in the second year, and regeneration of forest in eight subsequent years as described by Elliot et al. (1996). The biomass reduction due to harvest effects was described in the residue management and harvest index (harvest index = biomass removed/ biomass present) values in the management files. The values assumed are presented in Table 12.4. The climate for the simulations was stochastically generated with the CLIGEN generator (Flanagan and Livingston, 1995) from the Deadwood Dam, Idaho, climate statistics (mean annual precipitation = 830 mm [33 in.]).

An initial WEPP run was made with no disturbance. In this scenario, there was no runoff and no erosion. With the amount of residue cover and litter accumulation typical of forests, WEPP seldom predicts erosion. Our field observations generally

TABLE 12.3
Soil Properties Assumed for the WEPP Model
Computer Simulations

Soil property	Value	Units
Sand content	40	%
Silt content	45	%
Clay content	15	%
Interrill erodibility	2100	kg s m^{-4}
Rill erodibility	0.008	s m^{-1}
Critical shear	3	Pa
Saturated hydraulic conductivity		
Uncompacted	20	mm h^{-1}
Moderate compaction	15	mm h^{-1}
Severe compaction	8	mm h^{-1}
Hydrophobic	4	mm h^{-1}

confirm this, with most sediment from undisturbed watersheds coming from eroding ephemeral channels or landslides.

Tables 12.5 and 12.6 present the predicted runoff and erosion rates for different treatments. The WEPP predictions are generally logical. More compaction leads to greater runoff and greater erosion. The effect of removing greater amounts of vegetation also leads to greater erosion rates. The complete removal of biomass was modeled as removing 100% of the surface residue, which resulted in a small increase in runoff but a doubling of erosion rates. The role of surface residue is critical in controlling erosion in forests, just as it is in agriculture.

To compare the productivity impacts of soil erosion, we estimated the nitrogen losses associated with the above erosion rates. We assumed that the typical forest soil contains 4% organic matter and that organic matter is 2% nitrogen. The resulting nitrogen losses for 8 years of predicted erosion are presented in Table 12.7. The values in Table 12.7 can be compared to Table 12.2 to see that nutrient losses due to erosion are significant, greater than observed leaching losses, but not as great as losses due to vegetation removal. In a generally nutrient-deficient environment, such nitrogen losses will have a significant impact on future productivity.

TABLE 12.4
Values Describing the Effects of Timber Harvest in
the WEPP Model Computer Simulations

Treatment	Residue management	Harvest index
Complete biomass removal	100% surface residue removed	0.9
Bole and crown removed	No surface residue removed	0.8
Bole only removed	No residue management	0.4

TABLE 12.5
Predicted Average Annual Runoff (mm) from Rainfall from the WEPP Simulations for Five Simulated Forest Conditions

Treatment	Compaction (mm)		
	Moderate	Severe	None
Undisturbed	0.0	—	—
Complete biomass removal	12.8	18.8	35.6
Bole and crown removed	9.2	15.4	32.4
Bole only removed	9.1	16.1	32.7
Severe wildfire	65.0	—	—

Natural Fire Impacts

To model a severe fire, 100% of the residue was burned, and half of the remaining biomass was harvested in the autumn. This is generally much more severe than observed in the field but allows comparison of the extreme events. Generally, even "severe" fires do not remove more than 90–95% of the residue, and the remaining residue can reduce the predicted erosion rates by more than 90%. If the soil hydraulic conductivity remained unchanged, there was little difference in either runoff or erosion from the values predicted for the severe compaction, bole removal treatment. If the hydraulic conductivity was reduced to 4 mm/hr to reflect hydrophobic soil conditions that sometimes occur after severe fires, then the predicted runoff was doubled to 65 mm/year. The predicted erosion was 11.6 Mg ha^{-1}, greater than the bole and crown removal treatments but still somewhat less than the predicted rates on sites with complete biomass removal. As the soil hydrologically recovers following a severe fire, the runoff and erosion rates decline, a characteristic that WEPP is currently not capable of modeling continuously. Such a scenario could be developed with a series of 1-year runs with a different conductivity for each year.

TABLE 12.6
Predicted Average Annual Soil Loss (Mg ha^{-1}) from the WEPP Simulations for Five Simulated Forest Conditions

Treatment	Compaction (Mg ha^{-1})		
	Moderate	Severe	None
Undisturbed	0.0	—	—
Complete biomass removal	4.5	7.4	14.4
Bole and crown removed	2.0	3.3	7.2
Bole only removed	2.0	3.5	7.2
Severe wildfire	11.6	—	—

TABLE 12.7
Predicted Nitrogen Loss Due to Erosion in
the First 8 Years of Regrowth Following Harvest

Treatment	Compaction (kg ha^{-1})		
	None	Moderate	Severe
Undisturbed	0.0	—	—
Complete biomass removal	28.8	47.4	92.2
Bole and crown removed	12.8	21.1	46.1
Bole only removed	12.8	22.4	46.1
Severe wildfire	74.2	—	—

CONCLUSIONS

In our overview of the impacts of forest management activities on soil erosion and productivity, we show that erosion alone is seldom the cause of greatly reduced site productivity. However, erosion, in combination with other site factors, works to degrade productivity on the scale of decades and centuries. Extreme disturbances, such as wildfire or tractor logging, cause the loss of nutrients, mycorrhizae, and organic matter. These combined losses reduce long-term site productivity and may lead to sustained periods of extended erosion that could exacerbate degradation.

Managers should be concerned with harvesting impacts, site preparation disturbances, amount of tree that is removed, and the accumulation of fuel from fire suppression. On erosion-sensitive sites, we need to carefully evaluate such management factors.

Prescribed fire is generally an excellent tool in preparing sites for regeneration, for reducing fuel loads, and for returning sites to a more natural condition. Burning conducted under correct conditions will reduce the fire hazard, make planting easier, and retain the lower duff material to protect the mineral soil and conserve nutrients to sustain forest productivity.

The WEPP model can describe various impacts due to harvesting, but further work is required to model fire effects and the subsequent temporal and spatial variation in soil hydraulic conductivity and ground cover effects. From field observations and the modeling exercise, it appears that disturbances caused by harvest activities will lead to increases in erosion and runoff rates, much greater than natural conditions, even when extreme wildfire effects are considered.

REFERENCES

Agee, J. K. 1993. *Fire Ecology of Pacific Northwest Forests.* Island Press, Washington D.C., 493 pp.

Amaranthus, M.P., J.M. Trappe, and R.J. Molina. 1989. Long-term forest productivity and the living soil. In S.P. Gessel, D.S. Lacate, G.F. Weetman, and R.F. Powers (eds.).

Sustained Productivity of Forest Soils. Proc. 7th North American Forest Soils Conf. Faculty of Forestry Publ. 36–52. University of British Columbia, Vancouver.

Amaranthus, M.P., D.S. Page-Dumroese, A. Harvey, E. Cazares, and L.F. Bednar. 1996. Soil Compaction and Organic Matter Affect Conifer Seedling Nonmycorrhizal and Ectomycorrhizal Root Tip Abundance and Diversity. Research Paper PNW-RP-494. U.S. Department of Agriculture, Forest Service, Pacific Northwest Research Station, 12 pp.

Bilby, R.E., K. Sullivan, and S.H. Duncan. 1989. The generation and fate of road-surface sediment in forested watersheds in southwestern Washington. *J. For. Sci.* 35(2): 453–468.

Childs, S.W., S.P. Shade, D.W.R. Miles, E. Shepard, and H.A. Froehlich. 1989. Soil physical properties: importance to long-term productivity. In S.P. Gessel, D.S. Lacate, G.F. Weetman, and R.F. Powers (eds.). Sustained Productivity of Forest Soils. Proc. 7th North American Forest Soils Conf. Faculty of Forestry Publ. 53-67. University of British Columbia, Vancouver.

Corns, I.G.W. 1988. Compaction by forestry equipment and effects on coniferous seedling growth on four soils in the Alberta foothills. *Can. J. For. Res.* 18:75–84.

Dickerson, B.P. 1976. Soil compaction after tree-length skidding in northern Mississippi. *Soil Sci. Soc. Am. J.* 40:965–966.

Dissmeyer, G.E. and G.R. Foster. 1985. Modifying the Universal Soil Loss Equation for forest land. In S.A. El-Swaify, W.C. Moldenhauer, and A. Lu (eds.). *Soil Erosion and Conservation.* Soil and Water Conservation Society, Ankeny, IA, pp. 480–495.

Douglas, J.E. and O.C. Goodwin. 1980. Runoff and soil erosion from forest site preparation practices. In U.S. Forestry and Water Quality: What Course in the 80's? Proceedings. The Water Pollution Control Federation and Virginia Water Pollution Control Association, Richmond, pp. 50–74.

Elliot, W.J., R.B. Foltz, and M.D. Remboldt. 1994a. Predicting sedimentation from roads at stream crossings with the WEPP model. Presented at the 1994 ASAE International Winter Meeting. Paper No. 947511. ASAE, St. Joseph, MI, December 13–16.

Elliot, W.J., R.B. Foltz, and P.R. Robichaud. 1994b. A tool for estimating disturbed forest site sediment production. In Proc. Interior Cedar–Hemlock–White Pine Forests: Ecology and Management, Spokane, WA, Pullman, WA, March 2–4, 1993. Department of Natural Resource Science, Washington State University, pp. 233–236.

Elliot, W.J., C.H. Luce, and P.R. Robichaud. 1996. Predicting sedimentation from timber harvest areas with the WEPP model. In Proc. Sixth Federal Interagency Sedimentation Conf., Las Vegas, March 10–14, IX:46–53.

Flanagan, D.C. and S.J. Livingston (eds.). 1995. WEPP User Summary USDA–Water Erosion Prediction Project. National Soil Erosion Laboratory Report No. 11. West Lafayette, IN.

Froehlich, H.A. 1978. The Effect of Soil Compaction by Logging on Forest Productivity. Final Report, Contract No. 53500-CT4-5-5(N). Bureau of Land Management, Portland, OR, 19 pp.

Froehlich, H.A. 1979. Soil compaction from logging equipment: effects on growth of young ponderosa pine. *J. Soil Water Conserv.* 34:276–278.

Froehlich, H.A. and D.H. McNabb. 1984. Minimizing soil compaction in Pacific Northwest forests. In E.L. Stone (ed.). Forest Soils and Treatment Impacts. Proc. 6th American Forest Soils Conf., Knoxville, TN, pp. 159–192.

Geist, J.M. and P.H. Cochran. 1991. Influences of volcanic ash and pumice deposition on productivity of western interior forest soils. In A.E. Harvey and L.F. Neuenschwander (comps.). Proceedings — Management and Productivity of Western-Montane Forest

Soils. Gen. Tech. Rep. INT-GTR-280. U.S. Department of Agriculture, Forest Service, Intermountain Research Station, Ogden, UT, pp. 82–89.

Gent, J.A., Jr., R. Ballard, A.E. Hassan, and D.K. Cassel. 1984. Impact of harvesting and site preparation on physical properties of Piedmont forest soils. *Soil Sci. Soc. Am. J.* 48:173–177.

Greacen, E.L. and R. Sands. 1980. Compaction of forest soils — a review. *Aust. J. Soil Res.* 18:163–189.

Harvey, A.E., M.F. Jurgensen, and M.J. Larsen. 1979. Role of Forest Fuels in the Biology and Management of Soil. General Technical Report INT-65. USDA Forest Service, Intermountain Research Station, Ogden, UT, 8 pp.

Harvey, A.E., D.S. Page-Dumroese, M.P. Amaranthus, and G.I. McDonald. 1996. The Effects of Stump Removal and Simulated Harvest-Related Disturbances on Soil Properties, Ectomycorrhizal Development, Growth and Nutrition of Planted Western White Pine and Douglas-Fir in Northern Idaho. Research Paper. USDA Forest Service, Intermountain Research Station, Ogden, UT (in press).

Helms, J.A. and C. Hipkin. 1986. Effects of soil compaction on tree volume in a California ponderosa pine plantation. *West. J. Appl. For.* 1:121–124.

Incerti, M., P.F. Clinnick, and S.T. Willett. 1987. Changes in the physical properties of a forest soil following logging. *Aust. For. Res.* 17:91–98.

Jurgensen, M.F., A.E. Harvey, R.T. Graham, D.S. Page-Dumroese, J.R. Tonn, M.J. Larsen, and T.B. Jain. 1996. Impacts of timber harvesting on soil organic matter, nitrogen, productivity, and health of inland Northwest forests. *For. Sci.* (in press).

Knisel, W.G. 1980. CREAMS: A Field-Scale Model for Chemicals, Runoff, and Erosion from Agricultural Management Systems. Conservation Research Report No. 26. USDA, Washington, D.C., 643 pp.

Laflen, J.M., L.J. Lane, and G.R. Foster. 1991. WEPP: a new generation of erosion prediction technology. *J. Soil Water Conserv.* 46(1):34–38.

Liew, T.C. 1974. A note on soil erosion study at Tawau Hills Forest Reserve, Malay. *Nat. J.* 27:20–26.

Megahan, W.F. 1986. Recent studies on erosion and its control on forest lands in the United States. In F. Richard (ed.). Range Basin Sediment Delivery: Proceedings, Albuquerque, NM, August 1986. IAHS Publ. 159. Wallingford, Oxon, U.K., pp. 178–189.

Megahan, W.F. and W.J. Kidd. 1972. Effect of Logging Roads on Sediment Production Rates in the Idaho Batholith. Intermountain Research Station Research Paper INT-123. USDA Forest Service, Ogden, UT, 14 pp.

Miller, J.H. and D.L. Sirois. 1986. Soil disturbance by skyline yarding vs. skidding in a loamy hill forest. *Soil Sci. Soc. Am. J.* 50:462–464.

Miller, R.E., W.J. Stein, R.L. Heninger, W. Scott, S.M. Little, and D.J. Goheen. 1989. Maintaining and improving site productivity in the Douglas-fir region. In D.A. Perry, R. Meurisse, B. Thomas, R. Miller, J. Boyle, J. Means, C.R. Perry, and R.F. Powers (eds.). *Maintaining the Long-Term Productivity of Pacific Northwest Forest Ecosystems.* Timber Press, Portland, OR, pp. 98–136.

Moll, J.E. 1996. A Guide for Road Closure and Obliteration in the Forest Service. USDA Forest Service Technology and Development Program, Washington, D.C., 49 pp.

Mutchler, C.K., C.E. Murphree, and K.C. McGregor. 1994. Laboratory and field plots for erosion research. In R. Lal (ed.). *Soil Erosion Research Methods,* 2nd ed. Soil and Water Conservation Society, Ankeny, IA, pp. 11–37.

Page-Dumroese, D.S. 1996. Evaluating management impacts on long-term soil productivity: a research and national forest systems cooperative study — local results. In Proc. Western Regional Cooperative Soil Survey Conference (in press).

Pierce, F.J. 1991. Erosion productivity impact prediction. In R. Lal and F.J. Pierce (eds.). *Soil Management for Sustainability*. Soil and Water Conservation Society, Ankeny, IA, pp. 35–52.

Powers, R.F., D.H. Alban, R.E. Miller, A.E. Tiarks, C.G. Wells, P.E. Avers, R.G. Cline, R.O. Fitzgerald, and N.S. Loftus. 1990. Sustaining site productivity in North American forests: problems and perspectives. In S.P. Gessel, D.S. Lacate, G.F. Weetman, and R.F. Powers (eds.). Sustained Productivity of Forest Soils. Proc. 7th North American Forest Soils Conf. Faculty of Forestry Publ. 49-79. University of British Columbia, Vancouver.

Range, Air, Watershed and Ecology Staff Unit and Montana Cumulative Watershed Effects Cooperative. 1991. WATSED Water and Sediment Yields. Region 1, USDA Forest Service, Missoula, MT.

Reisinger, T.W., G.L. Simmons, and P.E. Pope. 1988. The impact of timber harvesting on soil properties and seedling growth in the South. *South. J. Appl. For.* 12(1):58–67.

Rice, R.M., J.S. Rothacher, and W.F. Megahan. 1972. Erosional consequences of timber harvesting: an appraisal. In Watersheds in Transition, American Water Resources Association Proceedings Series 14. Urbana, IL, pp. 321–329.

Robichaud, P.R. and T.A. Waldrop. 1994. A comparison of surface runoff and sediment yields from low- and high-severity site preparation burns. *Water Resour. Bull.* 30(1): 27–36.

Robichaud, P.R., R.T. Graham, and R.D. Hungerford. 1993a. Onsite sediment production and nutrient losses from a low-severity burn in the interior Northwest. In D.M. Baumgartner, J.E. Lotan, and J.R. Tonn (compilers). Interior Cedar–Hemlock–Whitepine Forests: Ecology and Management: Proceedings, Spokane, WA, March 1993, pp. 227–232.

Robichaud, P.R., C.H. Luce, and R.E. Brown. 1993b. Variation among different surface conditions in timber harvest sites in the Southern Appalachians. In International Workshop on Soil Erosion: Proceedings, Moscow, September 1993. The Center of Technology Transfer and Pollution Prevention, Purdue University, West Lafayette, IN, pp. 231–241.

Swanson, F.J., J.L. Clayton, W.F. Megahan, and G. Bush. 1989. Erosional processes and long-term site productivity. In D.A. Perry, R. Meurisse, B. Thomas, R. Miller, J. Boyle, J. Means, C.R. Perry, and R.F. Powers (eds.). *Maintaining the Long-Term Productivity of Pacific Northwest Forest Ecosystems*. Timbe Press, Portland, OR, pp. 67–82.

Swift, L.W., Jr. 1988. Forest access roads: design, maintenance, and soil loss. In W.T. Swank and D.A. Crossley, Jr. (eds.). *Forest Hydrology and Ecology at Coweeta*. Ecological Studies 66. Springer-Verlag, New York, pp. 313–324.

Wert, S. and B.R. Thomas. 1981. Effects of skid roads on diameter, height, and volume growth in Douglas-fir. *Soil Sci. Soc. Am. J.* 45:629–632.

Williams, J.R., C.A. Jones, and P.T. Dyke. 1984. A modeling approach to determining the relationship between erosion and soil productivity. *Trans. ASAE* 27:129–144.

Wischmeier, W.H. and D.D. Smith. 1978. Predicting Rainfall Erosion Losses — A Guide to Conservation Planning. USDA Agricultural Handbook No. 537.

Wronski, E.B. 1984. Impact of tractor thinning operations on soils and tree roots in a Karri forest, Western Australia. *Aust. For. Res.* 14:319–332.

Yoho, N.S. 1980. Forest management and sediment production in the South — a review. *South. J. Appl. For.* 4:27–36.

13 Rangeland Soil Erosion and Soil Quality: Role of Soil Resistance, Resilience, and Disturbance Regime

J.E. Herrick, M.A. Weltz, J.D. Reeder, G.E. Schuman, and J.R. Simanton

INTRODUCTION

The relationships between rangeland soil quality, soil resilience, and soil erosion depend on several interacting factors: (1) landscape and climate characteristics, (2) current disturbance regime, and (3) recent and evolutionary disturbance history. These factors tend to be more variable across rangelands than across agricultural lands. There are at least four specific relationships between soil quality and soil erosion which involve soil resistance or soil resilience. The first is the *historical resistance* of the soil to past disturbances, which can serve as an indicator of soil quality. Second, the *current resistance* of the soil to disturbance is related to soil erosion potential. The third relationship is the *current resilience* of the system following soil erosion. Finally, soil erosion is a *driver* in the system which determines soil quality. This final relationship illustrates the need to view the system dynamically: soil erosion both reflects and affects soil quality. These dynamic relationships depend, in turn, on the characteristics of historic and current disturbance regimes. Both ecosystems and species tend to evolve in response to dominant disturbance regimes, such as fire, drought, and grazing. The resistance and/or resilience of the system will tend to be higher for disturbance regimes which share key characteristics with historic and evolutionary patterns.

Over 30% of the U.S. land surface and 34% of the global land surface, exclusive of Antarctica, is classified as rangeland (World Resources Institute, 1992; National

1-57444-100-0/99/$0.00+$.50
© 1999 by Soil and Water Conservation Society

Research Council, 1994). Rangelands are arguably the most diverse of any class of productive land and are associated with infertile lowland soils throughout the humid tropics and with arid, semiarid, and steepland soils on nearly every continent. A common characteristic of most rangelands, however, is that they have some edaphic and/or climatic limitations which have prevented them from being used for intensive crop production. This functional definition of rangeland is implicit in the USDA land capability classification system (Dent and Young, 1981).

Soil quality can be generally defined as the long-term capacity of a soil to perform functions which sustain biological productivity and maintain environmental quality. This definition, similar to many of those listed in recent reviews by Doran and Parkin (1994) and the National Research Council (1993), explicitly does not favor one land use over another. Rangelands are valued for a wide variety of uses including food and fiber production, watershed protection, wildlife conservation, and recreation (National Research Council, 1994). A high-quality rangeland soil is one which will maintain its functional integrity and therefore sustain its many possible uses into the future. Consequently, the conservation of soil and water resources, or minimization of runoff and soil erosion, has emerged as a potentially key indicator of rangeland health, as well as soil quality. The relationships between soil erosion and soil quality, however, are not fully understood.

The objectives of this chapter are to define specific relationships between soil erosion and soil quality, to identify and describe several factors which determine the nature of these relationships for specific ecosystems, and to illustrate these relationships with examples from south-central New Mexico, southeastern Arizona, and northeastern Colorado. A brief discussion of the contribution of soil quality to rangeland health is also included. This chapter is designed to generate discussion relevant to assessing rangeland soil quality and its relationship to soil erosion. As such, it is not intended to serve as a review of the literature on rangeland soil erosion, for which the reader is referred to two edited volumes on the subject (Blackburn et al., 1994; Spaeth et al., 1996a).

RELATIONSHIPS BETWEEN SOIL EROSION AND SOIL QUALITY

Definitions

Disturbance, resistance, and resilience are interrelated terms which are critical to understanding relationships between soil erosion and soil quality. Definitions vary widely and frequently depend on both the author and the context in which the word is used. A disturbance is generally defined as any event which causes a significant change from the normal pattern in an ecosystem (Forman and Godron, 1986), where pattern includes both spatial and temporal distributions of plants, microtopographic features, and soil and plant community properties, processes, and functions. Changes caused by disturbances may be positive or negative. Whether or not an event is classified as a disturbance depends in part on the spatial and temporal scales of interest. The creation of a macropore by an earthworm at the base of a grass clump may be viewed as a disturbance at the individual plant scale during the course of a season. However, it would have little impact on hydrology at the watershed scale or at the plant scale over a period of several decades.

Whether or not an event is classified as a disturbance also depends on how resistant the system is to the particular event. Resistance is defined as the capacity of a system to continue to function without change through a disturbance (Pimm, 1984). The resistance of a system depends both on effects on individual system elements and the relationships between those elements and on the extent to which there is redundancy or overlap in ecosystem function. The recognition of functional redundancy among species has led ecologists to increasingly focus on groups of species which perform "keystone functions" rather than on individual keystone species (Mills et al., 1993). This paradigm can be broadened to include physical and chemical processes, such as macropore formation by soil biota versus shrinking and swelling.

The third term, resilience, has been defined in at least three very different ways. The most common definition is that resilience is proportional to the recovery of the functional integrity of a system following a disturbance (Pimm, 1984). Others have argued that the term is more useful if it is defined as a capacity of the system to recover following catastrophic disturbances (Holling and Meffe, 1996) or following several simultaneous and/or repeated catastrophic disturbances or stressors. While this third definition makes an already complex concept even more difficult to assess, it may serve to better identify the key periods when ecosystem thresholds are likely to be exceeded. All three definitions are useful. The first, most common definition will be applied in this chapter except where specified.

General Relationships

There are at least four specific relationships between soil erosion and rangeland soil quality (Figure 13.1). The first three include (1) the historic resistance of the soil to erosion, (2) the current resistance of the soil to erosion, and (3) the current resilience

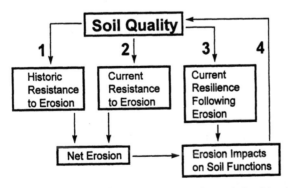

FIGURE 13.1   Conceptual framework illustrating the four relationships between soil erosion and rangeland soil quality.

of the system following erosion. These relationships reflect past and potential future responses of a system to erosion and can serve as indicators of soil quality. The fourth relationship is related to the third: changes in soil properties following erosional events not only serve as indicators of soil quality but also reflect the capacity of soil erosion to modify soil quality. Soil erosion, then, is a determinant of (#4), as well as a response to (#1–3), soil quality (Figure 13.1).

Historic Resistance

Soil properties which reflect the resistance of a system to soil erosion during the past several months or years are frequently suggested as indicators of soil quality (Arshad and Coen, 1992; Romig et al., 1995). The degree of pedestalling, heterogeneity of soil surface texture at the plant-interspace scale, the presence of rills, and signs of recent soil redeposition all indicate that some redistribution of soil resources has occurred. In linking soil loss to soil quality, the assumption is made that this loss results in a decline in the capacity of the ecosystem to fulfill one or more functions, such as water storage and nutrient supply. This assumption is frequently met in rangelands, which often occur on shallow soils.

While the relationship between the historic resistance of a system and past soil quality is relatively straightforward, the relationship between historic resistance and the indicators used to quantify it are not. The observed degree of pedestalling and the presence and characteristics of rills, for example, depend on a variety of factors. These factors include soil properties such as texture, the characteristics of the most recent storm(s), the time elapsed since the last storm, and the type and intensity of subsequent surface disturbances which could degrade or obscure the pedestals, rills, and depositional areas. Furthermore, identification of many of these features can be difficult without background knowledge of other processes occurring in the system. For example, soil accumulation around the bases of bunchgrasses in the Chihuahuan Desert is often attributed to pedestalling and/or deposition of material eroded from bare interspaces. While both processes can and do occur, much of the accumulation is frequently created by the activity of termites which bring soil to the surface and deposit it around standing dead vegetation. Consequently, indicators of historic resistance to soil erosion must be interpreted in the context of additional information and a knowledge of processes which may or may not be available for a specific site.

Current Resistance

The current resistance of a soil to erosion depends on both soil properties and vegetation characteristics. Soil properties such as aggregate stability, hydraulic conductivity, and ground cover can be directly related to soil quality. Karlen and Stott (1994) selected aggregate structure, surface sealing, and porosity as key soil quality indicators related to soil erosion by water. These properties have been measured and correlated with data from natural runoff and rainfall simulation plots for storms of different intensities and durations (e.g., Benkobi et al., 1993), and the results have been used to generate empirical relationships between specific soil properties and resistance to soil erosion (Weltz et al., 1996).

Vegetation cover is frequently the most important factor affecting site resistance to interrill erosion on arid and semiarid rangelands (Wood et al., 1987; Blackburn et al., 1992). Spaeth et al. (1996b,c) point out that while hydrologists have typically focused on quantitative soil factors, vegetation parameters such as cover, above- and belowground growth form, phenology, and spatial distribution can and should be used to enhance predictions of soil resistance to erosion and susceptibility to runoff. Plant community composition may be used as a surrogate for many of these attributes (Spaeth et al., 1996b,c). Attempts have been made to establish vegetation cover guidelines for resistance to soil erosion at the site level. Studies have identified minimum cover values ranging from 20% in Kenya (Moore et al., 1979) to near 100% in Australia, while typical recommendations for the United States range from 50 to 75% (Packer, 1951; Nobel, 1965; Orr, 1970; Gifford, 1984). This variability in cover values is a function of the interaction between the applied stress (rainfall intensity and duration) and the resistance of the soil (soil erodibility).

Correlations between vegetation characteristics and soil erosion are related to the effects of the vegetation on soil properties and raindrop impact and overland flow patterns. In the case of soil properties, vegetation measures can be simply viewed as surrogate indicators for soil quality: the soil affects, and is affected by, vegetation growing in it. The relationship is less clear for raindrop impact: Is a soil of higher quality because it supports higher plant cover and therefore is better protected from raindrops? Similarly, is a soil of higher quality if it supports a plant spatial distribution which increases the residence time of water on a slope by increasing the tortuosity of flow paths? For agronomic crops which are removed every year, this would not be true. According to this perspective, the assessment of soil quality should be based on soil characteristics alone or on crop productivity where production serves to reflect differences in soil quality. With the exception of this caveat, the crop is largely viewed as a secondary, independent factor. In rangelands, however, it is more difficult to separate the vegetation from the soil. Many rangelands are dominated by perennial vegetation, and the vegetative community which exists on a soil has, in many cases, developed with that soil (Blackburn et al., 1992). This also applies to annual grasslands which effectively reseed themselves with a similar suite of species year after year. In light of the overwhelming impact of vegetation on resistance to rangeland soil erosion and the posited key relationship between soil erosion and soil quality, we argue that vegetation characteristics should, at a minimum, be used to interpret and apply soil quality assessments.

Indicators of current resistance or its inverse, soil erodibility, have been widely employed in soil erosion models such as the Universal Soil Loss Equation (USLE), Revised Universal Soil Loss Equation (RUSLE), and Water Erosion Prediction Project (WEPP) (Weltz et al., 1996). Process-based erosion simulation models such as WEPP provide the potential to study the many interactive effects of management practices as they affect soil erosion.

Current Resilience

The resilience of a soil following erosion can be broadly defined in terms of the recovery of specific soil functions, such as infiltration and storage of water in the plant rooting zone (Figure 13.1, #3). More narrowly, however, it may be thought of in terms of the recovery of the resistance of the soil to future erosional events. In other words, to what extent does a single disturbance lead to increased susceptibility to future disturbances?

Resilience is one of the most critical issues to be addressed when assessing soil quality, yet it is also the most difficult to predict. Most studies have been designed to identify factors contributing to the resistance of systems to degradation. There are, however, some potentially useful, albeit largely untested, indicators of resilience following erosion. Soil depth is an obvious indicator of potential resilience, although the mere presence of a deep soil profile does not guarantee that surface soil structure will be regenerated following erosion, as illustrated by lateritic soils. Recent studies on the rate and extent of formation and reformation of soil aggregates in different soils (e.g., Chaney and Swift, 1986; Tisdall, 1996) and on soil organic matter regeneration (Reeder et al., in press) also provide some information on the relative importance of different soil constituents. However, vegetation and soil biotic indicators may be even more sensitive than any single soil chemical or physical parameter (Linden et al., 1994; National Research Council, 1994). This is particularly true for rangeland soils in which at least some of the vegetation remains intact or quickly regenerates following disturbance and where soil biotic communities are generally well established.

In addition to total plant cover, vegetation indicators of potential resilience might include the functional species composition and spatial distribution of the existing plant community, the soil seed bank, and the reservoir of seeds, including exotics, which are likely to disperse into a site following a catastrophic erosional event (National Research Council, 1994). Relevant questions include which species are likely to dominate following disturbance and what their likely impacts are on the regeneration of soil functional integrity and resistance to future disturbances. Although the vegetation recovery per se is not a direct indicator of soil resilience, the regeneration of soil functional integrity is inextricably linked to vegetation in most rangeland systems: the soil both affects and depends on the reestablishment of vegetative cover (DePuit and Redente, 1988).

Similar questions can be asked of the soil biotic community. In light of logistical difficulties in describing these communities, however, it is necessary to identify specific components which contribute directly to recovery processes, are relatively easy to census, and, most importantly, reflect the overall status of the living component of the soil. Recent attempts to isolate one or more key soil biotic indicators have yielded mixed results. In the case of ants, for example, studies in recovering Australian minelands on both basic biology (Majer, 1983) and community composition (Andersen, 1993) suggested that they should reflect soil biotic, if not physical, integrity. However, an extensive study recently completed in the Chihuahuan Desert yielded no significant relationships between ant community composition and site condition (W. Whitford, personal communication).

Soil Erosion as a Driver

Soil erosion can also be viewed as a *driver* in the system which determines soil quality (Figure 13.1, #4). Soil erosion is generally considered to be detrimental to soil quality at the field scale. With few exceptions, soil loss is associated with a reduction in the capacity of soils to perform ecosystem functions. The impacts of soil erosion are generally greater than the proportion of the soil profile removed would suggest due to the concentration of soil organic matter and nutrients near the soil surface. This generally negative view of soil erosion is less applicable at the watershed scale, at which processes of both erosion and deposition must be considered. Virtually the entire country of Bangladesh, for example, owes its relatively high-quality soils to deposition of sediment eroded from the mountains outside of its borders. This perspective can also be usefully applied at the microcatchment scale to predict potential future changes in soil quality based on current depositional patterns (Watters et al., 1996). Thus, soil quality assessments may be enhanced by including multiple scales and by quantifying the impacts of changes at one scale, or on one part of the landscape, on soil quality at other scales and in connected landscape units.

FACTORS INFLUENCING THE RELATIONSHIPS BETWEEN SOIL QUALITY AND SOIL EROSION

The nature of the four relationships between soil quality and soil erosion in each rangeland ecosystem depends on interactions among at least three sets of factors: (1) the characteristics of the disturbance regime associated with the current land use, (2) the characteristics of the disturbance regime(s) under which the current soil–landscape and soil–vegetation patterns developed, and (3) climate and landscape attributes, including degree of soil development and parent material. These three sets of factors largely determine the resistance of the soil to erosion, its resilience following erosion, and, ultimately, the impact of soil erosion on soil quality.

Disturbance Regime

The disturbance regime for an ecosystem can be defined by five attributes: disturbance type or types, spatial scale, intensity, frequency, and predictability. The type of disturbance can be simply defined in terms of the event which causes it, such as fire, logging, grazing, or vehicle traffic. In order to compare different disturbances, however, it is more useful to break each event down into individual components which affect soil processes using a disturbance matrix (Table 13.1). For example, fires, logging, and grazing all remove aboveground biomass, thereby affecting litter and vegetation cover and organic matter supply. Logging, grazing, and vehicle traffic compact the soil, which affects runoff, water availability, and aeration. Grazing and logging are differentiated by the component of the biomass removed, the nature of the compaction, and the form and distribution of nutrients returned to the system.

TABLE 13.1
Disturbance Matrix Illustrating Classification of Disturbance Events Based on Individual Components Which Affect Soil Processes

Disturbance event	Biomass removal	Soil compaction	Nutrient return	
			Form	Distribution
Fire	All	Diffuse[a]	Mineral	Follows vegetation
Logging	Woody	Linear	Unprocessed organic	Depends on practices
Grazing	Herbaceous	Linear and single point	Mineral and processed organic	Discrete, concentrated (dung); diffuse (trampled vegetation)
Vehicle traffic	All within wheel tracks	Linear		

[a] The effect here is indirect: fire increases the susceptibility of soil to crusting and compaction by removing the protective vegetation and litter layer.

The matrix in Table 13.1 can be expanded to include the scale of the disturbance. The impacts of a fire on runoff and soil erosion are much lower for a fire covering 10 m^2 than for one covering a section (square mile) or more. Similarly, the resilience of a system should be higher for the area covered by the smaller fire: this area would be immediately recolonized by organisms at the periphery. It would also be protected from future runoff events by the intact vegetation surrounding the affected area.

Disturbance frequency and intensity could also be added to the matrix in Table 13.1. The capacity of systems to recover from frequent, intensive compactive disturbances (such as a road or cattle path) is much lower than their capacity to recover from an occasional perturbation. In addition to the frequency, the timing of the disturbance can be very important. In the case of compaction, timing affects both resistance (which varies as a function of soil moisture) and resilience (which varies temporally with both biotic activity and physical processes such as frost heave).

Disturbance predictability affects plant and soil community composition. Long-lived perennial plants are often adapted to predictable disturbance. Annuals, conversely, tend to have large seed banks that allow them to quickly recover from a wide variety of disturbances (Barbour et al., 1987). Similarly, much of the soil biota in ecosystems dominated by predictable seasonal drought survive with well-adapted systems of timed reproduction. Populations of these same species, however, can be significantly reduced by a wet season drought (Steinberger and Whitford, 1984; Steinberger et al., 1984).

Disturbance History

The disturbance history of an ecosystem rarely can be used to predict the potential impact of a new disturbance regime. For example, the North American tallgrass prairie evolved under a disturbance regime which included high-intensity grazing by bison. Consequently, it should be both more resistant and more resilient under grazing than much of the Colorado Plateau. The current plant and soil communities of the Colorado Plateau may have never been impacted by large groups of large herbivores (Belnap, 1995). Identification of mechanisms of resistance to erosion can yield additional information on the potential resistance to new types of disturbances. Soils on the Colorado Plateau are relatively resistant to erosion, even after prolonged droughts, due to the stabilizing effects of soil surface cryptogams (primarily lichens). When the physical integrity of the soil surface is disrupted by new disturbances such as trampling, off-road vehicles, or mountain bikes, the resistance is lost (Anderson et al., 1982a). However, the resilience (rate of recovery) can be increased by carefully timing disturbances, such as grazing, to promote regeneration of the biologically active soil surface (Anderson et al., 1982b). This idea of timing disturbances to increase resilience is similar to more common recommendations to time grazing to coincide with periods of high resistance, such as when the soil is drier (Warren et al., 1986).

A combination of this historical perspective and a consideration of the attributes of each disturbance regime may be useful in resolving debates over grazing on public lands. Many of these debates are based on different perceptions of the impacts of grazing on soil quality and soil erosion. While the majority of studies have concluded that grazing generally increases runoff and erosion (Spaeth et al., 1996c), significant improvements under grazing in soil properties that are related to soil hydrology, such as organic matter content, have been recorded. Positive responses, such as that reported by Manley et al. (1995), are more frequently found for ecosystems with a history of large herbivore grazing. Contemporary attempts to improve soil quality and rangeland health under grazing in ecosystems in the southwestern United States (where historic grazing was probably intermediate between the Great Plains and the Colorado Plateau) are based on careful management of the intensity and frequency of fires and grazing with respect to precipitation.

Climate and Landscape

The ultimate impact of any disturbance regime on soil erosion and soil quality depends on interactions with inherent climate, soil, and landscape characteristics. Timing and characteristics of storms, slope, landscape position, topographic complexity, soil depth, and parent material affect resistance and resilience. Measurements of a soil's resistance to detachment by overland flow are more relevant if they are made during the season when precipitation events are likely to exceed infiltration capacity.

Factor-Based Approaches to Assessment

The above discussion suggests that rangeland soil quality assessments can be enhanced by considering the climate and landscape context together with the anticipated disturbance regime. An erosion-based assessment of soil quality should be completed in three interdependent stages. In the first stage, point and field-scale measurements of soil properties are scored using standard scoring functions and combined to generate a standard index (e.g., Karlen and Stott, 1994; Yakowitz et al., 1993). A typical disturbance regime and climate for the region are assumed for this stage. In the second stage, the scoring functions are modified according to local climate and landscape conditions, while the disturbance regime is held constant. In the third stage, the scoring functions are further modified to reflect different disturbance regime scenarios. At this stage, multiple assessments are possible, depending on the scenario selected. For example, a soil in an arid region which is protected from wind erosion by physical crusts may have high resistance to erosion under a disturbance regime of rainy season grazing (when physical crusts rapidly reform following rainstorms) but low resistance under a dry season or continuous grazing regime. Conversely, its long-term resilience may be lower under rainy season grazing due to the potentially negative impacts of growing season grazing on biomass production and subsequent soil carbon input.

CASE STUDIES

The Jornada Experimental Range, Walnut Gulch Experimental Watershed, and Central Plains Experimental Range (CPER) represent three distinct rangeland ecosystems (Figure 13.2). When compared with the global diversity of rangelands, these three sites appear relatively similar. They all lie within 1000 km of each other and

FIGURE 13.2 Locations of the three case studies: Central Plains Experimental Range (C), Jornada Experimental Range (J), and Walnut Gulch Experimental Watershed (W).

are classified as semiarid, with average annual precipitation ranging from 240 mm at the Jornada to 325 mm at the CPER. However, unique relationships between soil quality and soil erosion exist at each site. Post-European colonization changes in vegetation structure are believed to be associated with increased erosion and declines in soil quality at both the Jornada and Walnut Gulch. A change in disturbance regime is thought to be at least partially responsible in both cases. Water erosion has played a major role at the two locations, but much of the Jornada has also been severely altered by wind erosion. Unlike the Jornada or Walnut Gulch, the CPER continues to be dominated by grasses, although there is evidence that some changes in species composition have occurred with heavy grazing. With the exception of areas subjected to prolonged overgrazing coupled with drought and areas converted to croplands, the post-European colonization disturbance regime does not appear to have had a negative impact on soil quality or led to sharp increases in soil erosion on rangelands at the CPER.

Jornada Experimental Range

The Jornada Experimental Range is located in a closed basin in the northern Chihuahuan Desert. The elevation varies from 1190 to 1372 m. During the past 150 years, the plant community covering much of the basin, including most of the soils with a high sand content, has shifted from black grama- (*Bouteloua eriopoda*) dominated grassland (Figure 13.3a) to mesquite- (*Prosopis glandulosa*) dominated shrubland (Figure 13.3b). This shift reflects reductions in soil quality which have both led to and been reinforced by increased wind erosion (Buffington and Herbel, 1965; Gibbens et al., 1983). In many areas, eolian dunes up to several meters in height have developed around individual mesquite shrubs (Gould, 1982), leaving the wind-scalped interdunal areas largely devoid of perennial vegetation (Hennessy et al., 1983).

This transition is believed to be the result of an interaction between climate (drought) and a change in the disturbance regime (Nelson, 1934; Schlesinger et al., 1990). The historic disturbance regime consisted of unpredictable, large-scale droughts and more regular small-scale animal-induced soil surface and grazing disturbances. Antelope, deer, rabbits, prairie dogs, kangaroo rats, and a variety of other rodents all generated soil surface disturbances. All of these species are still present, with the exception of prairie dogs. Unlike the CPER (see below), there is little evidence to suggest that bison played a significant role in this system. With the introduction of cattle and water development, the intensity of grazing disturbances increased. This, together with the simultaneous invasion by mesquite (dispersed by the cattle) of the sandy grassland soils, is hypothesized to have caused the breakdown of the system during severe droughts (Nelson, 1934; Herbel et al., 1972). The system was not resistant to the new disturbance regime, nor was it resilient: grasses have not recovered even where cattle have been excluded for over 50 years (R.P. Gibbens, personal communication). In this case, the third definition of resilience discussed in the introduction is perhaps the most relevant insofar as it focuses on the capacity of the system to recover from multiple, simultaneous assaults.

(a)

(b)

FIGURE 13.3 Chihuahuan Desert grassland on the Jornada Experimental Range with soaptree yucca (*Yucca elata*) and invading mesquite (*Prosopis glandulosa*) in background (a) and former grassland now dominated by mesquite (b).

TABLE 13.2
**Average Soil Surface (0–5 mm) Structural Stability at Three Sites
as Measured by a Field Wet Aggregate Stability Test in Which
Air-Dry Soil Fragments Are Gently Sieved in Distilled Water
5 Minutes After Submersion**

Site	Grass cover (%)	Bare[a]	Grass[a]	Shrub[a]	Average (weighted by vegetative cover)
Mesquite dune	3.9	1.3	4.2	4.4	1.7
Grassland	23.4	1.3	5.1	4.8	2.2
Grassland — exclosure	25.3	2.4	5.0	5.6	3.4

Note: A value of 1 indicates that the visible structure of the fragment disintegrated within 5 sec of submersion, while a value of 6 was assigned to fragments which remained at least 75% intact after sieving. All sites are located within 750 m of each other and all contain 81–84% sand in the top 10 cm.

[a] de Soyza et al., in press.

From a soil quality perspective, three particularly significant changes have occurred. The first is that there has been a net loss of soil resources from the mesquite dune areas. Gibbens et al. (1983) recorded a net loss of 4.6 cm of soil from a 259-ha pasture over a period of 45 years. In this exclosure, the existing mesquite was killed and grasses were planted in the mid-1930s. By 1980, the shrubs had become reestablished and dunes had reformed. The second change is in soil texture. Based on soil samples taken from the same site, Hennessy et al. (1986) concluded that the material lost was confined to the silt and clay fractions. Many of these soils have sand contents well in excess of 80%, further magnifying the impacts of the loss of the fine fractions on soil aggregate stability and water and nutrient retention. Preliminary studies of soil aggregate stability on these sandy soils suggest that a reduction in soil surface structural stability occurs relatively early in the transition from grassland to shrubland (Table 13.2). This would suggest that soil resistance to disturbance in the grassland system is quite low and that it is only the protective grass cover which prevents catastrophic wind and water erosion from occurring (Figure 13.3b).

This net reduction in soil quality based on average site characteristics is reinforced by a third change: a redistribution of remaining resources within the site. Organic matter and associated nutrients tend to accumulate beneath shrubs (Schlesinger et al., 1990, 1996; Virginia et al., 1992). This tends to stratify the system into relatively high-quality microsites associated with shrubs and lower quality microsites in the interspaces. This is reflected in increased heterogeneity in soil structural stability (Table 13.2) and infiltration capacity (Table 13.3), with the highest stability and infiltration occurring beneath shrub canopies.

As a result of this heterogeneous distribution of resources, the interspace soil is both less resistant to erosion (Figure 13.1, #2) and less resilient after erosion has

TABLE 13.3
Relative Infiltration Capacity as Indicated by Time Required for 2.5 cm of Water to Infiltrate Saturated Soil from a 15-cm Ring Inserted to a Depth of 1.5 cm for Sites Listed in Table 13.2

Site	Min:sec Bare	Grass	Shrub	Average (weighted by vegetative cover)
Mesquite dune	8:38 (3:13)	2:48 (0:38)	2:12 (0:40)	7:18
Grassland	13:05 (3:14)	1:15 (0:33)	0:46 (0:10)	9:27
Grassland — exclosure	6:02 (1:04)	2:05 (0:38)	1:20 (0:28)	4:25

Note: Mean and standard deviation for $n = 3$.

occurred (Figure 13.1, #3). This then leads to a reduction in soil quality (Figure 13.1, #4) and results in a negative feedback loop of increasingly depleted interspaces and enriched shrub microsites (Schlesinger et al., 1990).

Walnut Gulch

The Walnut Gulch Experimental Watershed encompasses an area of 150 km² that surrounds Tombstone, in southeastern Arizona. Elevation of the watershed ranges from 1250 to 1585 m. The watershed is located primarily in a high foothill alluvial fan portion of the San Pedro River watershed (Renard et al., 1993; Weltz et al., 1996). Soils of the watershed reflect the parent material, with limestone-influenced alluvial fill as the dominant source. These soils are generally well-drained, calcareous, gravelly loams with a large percentage of rock and gravel on the soil surface (Breckenfield, 1996). Erosion pavement can exceed 70% on steep eroded hillslopes and typically ranges from 35 to 50%.

The watershed is in a transition zone between the Chihuahuan and Sonoran Desert plant communities (Figure 13.4). Historical records on plant community composition are limited but indicate that, like the Jornada, a larger percentage of the watershed was grass prior to European settlement in the late nineteenth century. Currently, the lower two-thirds of the watershed is dominated by shrubs that include creosote bush (*Larrea tridentata*), whitethorn (*Acacia constricta*), tarbush (*Fluorensia cernua*), burroweed (*Haplopappus tenuisectus*), and snakeweed (*Gutierrezia sarothrae*). The upper third of the watershed is dominated by desert grassland plant communities. Dominant grasses are black grama (*Bouteloua eriopoda*), blue grama (*B. gracilis*), and bush muhly (*Muhlenbergia porteri*).

The climate of Walnut Gulch watershed is classified as semiarid with frequent local droughts. Precipitation varies considerably both seasonally and annually. Average annual precipitation for the period 1956–90 was 312 mm with a standard deviation of 79 mm. Approximately two-thirds of the annual precipitation occurs as high-intensity, convective thunderstorms of limited areal extent during the summer monsoon period of June through September. Runoff and soil erosion on Walnut Gulch result almost exclusively from these summer convective storms.

(a)

(b)

FIGURE 13.4 Transition zone between Chihuahuan and Sonoran deserts at the Walnut Gulch Experimental Watershed. Photo point comparison shows an increase in burroweed and desert zinnia from 1967 (a) to 1994 (b).

The WEPP model was used to predict the impact of five management actions on soil erosion at the Lucky Hills watershed, a small instrumented catchment on Walnut Gulch (Renard et al., 1993). Lucky Hills is dominated by creosote bush and whitethorn, with an average canopy cover of 28% and ground cover of 56% (primarily rock and gravel cover). Little or no herbaceous vegetation exists in the watershed despite exclusion from grazing for 25 years. The five management scenarios included two types of disturbance and a range of disturbance intensities. Existing vegetation was maintained in two of the scenarios. In one of these, moderate grazing was imposed. In the other scenarios, three levels of grazing intensity (none, moderate, and heavy) were imposed on areas following herbicide-based shrub removal and grass reseeding.

The model simulation results show that grazing intensity increased water yield, sediment yield, and the magnitude of the 2-year frequency peak discharge over the 15-year simulation period. The most important effect of management was on hillslope sediment yield. Converting from shrub to grass with no grazing decreased hillslope sediment yield by 91%. However, this decrease translated into a much smaller (25%) decrease in reduction of sediment yield from the watershed. The sediment yield entering the channel decreased significantly, but runoff amounts and peak discharge rates did not decrease. This resulted in an increase in channel scour and a net decrease in watershed sediment yield of only 25%.

If on-site evaluation is limited to upland areas, then shifts in system stability related to hillslope, riparian corridor, and stream channel areas may go unrecognized. The major implication of this work is that the entire landscape must be evaluated for its resilience and resistance to soil erosion to avoid transferring the stress from one part of the landscape (hillslope) to another (channels) and destabilizing the entire landscape through complex feedback interactions.

A second study conducted at Walnut Gulch illustrates the importance of temporal variability in this ecosystem. Monthly evaluations of erosion rates using a rotating boom rainfall simulator (Swanson, 1965) were made on two treatments (natural and clipped) at three soil moisture contents on a Haplargid soil (Simanton et al., 1991) and compared to the soil erodibility factor (K) of the RUSLE model (Figure 13.5). The K factor of the RUSLE model is varied throughout the year and is a function of frost-free period and average annual erosivity (R: MJ*mm/ha*h*yr).

Measured erosion rates were lowest between May and July and highest in November. This is in complete contradiction to RUSLE model estimates (Figure 13.5). The discrepancy in the cycle of soil erodibility extremes may be due to the lack of freeze–thaw intensity in this Haplargid soil as compared to the soils from which the RUSLE K algorithm was developed (i.e., cropland soils from the east and midwestern United States) and needs to be modified to address rangeland conditions. Time-related changes in erosion rates associated with rangeland treatment need to be evaluated over a multiyear period using multiplot studies. Biotic factors, both flora and fauna, appear to significantly influence the temporal variability of soil quality and need to be considered before we can adequately define the interactions between soil quality and soil erosion in this ecosystem.

Central Plains Experimental Range

The CPER, established in 1937 to evaluate and develop improved management practices for fragile grasslands, is located in northeastern Colorado on the shortgrass steppe of the western Great Plains (Figure 13.6). The region is characterized by low but highly variable rainfall, frequent droughts, high evapotranspiration, and a short

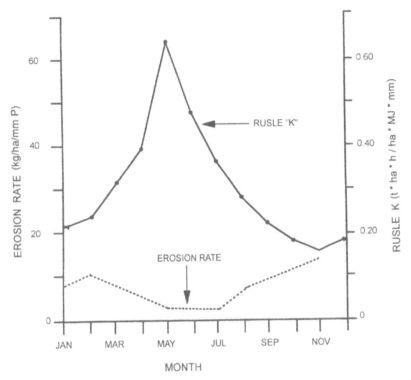

FIGURE 13.5   Monthly measured erosion rate versus RUSLE estimated monthly K from the Walnut Gulch Experimental Watershed.

FIGURE 13.6   Semiarid shortgrass rangeland at the Central Plains Experimental Range.

growing season (average 133 frost-free days). Annual precipitation averages 325 mm (range 109–580 mm, 80% in April through September, primarily as thunderstorms). Soils at the CPER are mostly sandy loams and loamy sands on a topography of rolling hills at elevations of 1600–1700 m. Vegetation is predominantly grasses (48–70% warm season and 8–10% cool season, as dry weight of peak standing crop), with shrubs (1–11%), forbs (5–9%), and plains prickly pear (10–24%) (Sims et al., 1978; ARS, unpublished data). Annual production averages 700 kg/ha, but varies from 300 to 1700 kg/ha, while carrying capacity averages 1.7 ha per animal unit month.

Grasses and forbs have been the dominant vegetation on the shortgrass steppe since the early Cenozoic, and herbivores have coevolved with the vegetation. The plant community has therefore experienced a long history of relatively heavy grazing pressure (Laurenroth and Milchunas, 1992). Fire has been a dominant force in maintaining the integrity of the prairie plant community (Reichman, 1987), and periodic extreme climatic events, such as floods or dust storms, have played a significant role in pedologic additions and translocations over time (Blecker et al., in press).

In addition to grazing by cattle and wildlife, extreme climatic events, and occasional fires, the current disturbance regime includes small surface disturbances caused by ants and small mammals. The effects of these small patch-producing disturbances depend on the frequency of occurrence of the disturbance (Coffin and Laurenroth, 1988). Tolerance of the shortgrass steppe to disturbance by grazing is well documented and is due in large part to the tolerance of the dominant grass species, blue grama, to heavy grazing (Klipple and Costello, 1960). A 12-year study at the CPER by Hyder et al. (1975) to evaluate repeated heavy grazing and N fertilization strategies suggested that, unlike other rangelands, the standing biomass of perennial and annual species of the shortgrass steppe tends to increase or decrease more in response to weather conditions than to heavy grazing.

The resistance of the soil to grazing disturbances largely depends on whether grazing results in changes in plant species composition and ground cover. In the long-term grazing pastures at the CPER, where a healthy stand of blue grama dominates the plant community irrespective of grazing intensity, soil erosion is minimal from normal high-intensity rainfall events. Studies by Frasier et al. (1995) indicated that soil loss was negligible when simulated rainfall was applied at rates of 55–110 mm/hr (a range of intensities common to thunderstorms in the area), although total runoff quantities and rates were higher from heavily grazed pastures than from lightly grazed pastures. The lower infiltration rates as a result of heavy grazing improved within 2 years after removing cattle from the heavily grazed pastures. Lower infiltration rates with heavy stocking rates also have been reported on mixed grass rangeland (Abdel-Magid et al., 1987).

Studies are currently under way at the CPER to evaluate effects of long-term grazing on other soil quality parameters such as organic matter content. The quantity and distribution of organic matter in a rangeland soil depend on the rooting characteristics of the plant community. Changes in plant species composition due to grazing pressure and consequent changes in total root biomass and distribution within the soil profile can alter soil organic matter concentration, composition, and distribution in the soil profile. Increases in surface soil organic carbon (C) have been reported for a mixed grass prairie as the result of grazing-induced reductions in needle-and-thread, a deep-rooted species, and increases in blue grama, a shallow-rooted species (Smoliak et al., 1972). Increases in mixed grass prairie soil C have also been attributed to grazing-enhanced decomposition of standing dead and surface litter (Manley et al., 1995). On the shortgrass prairie at the CPER, where blue grama has remained the dominant plant species irrespective of grazing intensity and where average annual production is low, preliminary investigations have revealed no significant differences in A-horizon organic C and N concentrations in 55-year-grazed versus 55-year-ungrazed pastures. However, because of higher soil bulk density with heavy grazing, the total organic C content of the A horizon in the heavily grazed pastures is significantly higher than the C content of nongrazed exclosures (G. Schuman and J. Reeder, ARS, unpublished data). Other studies conducted at the CPER have demonstrated that variability in soil organic matter content can be as high within a grazing treatment as between grazing treatments due to natural differences between bare soil and soil under plants (Hook et al., 1991; Burke et al., 1995).

Although the shortgrass prairie displays a high degree of resistance to grazing, it is not highly resilient where prolonged overgrazing coupled with drought conditions has degraded the plant community (Shoop et al., 1989). The soils of the region are highly erodible when not protected by vegetation, and attempts at renovation of damaged grasslands by seeding native species are largely unsuccessful because seed production by blue grama is low and variable (Coffin and Laurenroth, 1992), and the seedling morphology of blue grama is not well adapted to the low and sporadic precipitation common to the area (Hyder et al., 1975).

RANGELAND HEALTH AND SOIL QUALITY

While soil quality and soil health have emerged as new paradigms for assessing ecosystem condition in cultivated systems around the world, the term rangeland health has been proposed to refer to evolving approaches to the assessment and monitoring of noncultivated, nonforested lands. The National Research Council's Committee on Rangeland Classification (National Research Council, 1994) proposed three criteria for determining whether a rangeland is healthy, at risk, or unhealthy: "degree of soil stability and watershed function, integrity of nutrient cycles and energy flow, and presence of functioning recovery mechanisms." The first criterion is directly related to soil quality, while the latter two ultimately depend on soil stability and watershed function.

Tongway (1994) has proposed an approach which relies heavily on soil surface characteristics and other indicators of soil quality. This approach is being used to evaluate rangeland health in Australia. One characteristic of a soil quality–based approach to rangeland health assessment is that it depends on plant community structure rather than species composition. This approach has the potential to resolve contentious issues related to defining the preferred or ideal plant community for a site. In many areas, including most of the Chihuahuan Desert, soil quality is highly correlated with the desired potential natural community (PNC). However, exceptions do occur. For example, while buffalograss (*Buchloe dactyloides*) is a component of the native plant community in the Great Plains, and therefore is associated with good range condition, infiltration tends to be lower in buffalograss stands even when texture and organic matter are constant (Spaeth et al., 1996b). Where soil quality and comparisons with the PNC yield different conclusions, either one can be assigned precedence or the two can be combined using the weighting system discussed above (Karlen and Stott, 1994; Herrick and Whitford, 1995).

CONCLUSIONS

The link between soil quality and soil erosion has been clearly established in the popular press. In a recent *Los Angeles Times* article detailing the impacts of the mid-1990s drought in the Southwest, a social scientist with the National Center for Atmospheric Research was quoted as stating that "We shouldn't blame nature for destroying a lot of property. People keep moving in harm's way by building homes in fire-prone mountains, and farming and running cattle in regions of poor soil quality."

This intuitive link between soil erosion and soil quality is supported by four specific relationships. Soil quality is reflected in the historic and current resistance to erosion and in the current resilience following erosion. Erosion is also a key determinant of soil quality in many systems. The nature of these relationships is determined by the climate and the landscape together with the current disturbance regime and its similarity to the historic disturbance regime(s) under which the soil and biotic communities evolved.

These relationships between soil erosion and soil quality lead to a number of implications for the assessment of soil quality. Soil quality must be evaluated at a variety of spatial scales in order to incorporate the wide range of scales at which soil erosion processes occur. Ideally, the measurements at these scales should be linked to each other in parallel with hydrologic and eolian linkages in the landscape. Another implication of these relationships is that the current resistance to, and resilience following, erosion must be interpreted in the context of the current and historic disturbance regime. A soil which is highly resistant to erosion following drought may lose this resistance when the drought is combined with fire or intensive grazing. A third implication follows from the first two: the resistance of a soil to erosion ultimately depends on the spatial scale at which erosion is measured and on the intensity of the disturbance event which is assumed. Redeposition of soil within a landscape can disguise the magnitude of the potential sediment yield from a watershed until a 100-year storm occurs.

A final implication of the relationships between soil erosion and soil quality is that the resilience, as well as the resistance, of the soil must be incorporated into assessments if the long-term sustainability of the system is to be evaluated. While most models currently focus on the resistance of the system to soil erosion, many systems, such as soils on recent deep volcanic materials and alluvium, have such a high resilience that resistance is relatively less important. Conversely, many soils which are shallow, have lost most of their biota, or are in areas with severe climatic limitations have virtually no resilience. This last point illustrates the need to refine and improve definitions of "soil loss tolerance" as part of the effort to define soil quality and its relationship to soil erosion.

Future research on relationships between disturbance regimes, soil quality, and soil erosion should be conducted at three time scales: (1) immediately postdisturbance, for direct impacts on soil properties; (2) medium term, for impacts on soil biota and vegetation (including growth, growth form, and biomass allocation [root:shoot ratios]) which affect soil resistance and/or resilience; and (3) long term, for impacts on plant and soil biota community composition. Very few long-term disturbance studies exist, with the exception of some on grazing, and most of these do not include a hydrological component.

ACKNOWLEDGMENTS

L. Huenneke, A. deSoyza, M. Logan, and several anonymous reviewers provided useful comments on earlier versions. The preparation of this chapter was supported by a USDA-NRI grant to J. Herrick.

REFERENCES

Abdel-Magid, A.H., G.E. Schuman, and R.H. Hart. 1987. Soil bulk density and water infiltration as affected by grazing systems. *J. Range Mange.* 40:307–309.

Andersen, A.N. 1993. Ants as indicators of restoration success at a uranium mine in tropical Australia. *Restor. Ecol.* 1:156–167.

Anderson, D.C., K.T. Harper, and R.C. Holmgren. 1982a. Factors influencing the development of cryptogamic soil crusts in Utah deserts. *J. Range Mange.* 35:180–185.

Anderson, D.C., K.T. Harper, and S.R. Rushforth. 1982b. Recovery of cryptogamic soil crusts from grazing on Utah winter ranges. *J. Range Mange.* 35:355–359.

Arshad, M.A. and G.M. Coen. 1992. Characterization of soil quality: physical and chemical criteria. *Am. J. Altern. Agric.* 7:25–32.

Barbour, M.G., J.H. Burke, and W.D. Pitts. 1987. *Terrestrial Plant Ecology,* 2nd ed. Benjamin/Cummings, Menlo Park, CA, 634 pp.

Belnap, J. 1995. Surface disturbances: their role in accelerating desertification. *Environ. Monit. Assess.* 37:39–57.

Benkobi, L., M.J. Trlica, and J.L. Smith. 1993. Soil loss as affected by different combinations of surface litter and rock. *J. Environ. Qual.* 22:657–661.

Blackburn, W.H., F.B. Pierson, C.L. Hanson, T.L. Thurow, and A.L. Hanson. 1992. The spatial and temporal influence of vegetation on surface soil factors in semiarid rangelands. *Trans. ASAE* 35:479–486.

Blackburn, W.H., F.B. Pierson, Jr., G.E. Schuman, and R. Zartman (eds.). 1994. *Variability in Rangeland Water Erosion Processes.* Soil Science Society of America, Madison, WI, 106 pp.

Blecker, S.W., C.M. Yonker, C.G. Olson, and E.F. Kelly. In press. Paleopedologic and geomorphic evidence for Holocene climate variation, shortgrass steppe, Colorado. *Geoderma.*

Breckenfield, D.J. 1996. Soil Survey of Walnut Gulch Experimental Watershed, Arizona. Special Report. USDA-NRCS, USDA-ARS, and Arizona Agricultural Experiment Station, Tucson, AZ.

Buffington, L.C. and C.H. Herbel. 1965. Vegetational changes on a semidesert grassland range from 1858 to 1963. *Ecol. Monogr.* 35:139–164.

Burke, I.C., P.B. Hook, and W.K. Laurenroth. 1995. Effects of grazing and exclosure on soil organic matter pools and nitrogen availability in a shortgrass steppe. Abstract. In Central Plains Experimental Range Second Annual Symp., January 13, 1995. USDA-ARS and Colorado State University Long-Term Ecological Research Project, Fort Collins.

Chaney, K. and R.S. Swift. 1986. Studies on aggregate stability. I. Re-formation of soil aggregates. *J. Soil Sci.* 37:329–335.

Coffin, D.P. and W.K. Laurenroth. 1988. The effects of disturbance size and frequency on a shortgrass plant community. *Ecology* 69:1609–1617.

Coffin, D.P. and W.K. Laurenroth. 1992. Spatial variability in seed production of the perennial bunchgrass *Bouteloua gracilis* (Gramineae). *Am. J. Bot.* 79:347–353.

Dent, D. and A. Young. 1981. *Soil Survey and Land Evaluation.* Allen & Unwin, Winchester, MA, 278 pp.

DePuit, E.J. and E.F. Redente, 1988. Manipulation of ecosystem dynamics on reconstructed semiarid lands. In E.B. Allen (ed.). *The Reconstruction of Disturbed Arid Lands: An Ecological Approach.* Westview Press, Boulder, CO, 267 pp.

de Soyza, A.G., W.G. Whitford, and J.E. Herrick. In press. Sensitivity testing of indicators of ecosystem health. *Ecosystem Health.*

Doran, J.W. and T.B. Parkin. 1994. Defining and assessing soil quality. In J.W. Doran, D.C. Coleman, D.F. Bezdicek, and B.A. Stewart (eds.). *Defining Soil Quality for a Sustainable Environment.* SSSA Special Publication No. 35. Soil Science Society of America, Madison, WI, pp. 3–21.

Forman, R.T.T. and M. Godron. 1986. *Landscape Ecology.* John Wiley & Sons, New York, 619 pp.

Frasier, G.W., R.H. Hart, and G.E. Schuman. 1995. Rainfall simulation to evaluate infiltration/runoff characteristics of a shortgrass prairie. *J. Soil Water Conserv.* 50:460–463.

Gibbens, R.P., J.M. Tromble, J.T. Hennessy, and M. Cardenas. 1983. Soil movement in mesquite dunelands and former grasslands of southern New Mexico from 1933–1980. *J. Range Mange.* 36:145–148.

Gifford, G.F. 1984. Vegetation allocation for meeting site requirements. In *Developing Strategies for Rangeland Management: A Report by the Committee on Development Strategies for Rangeland Management, National Research Council, National Academy of Sciences.* Westview Press, Boulder, CO, pp. 35–116.

Gould, W.L. 1982. Wind erosion curtailed by controlling mesquite. *J. Range Manage.* 35: 563–566.

Hennessy, J.T., R.P. Gibbens, J.M. Tromble, and M. Cardenas. 1983. Vegetation changes from 1935 to 1980 in mesquite dunelands and former grasslands of southern New Mexico. *J. Range Manage.* 36:370–374.

Hennessy, J.T., B. Kies, R.P. Gibbens, and J.M. Tromble. 1986. Soil sorting by forty-five years of wind erosion on a southern New Mexico range. *Soil Sci. Soc. Am. J.* 50: 391–394.

Herbel, C.H., F.N. Ares, and R.A. Wright. 1972. Drought effects on a semidesert grassland range. *Ecology* 53:1084–1093.

Herrick, J.E. and W.G. Whitford. 1995. Assessing the quality of rangeland soils: challenges and opportunities. *J. Soil Water Conserv.* 50:237–242.

Holling, C.S. and G.K. Meffe. 1996. Command and control and the pathology of natural resource management. *Conserv. Biol.* 10:328–337.

Hook, P.B., I.C. Burke, and W.K. Laurenroth. 1991. Heterogeneity of soil and plant N and C associated with individual plants and openings in North American shortgrass steppe. *Plant Soil* 138:247–256.

Hyder, D.N., R.E. Bement, E.E. Remmenga, and D.F. Hervey, 1975. Ecological Responses of Native Plants and Guidelines for Management of Shortgrass Range. USDA Technical Bulletin 1053, 76 pp.

Karlen, D.L. and D.E. Stott. 1994. A framework for evaluating physical and chemical indicators of soil quality. In J.W. Doran, D.C. Coleman, D.F. Bezdicek, and B.A. Stewart (eds.), *Defining Soil Quality for a Sustainable Environment.* SSSA Special Publication No. 35. Soil Science Society of America, Madison, WI, pp. 53–72.

Klipple, G.E. and D.F. Costello. 1960. Vegetation and Cattle Responses to Different Intensities of Grazing on Short-Grass Ranges on the Central Great Plains. U.S. Department of Agriculture Technical Bulletin 1216. U.S. Government Printing Office, Washington, D.C.

Laurenroth, W.K. and D.G. Milchunas. 1992. Short-grass steppe. In R.T. Coupland (ed.). *Natural Grasslands. Introduction and Western Hemisphere.* Elsevier, New York, pp. 183–226.

Linden, D.R., P.F. Hendrix, D.C. Coleman, and P.C.J. van Vliet. 1994. Faunal indicators of soil quality. In J.W. Doran, D.C. Coleman, D.F. Bezdicek, and B.A. Stewart (eds.), *Defining Soil Quality for a Sustainable Environment.* SSSA Special Publication No. 35. Soil Science Society of America, Madison, WI, pp. 91–106.

Majer, J.D. 1983. Ants: bio-indicators of minesite rehabilitation, land use, and land conservation. *Environ. Manage.* 7:375–383.

Manley, J.T., G.E. Schuman, J.D. Reeder, and R.H. Hart. 1995. Rangeland soil carbon and nitrogen response to grazing. *J. Soil Water Conserv.* 50:294–298.

Moore, T.R., D.B. Thomas, and R.G. Barber. 1979. The influence of grass cover on runoff and soil erosion from soils in the Machakos area, Kenya. *Trop. Agric.* 50:333–339.

National Research Council. 1993. *Soil and Water Quality: An Agenda for Agriculture.* National Academy Press, Washington, D.C., 516 pp.

National Research Council. 1994. Rangeland Health: New Methods to Classify, Inventory, and Monitor Rangelands. National Academy Press, Washington, D.C., 180 pp.

Nelson, E.W. 1934. The Influence of Precipitation and Grazing on Black Grama Range. Technical Bulletin 409. U.S. Department of Agriculture. U.S. Government Printing Office, Washington, D.C.

Nobel, E.L. 1965. Sediment Reduction Through Watershed Rehabilitation. USDA Miscellaneous Publication 970:114–123. U.S. Government Printing Office, Washington, D.C.

Orr, H.K. 1970. Runoff and Erosion Control by Seeded and Native Vegetation on a Forest Burn: Black Hills, South Dakota. USFS Research Paper RM-60. U.S. Government Printing Office, Washington, D.C.

Packer, P.E. 1951. An approach to watershed protection criteria. *J. For.* 49:635–644.

Pimm, S.L. 1984. The complexity and stability of ecosystems. *Nature* 307:321–326.

Reeder, J.D., G.E. Schuman, and R.A. Bowman. In press. Soil carbon and nitrogen changes in CRP lands in the central Great Plains. *J. Soil Tillage Res.*

Reichman, O.J. 1987. *Konza Prairie. A Tallgrass Natural History.* University Press of Kansas, Lawrence, 226 pp.

Renard, K.G., L.J. Lane, W.E. Emmerich, J.J. Stone, M.A. Weltz, D.C. Goodrich, and D.S. Yakowitz. 1993. Agricultural impacts in an arid environment: Walnut Gulch case studies. *Hydrol. Sci. Technol.* 9:145–190.

Romig, D.E., M.J. Garlynd, R.F. Harris, and K. McSweeney. 1995. How farmers assess soil health and quality. *J. Soil Water Conserv.* 50:229–236.

Schlesinger, W.H., J.F. Reynolds, G.L. Cunningham, L.F. Huenneke, W.M. Jarrell, R.A. Virginia, and W.G. Whitford. 1990. Biological feedbacks in global desertification. *Science* 247:1043–1048.

Schlesinger, W.H., J.A. Raikes, A.E. Hartley, and A.F. Cross. 1996. On the spatial pattern of soil nutrients in desert ecosystems. *Ecology* 77:364–374.

Shoop, M., S. Kanode, and M. Calvert. 1989. Central Plains Experimental Range: 50 years of research. *Rangelands* 11:112–117.

Simanton, J.R., M.A. Weltz, and H. D. Larsen 1991. Rangeland experiments to parameterize the water erosion prediction project model: vegetation canopy cover effects. *J. Range Manage.* 44:276–282.

Sims, P.L., J.S. Singh, and W.K. Laurenroth. 1978. The structure and function of ten western North American grasslands. I. Abiotic and vegetational characteristics. *J. Ecol.* 66: 251–285.

Smoliak, S., J.F. Dormaar, and A. Johnston. 1972. Long-term grazing effects on *Stipa–Bouteloua* prairie soils. *J. Range Manage.* 25:245–250.

Spaeth, K.E., F.B. Pierson, M.A. Weltz, and R.G. Hendricks (eds.). 1996a. *Grazingland Hydrology Issues: Perspectives for the 21st Century.* Society for Range Management, Denver, CO.

Spaeth, K.E., F.B. Pierson, M.A. Weltz, and J.B. Awang. 1996b. Gradient analysis of infiltration and environmental variables as related to rangeland vegetation. *Trans. ASAE* 39:67–77.

Spaeth, K.E., T.L. Thurow, W.H. Blackburn, and F.B. Pierson. 1996c. Ecological dynamics and management effects on rangeland hydrologic processes. In K.E. Spaeth, F.B. Pierson, M.A. Weltz, and R.G. Hendricks (eds.). *Grazingland Hydrology Issues: Perspectives for the 21st Century.* Society for Range Management, Denver, CO, pp. 25–51.

Steinberger, Y. and W.G. Whitford. 1984. Spatial and temporal relationships of soil microarthropods on a desert watershed. *Pedobiologia* 26:275–284.

Steinberger, Y., D.W. Freckman, L.W. Parker, and W.G. Whitford. 1984. Effects of simulated rainfall and litter quantities on desert soil biota: nematodes and microarthropods. *Pedobiologia* 26:267–274.

Swanson, N.P. 1965. Rotating-boom rainfall simulator. *Trans. ASAE* 8:71–72.

Tisdall, J.M. 1996. Formation of soil aggregates and accumulation of soil organic matter. In M.R. Carter and B.A. Stewart (eds.). *Structure and Organic Matter Storage in Agricultural Soils.* Lewis Publishers, Boca Raton, FL, pp. 57–96.

Tongway, D. 1994. *Rangeland Soil Condition Assessment Manual.* CSIRO, Canberra, Australia.

Virginia, R.A., W.M. Jarrell, W.G. Whitford, and D.W. Freckman. 1992. Soil biota and soil properties associated with surface rooting zone of mesquite (*Prosopis glandulosa*) in historical and recently desertified Chihuahuan Desert habitats. *Biol. Fertil. Soils* 14:90–98.

Warren, S.D., T.L. Thurow, W.H. Blackburn, and N.E. Garza. 1986. Soil response to trampling under intensive rotational grazing. *J. Range Manage.* 50:491–495.

Watters, S.E., M.A. Weltz, and E.L. Smith. 1996. Evaluation of a site conservation rating system to describe soil erosion potential on rangelands. *J. Range Manage.* 49:277–284.

Weltz, M.A., H.D. Fox, S. Amer, F.B. Pierson, and L.J. Lane. 1996. Erosion prediction on range and grazinglands: a current perspective. In K.E. Spaeth, F.B. Pierson, M.A. Weltz, and R.G. Hendricks (eds.). *Grazingland Hydrology Issues: Perspectives for the 21st Century.* Society for Range Management, Denver, CO, pp. 97–116.

Wood, J.C., M.K. Wood, and J.M. Tromble. 1987. Important factors influencing water infiltration and sediment production on arid lands in New Mexico. *J. Arid Environ.* 12:111–118.

World Resources Institute. 1992. *World Resources: A Guide to the Global Environment.* Oxford University Press, New York.

Yakowitz, D.S., J.J. Stone, L.J. Lane, P. Heilman, J. Masterson, J. Abolt, and B. Imam. 1993. A decision support system for evaluating the effects of alternative farm management practices on water quality and economics. *J. Water Sci. Technol.* 28:47–54.

Section IV

Soil Erosion and Productivity

14 Relation Between Soil Quality and Erosion

R. Lal, D. Mokma, and B. Lowery

INTRODUCTION

Accelerated soil erosion is a global problem of modern times with severe economic (Pimentel et al., 1995) and environmental impacts (Lal, 1995). Economic impacts are due to decrease in crop yield, and environmental impacts are due to reduction in the soil's ability to regulate water and air qualities. Soil's environmental regulatory functions are also intimately related to its productivity, both being determined by soil quality (Bezdicek et al., 1996). The close interaction among soil quality (i.e., productivity and environmental regulatory capacity), soil degradative processes, and soil resilience (Lal, 1993, 1994a) is conceptualized in Figure 14.1. Soil resilience, soil's ability to restore its quality following a perturbation, depends on inherent soil properties and the net balance between soil formative and degradative processes (Lal, 1994a) as per Equation 14.1:

$$S_r = S_a + \int_{ti}^{tf} (S_n - S_d \pm I_m)dt \qquad (14.1)$$

where S_r is soil resilience, S_a is antecedent soil condition, S_n is the rate of new soil formation, S_d is the rate of soil degradation/depletion, I_m is the management input, and t is time. The magnitude and sign of the term $(S_n - S_d + I_m)$ are critical in determining soil resilience. The definition in Equation 14.1 can be applied to a specific soil property (e.g., rooting depth, topsoil thickness, soil organic carbon [SOC] content, plant-available water capacity [AWC], available nutrient capacity [ANC], etc.) as long as the time-dependent relation of that property is known. Soil resilience affects soil quality by mitigating the adverse effects of predominant degradative processes (Figure 14.1).

1-57444-100-0/99/$0.00+$.50
© 1999 by Soil and Water Conservation Society

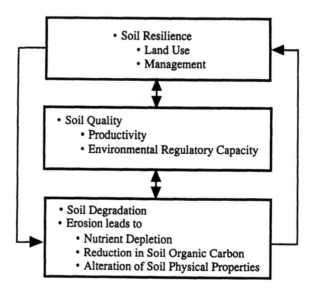

FIGURE 14.1   Interactive effects of soil resilience and degradative processes on soil quality.

Accelerated soil erosion is a predominant degradative process at the global scale (Oldeman, 1994). It causes a reduction in soil solum or depth to root or water-restricting layer and a decrease in plant-available water and nutrient capacities because of physical displacement of soil by forces of wind, water, and gravity. Therefore, soil erosion may be expressed as per Equation 14.2:

$$E = d(R_d, \text{AWC}, \text{ANC}, \text{SOC}) \,/\, dt \qquad (14.2)$$

where E is soil erosion rate, R_d is rooting/topsoil depth, and t is time. It is important to know soil-specific functions depicting temporal changes in parameters included in Equation 14.2. The value of E is negative for erosion and positive for deposition. Soil erosion may also be expressed as the rate of change of any of these parameters as per Equation 14.3:

$$E = d(S_n - S_d) \,/\, dt \qquad (14.3)$$

where S_n is the rate of accretion and S_d is the rate of decline of a specific soil characteristic (e.g., topsoil depth).

Soil erosion affects soil quality through its effects on inherent soil properties (i.e., R_d, AWC, ANC, clay content) and efficiency of use of external inputs (i.e., fertilizer, irrigation, tillage energy). Considering the principal determinants, soil quality (S_q) may be expressed as per Equation 14.4 (Lal, 1993):

$$S_q = f(P_s \times S_c \times R_d \times e_d \times N_c \times B_d)_t \qquad (14.4)$$

where P_s is soil productivity determined in appropriate units, S_c is index of soil structural characteristics including porosity and pore size distribution, R_d is rooting depth, e_d is charge density, N_c is nutrient reserve, B_d is a measure of biodiversity, and t is time. A sufficiency index of these parameters, determined for different land uses and farming systems, can be combined in a multiplicative manner as described by Lal (1994a). Accelerated erosion influences soil quality through adverse effects on the parameters listed in Equation 14.4.

EROSION–SOIL QUALITY INTERACTION

Accelerated soil erosion is driven by socioeconomic environments, and its magnitude is determined by management and key biophysical factors as shown in Figure 14.2. The impact of erosion on soil quality and vice versa is confounded by the interactive effects of biophysical factors and socioeconomic (which includes management) environments. Soil quality affects the magnitude of erosion and is in turn affected by its ramifications. These interactive effects are determined by inherent soil properties, their critical limits, and spatial and temporal changes due to management. Important among these are the following.

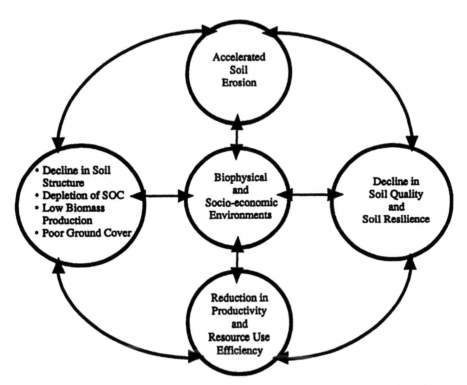

FIGURE 14.2 The complex relationship between soil quality and accelerated erosion as affected by biological and socioeconomic factors.

Solum or Rooting Depth

Erosion effects on soil quality depend on the minimum (critical) soil depth required to sustain productivity and maintain environmental regulatory capacity (Lal, 1994b). The magnitude of critical soil depth may depend on land use, farming systems, management, climate, and soil resilience. Erosion effects on soil quality and productivity are minimal for a soil with a deep profile of uniformly favorable properties (e.g., alluvial or loess deposits).

Soil Life

The duration for which soil productivity and environmental capacity are maintained at a satisfactory level has been defined as the soil life (Biot et al., 1989). Soil life depends on soil resilience, land use, farming system, and management. The time required for soil quality decline to reduce its productivity to half its antecedent level is called half-life.

Spatial Variability in Soil Quality

Erosional processes are highly variable over a watershed, hillside, or plot. Some soilscape or landscape units are more susceptible to erosion and erosion-induced degradation than others. Ratio of areas of susceptible to unsusceptible or less susceptible soilscape units is an important factor determining the mean soil quality index over the watershed in relation to its interaction with erosional processes. Indicators of susceptibility of soilscape units to erosional processes are related to soil quality and vary among soils.

Soil Physical Quality

Determinants of soil physical quality (Figure 14.3) are adversely affected by soil erosion. The most reactive portion of the soil is the organic and clay fractions. Accelerated erosion is a selectively destructive process (McDaniel and Hajek, 1985; White et al., 1985; Thomas et al., 1985), because it removes one of the key reactive fractions of soil (Lowery et al., 1995) and leads to degradation in soil physical properties (Table 14.1). As erosion becomes more severe, the composition of lower horizons increasingly determines the soil physical quality. Increase in soil erosion can cause a soil to shift from one erosion phase to another with an attendant decrease in soil quality. This shift is attributed to unfavorable changes in surface soil properties (Lowery and Larson, 1995).

Soil structure is an important attribute of soil physical quality, and its maintenance and improvement are essential to reducing soil erosion risks. Soil erosion rate is determined by the ability of soil aggregates to withstand the forces of raindrop impact and surface flow (Meyer, 1981). The SOC is one of the soil properties that affects soil detachment (Young and Onstad, 1978). Soil aggregation decreases with increasing severity of erosion (Gollany et al., 1991). Soil aggregates are formed by biological, chemical, and physical processes, but are stabilized primarily by

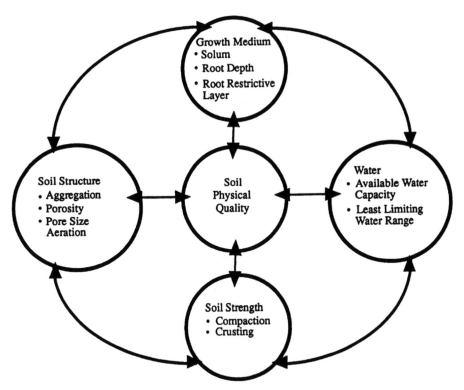

FIGURE 14.3   Determinants of soil physical quality. Accelerated erosion reduces soil physical quality through effects on these parameters.

SOC (Harris et al., 1966). The SOC can be divided into three major components depending on their capacity to stabilize aggregates (Tisdall and Oades, 1982). Humic substances associated with polyvalent metal cations are the most persistent stabilizing agents and are associated with microaggregation. Roots and fungal hyphae are temporary stabilizing agents and are primarily associated with macroaggregation. Plant and microbial polysaccharides and gums and more complex soil polysaccharides are the transient stabilizing agents and contribute to micro- and macroaggregation. These transient agents are effective for short periods, from weeks to a few months. Aggregation also controls soil porosity, pore size distribution, and their ability to transmit fluids. Coarse pores control infiltration and drainage rates, whereas fine pores control plant-available water and water storage capacity. Because of its strong impact on aggregation and soil quality, it is essential that SOC levels be maintained or increased.

Soil aggregation also affects crusting and surface soil formation. Crusts form when surface aggregates break down, filling pores with fine particles. Crusts impede water infiltration, which leads to increased runoff and erosion. Strong crusts contain less SOC than weak crusts (Arshad and Schnitzer, 1987), and eroded soils have

TABLE 14.1
Soil Erosion Effects on Soil Physical Properties of Ap Horizon

State	Soil	Bulk density (mg m⁻³)			0.03 MPa water content (%)			Hydraulic conductivity (cm h⁻¹)		
		SL	M	SV	SL	M	SV	SL	M	SV
Michigan	Marlette	1.37	1.48	1.50	27.7	24.6	23.7	—	—	—
Wisconsin	Dubuque	1.34	1.36	1.37	32.0	—	43.0	41.8	15.6	17.9
Ohio	Miamian	1.44	1.47	1.51	41.0	39.0	37.0	6.11	2.82	1.45

Note: SL = slight erosion, M = moderate erosion, and SV = severe erosion.

Modified from Lowery et al., 1995; Fahnestock et al., 1996a.

stronger crusts and more runoff than uneroded soils (Miller et al., 1988). In addition to low aggregate stability, the exchangeable sodium percentage (ESP) of a soil also affects crust formation (Agassi et al., 1985). Rainwater falling on sodic soils causes chemical dispersion of clay particles which move with the infiltrating water to regions of less porosity, where they lodge and clog the conducting pores. Soil loss is closely related to ESP in most sodic soils (Levy et al., 1994).

Clay mineralogy is an important factor in aggregate formation and stabilization. Expandable clays, smectite and vermiculite, may cause dry aggregates in the surface to break down as raindrops suddenly wet them. As the dispersed clay is dehydrated, it forms a dense and hard crust. Dehydration of flocculated clays produces a loose assemblage of small aggregates. Aggregates of nonexpanding clays, mica and kaolinite, are less susceptible to disruption during hydration. Nature and amount of exchangeable cations affect how the clay minerals respond to hydration and the strength of crusts. Soil erosion may not alter the clay mineralogy of the plow layer, but it may modify the amount and proportion of cations, thereby altering the stability of aggregates.

Finely divided Fe oxides are binding agents that cement primary soil particles into aggregates. In the cementing process, the Fe oxides fill in significant proportions of pores between the particles. The aggregates form more through an attraction between positively charged Fe oxides and negatively charged clay minerals rather than through crystal growth. The charge of Fe oxide particles is pH dependent; therefore, their effect on aggregation is pH dependent. Soil erosion may increase the amount of Fe oxides in the plow layer through the incorporation of subsoil into the plow layer. Erosion alters the pH of the plow layer, thereby affecting the stability of the aggregates bound by Fe oxides.

Soil Chemical and Mineralogical Quality

Soil chemical quality is determined by a wide range of properties (Figure 14.4), including cation exchange capacity (CEC), pH, plant-available nutrients (N, P, K),

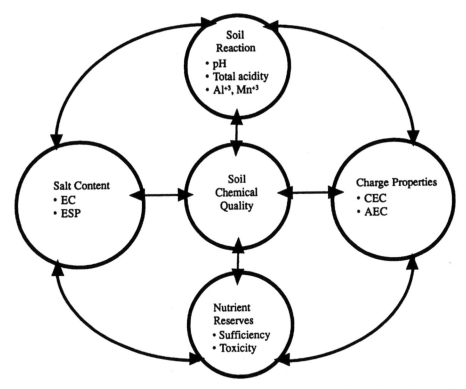

FIGURE 14.4 Determinants of soil chemical quality. Accelerated erosion reduces soil chemical quality through effects on these parameters.

salinity/conductivity, and elemental toxicities (Doran and Parkin, 1994; Karlen and Stott, 1994). Of these properties, soil erosion directly or indirectly affects CEC, pH (Table 14.2), and nutrient availability.

The CEC of a soil is reduced by soil erosion through the loss of SOC even though the clay content increases in soils that have Bt horizons, because SOC has a much greater CEC than clay. Reduction in CEC also has adverse effects on ANC, plant growth, and productivity.

Soil pH is also related to the amount and kind of cations or the ANC. The availability or solubility of plant nutrients is related to pH. Thus, pH influences plant growth and subsequently the amount of organic matter added to the soil. Even though nutrient loss through soil erosion can be corrected relatively easily by addition of fertilizers, nutrient availability and ANC are important to sustaining plant growth. Soil quality is adversely affected by loss of nutrients. Ultisols and Ultic subgroups of Alfisols have acid subsoils. Erosion of surface layers of these soils may cause some of the acidic subsoil material to be incorporated into the plow layer. The resulting high Al concentrations result in reductions in root growth and may limit crop production. Consequently, it is often difficult to alleviate crop

TABLE 14.2
Erosion Effects on Soil Chemical Properties of Ap Horizon

State	Soil	CEC (cmol kg⁻¹)			pH		
		SL	M	SV	SL	M	SV
Nebraska	Sharpsburg	16.3	18.0	22.2	5.1	5.7	5.5
Ohio	Miamian	9.2	13.6	17.0	5.6	5.9	5.7
Ohio	Strawn	14.0	14.0	20.0	6.5	6.6	7.6
Michigan	Marlette	6.8	9.2	9.5	6.4	6.7	7.4
Minnesota	Ves	18.0	18.0	13.2	5.8	6.2	8.0
South Dakota	Beadle	29.9	31.6	22.7	6.8	7.1	7.5

Note: SL = slight erosion, M = moderate erosion, and SV = severe erosion.

Modified from Ebeid et al., 1995; Fahnestock et al., 1996a; Mokma et al., 1996.

yield reductions of eroded Ultisols without addition of fertilizers, lime, and organic amendments.

Soil Biological Quality

Erosion affects soil biological quality through reduction in quality and quantity of SOC content (Figure 14.5 and Table 14.3). Erosion also affects biomass carbon, SOC (Dzhadan et al., 1975), and soil biodiversity (Lal, 1991a). Soil is a living entity and a habitat for a vast number of diverse species of animals and microorganisms. Activity of soil biota produces organic polymers that bind the clay particles into aggregates (Lynch and Bragg, 1985). When devoid of its biota, through erosion and related degradative processes, the uppermost layer of earth ceases to be a soil.

TEMPORAL CHANGES IN SOIL QUALITY IN RELATION TO EROSION

Erosion effects on temporal changes in soil quality depend on many factors, including land use and management.

Land Use

The magnitude of erosion and, therefore, its effect on soil quality vary with land use. Land recently converted from permanent vegetation usually has favorable soil structure, fine roots, and high SOC content. Intensive cultivation can cause adverse changes in these properties and reduce soil quality. The amount and stability of aggregates are functions of climate, soil type, SOC content, and fine roots. The greatest rates of change in SOC usually occur in the first 20 years of cultivation (Mann, 1986). As the intensity of land use increases, the amount of ground cover and associated root contributions decreases, the quantity and quality of SOC decrease, and soil quality is reduced.

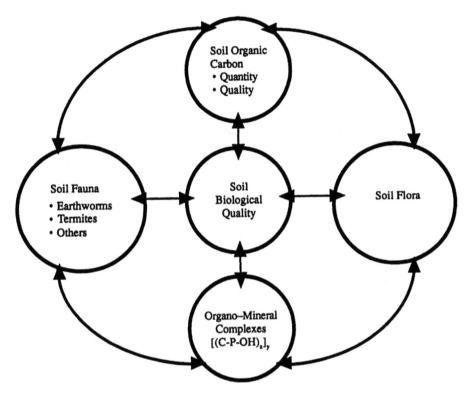

FIGURE 14.5 Determinants of soil biological quality.

Crop Residue Management

Crop residues returned to the soil contribute to the maintenance of and increase in SOC content and soil fertility. Because of its importance, the USDA (1992) pro-

TABLE 14.3
Erosion Effects on SOC Content of Ap Horizon

State	Soil	Erosion phases		
		SL	M	SV
Nebraska	Sharpsburg	2.00	1.60	1.40
Ohio	Miamian	1.04	1.00	1.06
Ohio	Strawn	1.09	1.03	0.94
Michigan	Marlette	1.10	0.90	0.70
Minnesota	Ves	2.00	1.80	1.40
South Dakota	Beadle	2.0	1.8	1.6

Note: SL = slight erosion, M = moderate erosion, and SV = severe erosion.

Modified from Ebeid et al., 1995; Fahnestock et al., 1996a; Mokma et al., 1996.

posed a soil quality rating index based on crop residue management as per Equation 14.5:

$$S_q R = OM + TP + E \qquad (14.5)$$

where $S_q R$ is the soil quality rating, OM is the amount of organic matter that must be returned to the soil to maintain the desired level of SOC content, TP is the subfactor related to all field operations which break down residues and aerate the soil including tillage, and E is the erosion subfactor which relates productivity decline to soil erosion as predicted by the Universal Soil Loss Equation.

Cropping Systems and Cover Crops

Cropping systems may enhance or diminish SOC content depending on soil and crop management (Robinson et al., 1996). Inadequately fertilized monocrop systems are the most detrimental to SOC, whereas rotations that include meadow crops conserve and even enhance SOC content. Inclusion of uncropped fallow into a crop rotation decreases SOC content within the first 20 years (Bremer et al., 1995). The negative impact is thought to be the result of increased decomposition of SOC during the fallow periods. Loss of plant nutrients and fertility depletion reduce the amount of biomass produced, decrease SOC content, and reduce associated microbial processes. Soil erosion increases as a result of adverse changes in these soil properties. In contrast, use of fertilizers, manure, and cover crops in rotation usually increases SOC contents (Janzen, 1987; Angers, 1992), decreases erosion, and improves soil quality. Judicious use of fertilizers to increase soil fertility improves biomass production and SOC content. The efficient use of N from diverse sources is an essential component of sustainable agricultural practices. Even with the best production practices, only 40–60% of the applied organic and inorganic nutrients are recovered by plants. Nutrient use efficiency drastically decreases with increase in soil erosion.

Aggregate stability increases significantly under alfalfa and pastures but not under cereals and uncropped fallow (Angers and Mehuys, 1988; Angers, 1992; Perfect et al., 1990). The greatest improvement in soil structure occurs under grasses, an intermediate improvement under legumes, and no improvement or even deterioration under continuous monoculture cereals (e.g., corn) (Perfect et al., 1990). Soil aggregation and associated porosity and infiltration capacity are improved with increase in SOC (Pikul and Zuzel, 1994). Infiltration rates are usually favorable in systems with rather than without cover crops. The cover crop intercepts raindrops, absorbs their energy, and reduces crust formation. Incorporation of cover crops into the soil increases SOC, thereby improving aggregate stability, porosity, and soil physical quality.

Tillage

The kind and amount of tillage influence soil erosion and soil quality. Conservation tillage systems, including no-tillage systems, reduce soil erosion. This reduction is attributed primarily to residue cover on the soil surface that protects the soil from raindrop impact and reduces crust formation. Residues returned to the soil increase SOC content, improve aggregate stability (West et al., 1992), and minimize crusting (Wagger and Denton, 1992; Zobeck and Popham, 1990). Conversion of moldboard plowing to no-tillage can drastically alter soil properties (Pierce et al., 1994). Within 4–5 years, the effects of moldboard plowing may be dissipated and soil properties are similar to those with continuous no-tillage management. Timing of tillage in conventional systems also affects soil erosion, because soil is most erodible immediately after tillage. Erodibility decreases after initial tillage as soil consolidates through wetting–drying cycles.

MECHANISMS OF EROSION IMPACT ON SOIL QUALITY

Accelerated erosion adversely affects soil quality through several mechanisms, including reduction in solum depth, soil productivity, and nutrient imbalance.

Decline in Solum Depth

Soil erosion adversely affects soil physical quality, especially rooting depth and AWC (Thomas and Cassel, 1979; Swan et al., 1987; Andraski and Lowery, 1992). Soil bulk density increases with erosion (Frye et al., 1982), soil structure and aggregation are reduced, and risks of crusting and compaction are increased because of the loss of SOC content and reduction in soil biodiversity.

Soil erosion removes topsoil and subsequent tillage incorporates subsoil into the plow layer. As a result, the properties of the surface soil are altered. The SOC, N, and extractable P contents decrease and clay content increases if the Ap horizon is underlain with a Bt horizon, bulk density increases slightly, and saturated hydraulic conductivity decreases (Olson and Nizeyimana, 1988; Mokma and Sietz, 1992; Bauer and Black, 1994; Cihacek and Swan, 1994; Lowery et al., 1995; Mokma et al., 1996). These properties negatively impact soil quality by increasing crusting, reducing water infiltration and AWC, loss of nutrients and ANC, and decrease in soil biodiversity. Swan et al. (1987) observed that corn yield increased as depth to a clay residuum increased. The clay residuum has a low plant-available water capacity; thus, as the depth to the clay increases, the amount of plant-available water and yield increase. Swan et al. suggested that the depth of soil is partially a function of soil erosion (i.e., soils shallow to clay have been eroded more). In general, eroded soils are of low quality because the organic-enriched layers have been removed and the water storage capacity reduced. Decrease in rooting depth by 0–79 cm for some soils of the north central region decreased corn grain yield by as much as 25% (Figure 14.6).

Root-Restrictive Subsoil

The magnitude of yield reduction by erosion depends on soil quality, especially the subsoil. Root-restricting layers limit the amount of water available to plants, because

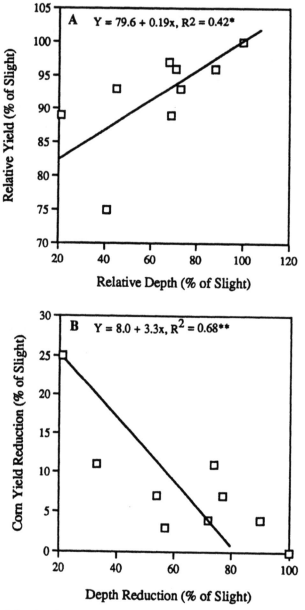

FIGURE 14.6 Solum/rooting depth effects on corn grain yield. (Adapted from Mokma et al., 1996.)

erosion reduces the thickness of the rooting zone. Drastic reductions in AWC occur with increasing severity of erosion (Table 14.4). Severe erosion on soils with root-restricting layers can also adversely affect crop stand (Olson and Carmer, 1990;

TABLE 14.4
Effect of Erosion on Weighted Mean Available Water
Capacity in the Upper 100 cm of the Profile for Some
Soils of the North Central Region

State	Soil	Erosion phase		
		SL	**M**	**SV**
South Dakota	Beadle	17.0	15.6	13.0
South Dakota	Egan	16.5	12.2	11.4
Nebraska	Sharpsburg	31.4	29.4	28.3
Nebraska	Wymore	21.2	23.3	23.6

Note: SL = slight erosion, M = moderate erosion, and SV = severe erosion.

Modified from Lowery et al., 1995.

Mokma and Sietz, 1992). Corn grain yield reductions over a 10-year period for three soils in Indiana ranged from 9 to 18% as the depth to compacted, calcareous glacial till decreased from slightly eroded phase to severely eroded phase (Weesies et al., 1994). Soybean yield reductions for the same three soils ranged from 17 to 24%. Corn grain yields for two Kentucky soils were 12 and 21% less on eroded phases compared to slightly eroded phases (Frye et al., 1982). In Ohio, Fahnestock et al. (1996a) observed that corn grain yield reduction was 20% on severely and 6% on moderately eroded phases compared with slight erosion phases. Similarly, soybean grain yield in Ohio was reduced by 27% on severely eroded phase and 12% on moderately eroded phase.

Nutrient Imbalance in Subsoil

Accelerated erosion can also cause severe yield reductions when subsoil horizons have drastic nutrient deficiency or toxicity problems. Because erosion causes a considerable loss of plant nutrients (Thomas and Cassel, 1979), nitrogen deficiency is commonly observed on severely eroded soils. In addition, data in Table 14.5 from some soils in the north central region also show deficiency of zinc. Because of decrease in SOC, severe erosion may lead to deficiency of other micronutrients as well. Erosion effects on nutrient depletion can be predicted from information on soil quality changes (Rosenberry et al., 1980). Soils in the southern Piedmont generally are infertile with low base saturation, acid subsoils, and high Al concentrations. Corn yield reductions for several soils in the southern Piedmont ranged from 23 to 42% when severely eroded phases were compared to slightly eroded phases (Langdale et al., 1979; Bruce et al., 1983). Yield reductions for soybean on similar soils ranged from 31 to 50% (Bruce et al., 1983, 1984; Thomas et al., 1985; White et al., 1985).

Daniels et al. (1989) believe that results based on soil analyses overestimate the effect of erosion on productivity because of the confounding effect of the landscape position. Effects of landscape position and degree of erosion on productivity may be

TABLE 14.5

Effect of Erosion on Nitrate–N in the Upper 60 cm of the Profile and DTPA-Extractable Zn in the Ap Horizon for Some Soils from the North Central Region

State	Soil	Erosion phase	kg ha^{-1}	
			N	Zn
South Dakota	Beadle	Slight	96	1.1
		Moderate	132	0.9
		Severe	125	0.7
South Dakota	Egan	Slight	131	2.0
		Moderate	124	0.8
		Severe	120	0.6
Minnesota	Ves	Slight	111	0.5
		Moderate	103	0.4
		Severe	74	0.3
North Dakota	Barnes	Slight	49	1.1
		Severe	27	0.5

Modified from Cihacek and Swan, 1994.

difficult to separate (Stone et al., 1985). In some studies, crop yields of moderately eroded phases equaled or exceeded those of slightly eroded phases (McDaniel and Hajek, 1985; Stone et al., 1985; Ebeid et al., 1995; Fahnestock et al., 1996b). Crop yields on depositional sites are often superior to other erosional phases during seasons with below-normal rains. Soybean grain yield in Ohio was 1.5 mg ha^{-1} for slight erosion, 1.4 mg ha^{-1} for moderate erosion, 1.6 mg ha^{-1} for severe erosion, and 3.1 mg ha^{-1} for deposition site (Table 14.6) (Ebeid et al., 1995). Because of favorable soil quality, crop growth and yields are generally superior on depositional sites

TABLE 14.6

Total Biomass, Grain Yield, Harvest Index, and Weight of 100 Grains of Soybeans Harvested from Each Erosion Class

Erosion class	Total biomass (kg/ha)	Grain yield (kg/ha)	Harvest index (%)	100 grain weight (g)
Slight	3135	1504	47.9	17.4
Moderate	3316	1395	41.0	17.2
Severe	4639	1631	35.0	25.2
Deposition	7194	3080	44.2	32.8
Least significant difference (0.05)	1367	669	7.0	2.3

Note: Biomass and grain yield may be biased because of the exceptionally dry season and the small area harvested.

Modified from Ebeid et al., 1995.

TABLE 14.7
Effect of Erosion Phase on Soybean Yields Evaluated on Sites A, B, and C in 1992 and 1993

Soybean yield	Erosion phase				
	Slight	Moderate	Severe	Deposition	Least significant difference (0.05)
Site A, 1993					
Grain (mg ha^{-1})	3.3	2.9	2.4	4.2	NS[a]
Straw (mg ha$^{-1)}$	7.4	7.1	5.8	8.2	NS
Harvest index (%)	31.2	28.7	29.6	33.5	NS
Site B, 1993					
Grain (mg ha^{-1})	3.4	3.3	3.6	5.6	1.1
Straw (mg ha^{-1})	7.1	7.5	7.5	10.6	1.9
Harvest index (%)	32.9	30.4	31.8	34.8	3.8
Site C, 1992					
Grain (mg ha^{-1})	4.0	3.6	2.1	3.4	0.6
Straw (mg ha^{-1})	7.6	6.8	3.7	5.6	1.8
Harvest index (%)	34.6	34.2	36.4	35.2	NS

[a] Not significant.

Modified from Fahnestock et al., 1996.

and often better than those on slightly eroded sites (Table 14.7). High yields on depositional sites may compensate for the adverse yield effects on severely eroded sites.

SOIL QUALITY MANAGEMENT OF ERODED SOILS

Strategies of soil quality management of eroded soils include SOC enhancement, water management, and improved cropping systems.

Improvement of SOC Content

The addition of organic matter from crop residue or waste products such as animal manure, sludge, and compost can increase SOC content of eroded soil (Karlen et al., 1994; Touray, 1994). Chang et al. (1991) found an increase in SOC in a clay loam soil from 2% to 4.5% with the application of cattle manure over 11 years. The depth of increase in SOC ranged from 30 cm with no irrigation to 60 cm with irrigation. It has been shown that a continuous application of manure is better than a single application (Mathers and Stewart, 1984). However, Wei et al. (1985) reported a 13% increase in SOC 5 years after a single application of 112 mg ha^{-1} of dry sewage sludge to a clay loam soil. Although this soil did not show signs of severe erosion, the texture of its surface is typical of an undesirable (low-quality) subsurface horizon of many key agricultural soils (such as Dubuque silt loam) that would be exposed following severe erosion. Wei et al. found significant improvement in many soil physical properties, including bulk density, infiltration, and aggregate stability, with increasing SOC content. They applied 0, 11.2, 22.4, 44.8, and 112.0 mg ha^{-1} dry dewatered sludge in 1977 and a yearly application from 1977 to 1982 which totaled 134.4 mg ha^{-1}. The SOC content can also be improved by use of crop residue mulch, conservation tillage, and incorporation of cover crops in a rotation cycle.

Water Management

Low AWC is a principal cause of low yields from eroded soils, and overland flow further depletes the water resources on sloping lands (Lowery et al., 1982). With adequate fertility and a good level of management, productivity is largely determined by the capacity of the soil to store and supply water to plants (Leeper et al., 1974). Andraski and Lowery (1992) found that the total quantity of plant-extractable water stored in the upper 1 m of slightly eroded Dubuque silt loam soil was 181 mm, which is 7% more than that for moderately eroded Dubuque soil. In this study, the moderately eroded soil had 14% more water stored (169 mm) than the severely eroded soil (159 mm). Swan et al. (1987) found a significant interaction effect on corn yield between climate and soil water-holding capacity. In addition to the use of cultural practices that improve *in situ* conservation of soil water, judicious use of supplemental irrigation through appropriate methods of water delivery is another viable option of improving the AWC of eroded soils. Enhancement of SOC content improves AWC (Hudson, 1994).

Improved Systems

The relationship between soil erosion and soil productivity is complex. The degree to which yield losses occur for different soils is a function of numerous interacting factors, including soil physical, chemical, and biological qualities; landscape position; crop being grown; crop and soil management practices used; and weather conditions before and during the growing season. Reduction in crop yield with increasing erosion is well documented (Langdale et al., 1979; Alberts and Spomer, 1987; Swan et al., 1987; Andraski and Lowery, 1992; Weesies et al., 1994; Changere and Lal, 1995; Fahnestock et al., 1996b). Experiments conducted prior to 1950 focused on the depletion of ANC in surface horizons as the cause of yield reduction. Nitrogen was frequently identified as limiting nutrients because the source of nearly all plant-available N was the SOC. A North Dakota study found that loss of productivity associated with depletion of SOC by soil erosion was primarily a consequence of the concomitant loss of plant nutrients (Bauer and Black, 1994). However, use of new crop varieties, improved fertilizer technology, and scientific management practices frequently mask the impact that erosion has on crop yield (Krauss and Allmaras, 1982; Langdale and Shrader, 1982). Although increased use of fertilizer may overcome the depletion of plant nutrients, continued loss of SOC through soil erosion may impact other soil properties that reduce productivity. Langdale et al. (1979) found that even though corn yields in the southern Piedmont had more than doubled since about 1949, yield reductions due to soil erosion had remained near 40%. Experiments conducted since the 1980s have focused on the relationship of productivity to differences in physical properties as they impact AWC, in addition to ANC and chemical properties, resulting from soil erosion. Soil erosion modifies soil properties that have been identified as indicators of soil quality (Doran and Parkin, 1994; Karlen and Stott, 1994).

INDICATORS OF SOIL QUALITY IN RELATION TO EROSION

The depth of a soil solum has a profound influence on soil quality. Soils which are deep to a root or water-restricting layer are considered to be of good quality, while those shallow to such a layer are in general considered to be of poor quality with respect to plant growth under rain-fed conditions. For the Palouse and Naff soils in northwestern Idaho, Bramble-Brodahl et al. (1985) observed that wheat yield declined exponentially with decrease in depth of the Mollic epipedon as per Equation 14.6:

$$Y = 5050 + 347(1 - e^{-0.019D}) \qquad (14.6)$$

where Y is wheat yield (in kg ha^{-1}) and D is depth of the Mollic epipedon (in cm). Therefore, as a soil shallow to a restrictive layer is eroded, its AWC, ANC, soil quality, and crop yield are reduced.

In addition to the depth of soil solum, the SOC content is an important indicator of soil quality. As SOC is increased, soil physical properties are improved (Wei et al., 1985). The bulk density of a compacted soil is reduced, the hydraulic conductivity and infiltration rates are increased, and aggregate stability is increased as the SOC content is increased from improved cultural practices including application of sewage sludge, animal manure, residue management, or improved cropping systems. Touray (1994) found increases in hydraulic conductivity of eroded Dubuque soil with application of animal manure. The greatest increase was noted in the moderately eroded plots.

The data presented in this review show that soil quality indicators in relation to erosion include those indicated in Equation 14.7:

$$S_q = f(R_d, \text{LLWR}, N_c, \text{SOC}, S_c, B_d) \qquad (14.7)$$

Most important determinants of soil quality are depth to root or water-restricting layer (R_d), least limiting water range (LLWR) (Thomasson, 1978; Letey, 1985; da Silva et al., 1994), nutrient capacity (N_c), SOC, structural characteristics (S_c) including aggregation and water retention/transmission properties, and soil biodiversity (B_d) including biomass carbon and activity and species diversity of soil fauna (e.g., earthworms). Rating system and methods of evaluation of these indicators are described by Lal (1994c). Pedotransfer functions relating crop yield to soil proper-

TABLE 14.8
Pedotransfer Functions Relating Crop Yield to Erosion-Induced Changes in Soil Properties

Crop	Soil	Pedotransfer function	R^2
Corn	Miamian	$y = 10.0 + 0.05\ I + 2.8\ SOC + 0.11\ CL - 7.5\ \rho + 0.6\ MWD$	0.97[a]
Corn	Strawn	$y = -15.1 - 0.19\ SL + 19.4\ \rho_b + 5.5\ AWC - 0.49\ CEC$	0.95[a]
Soybean	Strawn	$y = 7.4 - 0.1\ CL + 2.0\ SOC - 7.2\ AWC - 0.4\ pH - 0.03\ P$	0.91[a]
	Strawn	$y = 6.6 + 5.4\ SOC - 23.6\ AWC - 0.6\ pH - 0.003\ P$	0.97[a]
	Miamian	$y = 9.0 + 0.15\ SL - 5.4\ \rho_b - 0.5\ pH + 0.005\ K$	0.94[a]

Note: y is yield (mg ha^{-1}), I is cumulative infiltration (cm in 3 hr), CL is clay (%), SL is silt (%), SOC is soil organic carbon (%), ρ_b is soil bulk density (mg m^{-3}), AWC is available water capacity (%), CEC is in cmol kg^{-1}, and P and K are in kg ha^{-1}.

[a] Denotes significance at 0.01 level of probability.

Modified from Changere and Lal, 1995; Fahnestock et al., 1996b.

ties for different erosion phases have been reported by Changere and Lal (1995) and Fahnestock et al. (1996a) and are shown in Table 14.8.

CONCLUSIONS

Soil quality and soil erosion are strongly interrelated phenomena. Soil quality affects the rate of soil erosion and is in turn affected by it. Erosion effects on soil quality are determined by soil resilience, soil half-life, spatial variability in erosional processes over the soilscape, land use, farming system and management, and inherent soil properties. Principal determinants of soil physical quality are rooting depth or solum thickness, available water-holding capacity, least limiting water range, and soil structural attributes. Soil chemical quality is determined by pH, CEC, and nutrient availability including deficiency and toxicity. Soil biological quality is affected by SOC content, biomass carbon, and soil biodiversity. Temporal changes in soil quality in relation to erosion are determined by land use, crop residue management, cropping system, and tillage methods. Principal mechanisms of erosion impact on soil quality involve reduction in rooting/solum depth, decrease in available water and nutrient retention capacity, presence of root-restrictive horizon at shallow depth, and nutrient imbalance. Therefore, erosion-influenced determinants of soil quality are rooting depth, least limiting water range, SOC, structural attributes, plant nutrient factors, and soil biodiversity.

REFERENCES

Agassi, M., J. Morin, and I. Shainberg. 1985. Effect of raindrop impact energy and water salinity on infiltration rates of sodic soils. *Soil Sci. Soc. Am. J.* 49:186–190.

Alberts, E.E. and R.G. Spomer. 1987. Corn grain yield response to topsoil depth on deep loess soil. *Trans. ASAE* 30:977–981.

Andraski, B.J. and B. Lowery. 1992. Erosion effects on soil water storage, plant water uptake and corn growth. *Soil Sci. Soc. Am. J.* 56:1911–1919.

Angers, D.A. 1992. Changes in soil aggregation and organic carbon under corn and alfalfa. *Soil Sci. Soc. Am. J.* 56:1244–1249.

Angers, D.A. and G.R. Mehuys. 1988. Effects of cropping on macro-aggregation of a marine clay soil. *Can. J. Soil Sci.* 68:723–732.

Arshad, M.A. and M. Schnitzer. 1987. Characteristics of the organic matter in a slightly and in a severely crusted soil. *Z. Pflanzenernaehr. Bodenkde.* 50:412–416.

Bauer, A. and A.L. Black. 1994. Quantification of the effect of soil organic matter content on soil productivity. *Soil Sci. Soc. Am. J.* 58:185–193.

Bezdicek, D.F., R. Papendick and R. Lal. 1996. Introduction. In J. Doran and A. Jones (eds.). *Handbook of Methods for Assessing Soil Quality.* SSSA Special Publication No. 49, Soil Science Society of America, Madison, WI, pp. 1–8.

Biot, Y., M. Sessay, and M. Stocking. 1989. Assessing the sustainability of agricultural land in Botswana and Sierra Leone. *Land Degrad. Rehabil.* 1:263–268.

Bramble-Brodahl, M., M.A. Fosberg, D.J. Walker, and A.L. Falen. 1985. Changes in soil productivity related to changing topsoil. In *Erosion and Soil Productivity.* ASAE Publ. 8-85. ASAE, St. Joseph, MI, pp. 18–21.

Bremer, E., B.H. Ellert, and H.H. Janzen. 1995. Total and light-fraction carbon dynamics during four decades after cropping changes. *Soil Sci. Soc. Am. J.* 59:1398–1403.

Bruce, R.R., A.W. White, Jr., A.W. Thomas, and G.W. Langdale. 1983. Effect of water erosion upon physical character of rooting volume of Typic Hapludults. *Agronomy Abstracts.* American Society of Agronomy, Madison, WI, p. 196.

Bruce, R.R., A.W. White, Jr., A.W. Thomas, G.W. Langdale, and H.F. Perkins. 1984. Describing the effects of soil erosion on crop productivity of Cecil–Pacolet soils. *Agronomy Abstracts.* American Society of Agronomy, Madison, WI, p. 245.

Chang, C., T.G. Sommerfeldt, and T. Entz. 1991. Soil chemistry after eleven annual applications of cattle feedlot manure. *J. Environ. Qual.* 20:475–480.

Changere, C. and R. Lal. 1995. Soil degradation by erosion of a Typic Hapludalf in central Ohio and its rehabilitation. *Land Degrad. Rehabil.* 6:223–238.

Cihacek, L.J. and J.B. Swan. 1994. Effects of erosion on soil chemical properties in the north central region of the United States. *J. Soil Water Conserv.* 49:259–265.

Daniels, R.B., J.W. Gilliam, D.K. Cassel, and L.A. Nelson. 1989. Soil erosion has limited effect on field scale crop productivity in the southern Piedmont. *Soil Sci. Soc. Am. J.* 53:917–920.

da Silva, A.P., B.D. Kay, and E. Perfect. 1994. Characterization of the least limiting water range of soils. *Soil Sci. Soc. Am. J.* 58:1775–1781.

Doran, J.W. and T.B. Parkin. 1994. Defining and assessing soil quality. In J.W. Doran, D.C. Coleman, D.F. Bezdicek, and B.A. Stewart (eds.). *Defining Soil Quality for a Sustainable Environment.* SSSA Special Publication No. 35. Soil Science Society of America, Madison, WI, pp. 3–21.

Dzhadan, G.I., M.K. Demidenko, and G.N. Chabanov. 1975. Effect of the degree of erodibility of soils on their agrochemical properties and crop grain yield. *Soviet Soil Sci.* 7:579–597.

Ebeid, M.M., R. Lal, G.F. Hall, and E. Miller. 1995. Erosion effects on soil properties and soybean yield of a Miamian soil in western Ohio in a season with below normal rainfall. *Soil Technol.* 8:97–108.

Fahnestock, P., R. Lal, and G.F. Hall. 1996a. Land use and erosional effects on two Ohio Alfisols. I. Soil properties. *J. Sustain. Agric.* 7:63–84.

Fahnestock, P., R. Lal, and G.F. Hall. 1996b. Land use and erosional effects on two Ohio Alfisols. II. Crop yields. *J. Sustain. Agric.* 7:85–100.

Frye, W.W., S.A. Ebelhar, L.W. Murdock, and R.L. Blevins. 1982. Soil erosion effects on properties and productivity of two Kentucky soils. *Soil Sci. Soc. Am. J.* 46:1051–1055.

Gollany, H.T., T.E. Schumacher, P.D. Evenson, M.J. Lindstrom, and G.D. Lemme. 1991. Aggregate stability of an eroded and desurfaced Typic Argiustoll. *Soil Sci. Soc. Am. J.* 55:811–816.

Harris, R.F., G. Chesters, and O.N. Allen. 1966. Dynamics of soil aggregation. *Adv. Agron.* 18:107–169.

Hudson, B.D. 1994. Soil organic matter and available water capacity. *J. Soil Water Conserv.* 49:189–194.

Janzen, H.H. 1987. Effect of fertilizer on soil productivity in long-term spring wheat rotations. *Can. J. Soil Sci.* 67:165–174.

Karlen, D.L. and D.E. Stott. 1994. A framework for evaluating physical and chemical indicators of soil quality. In J.W. Doran, D.C. Coleman, D.F. Bezdicek, and B.A. Stewart (eds.). *Defining Soil Quality for a Sustainable Environment.* SSSA Special Publication No. 35. Soil Science Society of America, Madison, WI, pp. 53–72.

Karlen, D.L., N.C. Wollenhaupt, D.C. Erbach, E.C. Berry, J.B. Swan, N.S. Eash, and J.L. Jordahl. 1994. Crop residue effects on soil quality following 10 years of no-till corn. *Soil Tillage Res.* 31:149–167.

Krauss, H.A. and R.R. Allmaras. 1982. Technology masks the effects of soil erosion and wheat yields — a case study in Whitman County, Washington. In B.L. Schmidt, R.R. Allmaras, J.V. Mannering, and R.I. Papendick (eds.). *Determinants of Soil Loss Tolerance.* ASA Special Publication No. 45. American Society of Agronomy, Madison, WI, pp. 75–86.

Lal, R. 1991a. Soil conservation and biodiversity. In D.L. Hawksworth (ed.). *The Biodiversity of Micro-organisms and Invertebrates: Its Role in Sustainable Agriculture.* CAB International, Wallingford, U.K., pp. 89–104.

Lal, R. 1991b. Expectations of cover crops for sustainable agriculture. In W.L. Hargrove (ed.). *Cover Crops for Clean Water.* Soil and Water Conservation Society, Ankeny, IA, pp. 1–11.

Lal, R. 1993. Tillage effects on soil degradation, soil resilience, soil quality and sustainability. *Soil Tillage Res.* 27:1–8.

Lal, R. 1994a. Sustainable land use systems and soil resilience. In D.J. Greenland and I. Szabolcs (eds.). *Soil Resilience and Sustainable Land Use.* CAB International, Wallingford, U.K., pp. 41–67.

Lal, R. 1994b. Soil conservation. In S.M. Virmani, J.C. Katyal, H. Eswaran, and I.P. Abrol (eds.). *Stressed Ecosystems and Sustainable Agriculture.* Oxford and IBH Publishing, New Delhi, pp. 267–279.

Lal, R. 1994c. Methods and Guidelines for Assessing Sustainable Use of Soil and Water Resources in the Tropics. SMSS Tech. Monograph 21, Washington, D.C., 78 pp.

Lal, R. 1995. Global soil erosion by water and carbon dynamics. In R. Lal, J. Kimble, E. Levine, and B.A. Stewart (eds.). *Soils and Global Change.* Lewis Publishers, Boca Raton, FL, pp. 131–140.

Lal, R., E. Regnier, D.J. Eckert, W.M. Edwards, and R. Hammond. 1991. Expectations of cover crops for sustainable agriculture. In W.L. Hargrove (ed.). *Cover Crops for Clean Water.* Soil and Water Conservation Society, Ankeny, IA, pp. 1–11.

Langdale, G.W. and W.D. Shrader. 1982. Soil erosion effects on soil productivity of cultivated cropland. In B.L. Schmidt, R.R. Allmaras, J.V. Mannering, and R.I. Papendick (eds.). *Determinants of Soil Loss Tolerance.* ASA Special Publication No. 45. American Society of Agronomy, Madison, WI, pp. 41–51.

Langdale, G.W., J.E. Box, Jr., R.A. Leonard, A.P. Barnett, and W.G. Fleming. 1979. Corn yield reduction on eroded southern Piedmont soils. *J. Soil Water Conserv.* 34:226–228.

Leeper, A.A., E.C.A. Runge, and W.M. Walker. 1974. Effect of plant-available stored soil moisture on corn yields. I. Constant climatic conditions. *Agron. J.* 66:723–727.

Letey, J. 1985. Relationship between soil physical properties and crop production. *Adv. Soil Sci.* 1:277–294.

Levy, G.J., J. Levin, and I. Shainberg. 1994. Seal formation and interrill soil erosion. *Soil Sci. Soc. Am. J.* 58:302–209.

Lowery, B. and W.E. Larson. 1995. Symposium: erosion impact on soil productivity; preamble. *Soil Sci. Soc. Am. J.* 59:647–648.

Lowery, B., G.F. Kling, and J.A. Vomocil. 1982. Overland flow from sloping land: effects of perched water tables and subsurface drains. *Soil Sci. Soc. Am. J.* 46:93–99.

Lowery, B., J. Swan, T. Schumacher, and A. Jones. 1995. Physical properties of selected soils by erosion class. *J. Soil Water Conserv.* 50:306–311.

Lynch, J.M. and E. Bragg. 1985. Microorganisms and soil aggregate stability. *Adv. Soil Sci.* 2:133–171.

Mann, L.K. 1986. Changes in soil carbon storage after cultivation. *Soil Sci.* 142:279–288.

Mathers, A.C. and B.A. Stewart. 1984. Manure effects on crop yields and soil properties. *Trans. ASAE* 27:1022–1026.

McDaniel, T.A. and B.F. Hajek. 1985. Soil erosion effects on crop productivity and soil properties in Alabama. In *Erosion and Soil Productivity.* Publ. 8-85. ASAE, St. Joseph, MI, pp. 48–58.

Meyer, L.D. 1981. How rain intensity affects interrill erosion. *Trans. ASAE* 24:1472–1475.

Miller, W.P., C.C. Truman, and G.W. Langdale. 1988. Influence of previous erosion on crusting behavior of Cecil soils. *J. Soil Water Conserv.* 43:338–341.

Mokma, D.L. and M.A. Sietz. 1992. Effects of soil erosion on corn yields on Marlette soils in south-central Michigan. *J. Soil Water Conserv.* 47:325–327.

Mokma, D.L., T.E. Fenton, and K.R. Olson. 1996. Effect of erosion on morphology and classification of soils in the north central United States. *J. Soil Water Conserv.* 51:171–175.

Oldeman, L.R. 1994. The global extent of soil degradation. In D.J. Greenland and I. Szabolcs (eds.). *Soil Resilience and Sustainable Land Use.* CAB International, Wallingford, U.K., pp. 99–117.

Olson, K.R. and S.G. Carmer. 1990. Corn yield and plant population differences between eroded phases of Illinois soils. *J. Soil Water Conserv.* 45:562–566.

Olson, K.R. and E. Nizeyimana. 1988. Effects of soil erosion on corn yields of seven Illinois soils. *J. Prod. Agric.* 1:13–19.

Perfect, E., D.D. Kay, W.K.P. van Loon, R.W. Sheard, and T. Pojasok. 1990. Rates of change in soil structural stability under forages and corn. *Soil Sci. Soc. Am. J.* 54:179–186.

Pierce, F.J., M.C. Fortin, and M.J. Staton. 1994. Periodic plowing effects on soil properties in a no-till farming system. *Soil Sci. Soc. Am. J.* 58:1782–1787.

Pikul, J.L., Jr. and J.F. Zuzel. 1994. Soil crusting and water infiltration affected by long-term tillage and residue management. *Soil Sci. Soc. Am. J.* 58:1524–1530.

Pimentel, D., C. Harvey, P. Resosudarmo, K. Sinclair, D. Kurz, M. McNair, S. Crist, L. Shpritz, L. Fitton, R. Saffouri, and R. Blair. 1995. Environmental and economic costs of soil erosion and conservation benefits. *Science* 267:1117–1123.

Robinson, C.A., R.M. Cruse, and M. Ghaffarzadeh. 1996. Cropping system and nitrogen effects on Mollisol organic carbon. *Soil Sci. Soc. Am. J.* 60:264–269.

Rosenberry, P., R. Knutson, and L. Harmon. 1980. Predicting the effects of soil depletion from erosion. *J. Soil Water Conserv.* 35:131–134.

Stone, J.R., J.W. Gilliam, D.K. Cassel, R.B. Daniels, L.A. Nelson, and H.J. Kleiss. 1985. Effect of erosion and landscape position on the productivity of Piedmont soils. *Soil Sci. Soc. Am. J.* 49:987–991.

Swan, J.B., M.J. Shaffer, W.H. Paulson, and A.E. Peterson. 1987. Simulating the effects of soil depth and climatic factors on corn yield. *Soil Sci. Soc. Am. J.* 51:1023–1032.

Thomas, A.W., R.L. Carter, and J.R. Carreker. 1968. Soil and water nutrient losses from Tifton loamy sand. *Trans. ASAE* 11:677–679.

Thomas, D.J. and D.K. Cassel. 1979. Land forming Atlantic Coastal Plain soils: crop yield relationships to soil physical and chemical properties. *J. Soil Water Conserv.* 34:20–24.

Thomas, P.J., J.C. Baker, and T.W. Simpson. 1985. Effects of erosion on productivity of Cecil–Pacolet soils in Virginia. *Agronomy Abstracts.* American Society of Agronomy, Madison, WI, p. 213.

Thomasson, A.J. 1978. Towards an objective classification of soil structure. *J. Soil Sci.* 29:38–46.

Tisdall, J.M. and J.M. Oades. 1982. Organic matter and water-stable aggregates in soil. *J. Soil Sci.* 33:141–163.

Touray, K.S. 1994. Erosion and Organic Amendment Effects on the Physical Properties and Productivity of a Dubuque Silt Loam Soil. Ph.D. thesis. University of Wisconsin, Madison.

USDA. 1992. Proceedings of the Soil Quality Standards Symposium, San Antonio, Texas. USDA Forest Service, Washington, D.C., 80 pp.

Wagger, M.G. and H.P. Denton. 1992. Crop and tillage rotations: grain yield, residue cover, and soil water. *Soil Sci. Soc. Am. J.* 56:1233–1237.

Weesies, G.A., S.J. Livingston, W.D. Hosteter, and D.L. Schertz. 1994. Effect of soil erosion on crop yield in Indiana: results of a 10 year study. *J. Soil Water Conserv.* 49: 597–600.

Wei, Q.F., B. Lowery, and A.E. Peterson. 1985. Effect of sludge application on physical properties of a silty clay loam soil. *J. Environ. Qual.* 14:178–180.

West, L.T., W.P. Miller, R.R. Bruce, G.W. Langdale, J.M. Laflen, and A.W. Thomas. 1992. Cropping system and consolidation effects on rill erosion in the Georgia Piedmont. *Soil Sci. Soc. Am. J.* 56:1238–1243.

White, A.W., Jr., R.R. Bruce, A.W. Thomas, G.W. Langdale, and H.F. Perkins. 1985. Characterizing productivity of eroded soils in the southern Piedmont. In *Erosion and Soil Productivity.* Publ. 8-85. ASAE, St. Joseph, MI, pp. 83–95.

Young, R.A. and C.A. Onstad. 1978. Characterization of rill and interrill eroded soil. *Trans. ASAE* 21:1126–1130.

Zobeck, T.M. and T.W. Popham. 1990. Dry aggregate size distribution of sandy soils as influenced by tillage and precipitation. *Soil Sci. Soc. Am. J.* 54:198–204.

15 Erosion Impacts on Crop Yield for Selected Soils of the North Central United States

*K.R. Olson, D.L. Mokma, R. Lal,
T.E. Schumacher, and M.J. Lindstrom*

INTRODUCTION

The relationship between soil productivity and erosion is complex, and soils are not the sole factors controlling crop yields. The degree to which yield losses occur for various soils is a function of interacting factors including soil physical, chemical, and biological properties; landscape position; the crop grown; management practices used; and weather conditions before and during the growing season. Results from work before 1950 differ from current results as a consequence of technology advances. More efficient agronomic practices frequently mask the effects of erosion on yield as a result of new crop varieties, management practices, and fertilizer technology. The objectives of this study were to measure the relationship of corn grain yields to soil physical properties and climate under conditions of high input management systems and to determine the effects of erosion on the quality and productivity of important regional soils. The project was designed to minimize the effects of landscape position and agronomic practices on data interpretation. Soil, climate, and agronomic data were collected according to standardized procedures on eroded soil phases located in similar landscapes. Relative corn grain yield loss averaged 18% for severely eroded soil phases, when compared to less eroded soil phases, for soils with root-restricting subsoils and 0% for soils without root-restricting subsoils. For soils

1-57444-100-0/99/$0.00+$.50
© 1999 by Soil and Water Conservation Society

with root-restricting subsoils, the long-term productivity, measured by corn grain yield loss, was greater than the 15% threshold value, suggesting a major degradation of soil quality. The changes in topsoil properties which appeared to have the greatest impact on soil productivity and quality were an increase in clay content (9 of 12 soils) and a decrease in organic C (10 of 12 soils) for the severely eroded soil phases. For selected fine-texture soils, this resulted in a significant reduction in plant population, which affected corn yields. Plant population differences between eroded phases of soils with favorable soils for rooting were not significant. With increased erosion, the subsoil properties which changed the most dramatically were rooting depth and water storage capacity for soils with root-restricting subsoils.

The degree to which yield losses occur for various soils is a function of interacting factors including soil physical, chemical, and biological properties; landscape position; the crop grown; management practices used; and weather conditions before and during the growing season of interest. The relationship between soil productivity and erosion has been studied for more than six decades. Results from the early work (before 1950) differ from current findings as a consequence of technology advances (Krauss and Allmaras, 1982). More efficient agronomic practices frequently mask the effects of erosion on yield as a result of new crop varieties, management practices, and fertilizer technology.

Soil erosion–productivity relationships are complex, as the vast array of soils with associated soil properties are not the sole factors controlling crop yields. Crop yields have been found to vary with soil series (Pierce et al., 1983) primarily as a consequence of one or more limiting soil properties. Physical properties controlling water availability, soil aeration, and root growth are important determinants of yield on soils that have dense subsoils or shallow root-restricting layers (Mokma and Sietz, 1992). There are many soils in the north central United States that either have dense subsoils (fragipans or clay pans) or have developed over dense glacial-till parent material.

Climatic factors, such as growing season precipitation, evapotranspiration, growing season length, and temperature, also have a major modifying influence on erosion–productivity relationships (Cassel and Fryrear, 1990; Shaffer et al., 1994). Climatic factors are considered a major limiting parameter for crop yields. An examination of the effects of erosion on soil productivity for a large region over several growing seasons must include some account of the effects of climate on productivity. In some studies where landscape position varies, water movement is controlled by landscape position, which can affect soil development as well as influence soil productivity separately from soil erosion (Stone et al., 1985).

Many studies, particularly from the north central, northwest, and southeast regions, have implicated a loss of nutrients, particularly nitrogen, as a major factor contributing to a reduction in soil productivity as a result of erosion (Cassel and Fryrear, 1990; Cihacek and Swan, 1994; Krauss and Allmaras, 1982). Productivity losses from erosion-induced loss of nitrogen are hidden and many times can be corrected by the application of nitrogen-containing fertilizers or manure.

Methods of soil and crop management have a profound effect on soil properties with or without erosion (Langdale and Shrader, 1982). Soil properties sensitive to cultivation per se include organic matter content, pH, percent aggregation, size distribution and stability of aggregates, infiltration rate, soil water retention, rooting, porosity, and pore size distribution (Lowery et al., 1995). Invariably, there are also other causes involved which make it difficult to attribute varying crop yield responses to differences in past erosion.

Methods to Determine the Effects of Soil Erosion on Soil Properties and Crop Yields

Topsoil removal and additions have been used to create various topsoil thicknesses in a randomized complete block statistical design (Gollany et al., 1992; Stallings, 1950). Such studies are usually conducted on nearly uniform, level ridgetops, which eliminates landscape position variability. In a natural setting, topsoil is seldom completely eroded, with the associated loss of most of the organic C, microorganisms, nutrients, and aggregation. Many researchers (Gollany et al., 1992; Olson et al., 1994a) believe this approach is not a valid comparison as even slightly eroded soils have a mixture of topsoil and subsoil in the plow layer.

Paired comparisons between selected eroded soil phases within a field have been used by many researchers within the last 10 years (Chengere and Lal, 1995; Mokma et al., 1996; Mokma and Sietz, 1992; Olson and Carmre, 1990; Pierce et al., 1983; Stone et al., 1985) to evaluate erosion effects. This approach appears to work best in relatively young landscapes with documented cropping histories (the past 50–150 years) where it is possible to determine estimates of past erosion. In some cases, uncultivated landscapes are available for comparison purposes. Researchers (Olson and Nizeyimana, 1988; Olson and Carmer, 1990; Schumacher et al., 1994) have attempted to reduce slope and landscape effects between moderately and severely eroded phases of a soil by locating comparable plots on the same landscape position with similar slope characteristics.

Simulation models, such as Erosion Productivity Impact Calculator (EPIC) (Williams et al., 1983), Productivity Index Model (Lindstrom et al., 1992), and Nitrogen-Tillage Residue Management Model (NTRM) (Shaffer et al., 1983, 1994), can be used to identify soil, climatic, and crop factors that limit crop production at specific sites. One model, the Productivity Index (PI) (Pierce et al., 1983), assumes loss of potential crop production from soil erosion to be associated with changes in profile water-holding capacity, bulk density, and pH. The extent to which removal of topsoil by erosion improves or worsens those soil parameters dictates if and how much erosion increases or decreases the crop production potential. There is a need for soil and yield data to test these models.

The interaction of erosion-modified physical properties of soil which appear to influence production on eroded soils can be assessed using simulation models. The NTRM has been used in Minnesota to better quantify the interaction of climate, erosion-modified soil properties, and crop production. This modeling effort has identified restoration alternatives that may contribute directly to improved production. The effectiveness of models, however, is dependent on the validity of the assumptions and the reliability of the data on which they are based.

Each erosion–productivity method has inherent strengths, weaknesses, and biases which can result in the measured soil productivity response attributed to erosion being potentially confounded with other variables such as landscape position, soil formation, and management (Olson et al., 1994a). A recommended method used by the NC-174 Committee on Soil Erosion–Productivity to determine the effects of erosion on soil properties and crop yields of cultivated sites included the following: (1) locate at least three plots on each slightly, moderately, and severely eroded phase of a soil series in the cultivated field on the same landscape position with similar slope characteristics and (2) compare the mean crop yields under uniform management values for each slightly, moderately, and severely eroded phase of a soil series in the cultivated field to determine the effects of erosion on soil properties and crop yields. This appears to be a logical best approach to determine the effect of erosion on soil properties and productivity (Olson et al., 1994a). Landscape component differences, such as surface shape, or subsurface drainage differences can be minimized but may still be a co-variable. Landscape position differences can be eliminated. Past and present management variability between eroded phases of a soil can be minimized.

Effects of Soil Erosion on Soil Properties

The effects of erosion on soil properties vary by soil series, management, and location. In general, soil erosion affects the chemical properties of soils by (1) loss of organic matter, (2) loss of soil minerals containing plant nutrients, and (3) exposure of subsoil materials with low fertility or high acidity. Soil erosion causes changes in physical properties of soils, such as structure, texture, bulk density, infiltration rate, depth for favorable root development, and available water-holding capacity (Frye et al., 1982). Soil erosion causes changes in mineralogical properties of soils by thinning the plow layer (Ap horizon) and subsequent mixing of the subsoil (B horizon) into the Ap horizon by tillage equipment. The clay mineralogy of a soil depends on the parent materials present, weathering, and clay redistribution (Nizeyimana and Olson, 1988).

Numerous factors such as weather and plant genetic potential control the overall production of crops in a given geographic area; however, the soil system remains a major determinant of yields because of the environment it provides for root growth. Researchers (Nizeyimana and Olson, 1988) have studied the effects of degree of erosion on the chemical, mineralogical, and physical properties of seven soils. For most soils studied, degree of erosion significantly reduced the organic C and water storage porosity values of the Ap horizons. Clay mineral type estimates of the Ap horizons of severely eroded soils change measurably as a result of the mixing of the underlying Bt (argillic) horizon materials, higher in hydrous mica or smectite, into the residual topsoil by tillage. With increased erosion, pH, cation exchange capacity, potassium, and base saturation value trends varied with soil series. Erosion of soils with root-restricting layers, such as dense subsoils, leads to these layers occurring closer to the surface and reduced water storage capacities.

Chengere and Lal (1995) used stepwise regression analysis to relate crop yield to soil properties. Corn grain yield for Miamian soil in 1990 was shown to correlate highly ($R^2 = 0.97$) with soil organic carbon, mean aggregate diameter, bulk density, cumulative infiltration, and clay content. In 1993, Fahnestock et al. (1995) found corn grain yield correlated highly ($R^2 = 0.88$) with soil organic carbon, available water capacity, pH, and P.

Organic matter is important in both the development and stability of soil aggregates (Andraski and Lowery, 1992). Aggregates stabilized by humus are more stable than those bound by clay. Subsurface aggregates low in humus are more easily broken down by the impact of raindrops and increase the rate of runoff. Reduced soil water recharge potential can result in reduced productivity.

When clayey subsoil materials of an eroded soil are incorporated into the plow layer by tillage, the moisture range at which the soil can be easily and safely tilled becomes narrower (Frye et al., 1985). If the soil is tilled wet, soil structure tends to break down, leading to decreased pore space, aeration, infiltration, and percolation. If tilled dry, the clayey subsoil becomes cloddy and difficult to work, producing a poor seedbed, lowering plant emergence, and raising energy costs.

As a field erodes to varying degrees, one area needs much higher inputs for optimum economic yield as compared to other areas. As a consequence, management decisions and field operations become more difficult. Differential erosion can also result in variations in water infiltration and storage. Water availability usually becomes the major limiting factor as erosion occurs. Supplying water by irrigation is difficult and expensive in variably eroded and sloping fields.

Effects of Soil Properties on Crop Yields

Erosion usually reduces the immediate and long-term crop production potentials (American Society of Agricultural Engineers, 1985; Follett and Stewart, 1985; Hammel et al., 1985; Lindstrom et al., 1992; Olson et al., 1994a; Pierce et al., 1983; Schumacher et al., 1994; Shaffer et al., 1994). The severity of the impact appears to depend upon numerous characteristics of topsoil and subsoil, and in some cases all factors have not been identified. The crop yield response to changes in soil properties varies by soil series, landscape, hydrology, climate, and management practices.

Many early studies evaluating the effects of erosion on crop yields were made by removing topsoil to simulate severe erosion. In Ohio, researchers (Stallings, 1950; Uhland, 1949) found that yields of corn (*Zea mays* L.), wheat (*Triticum aestivum* L.), and oat (*Avena sativa* L.) grown on exposed subsoil plots were less than half those on the topsoil plots. Corn and wheat yields were considerably greater on plots with double thickness of topsoil. Other researchers (Copley et al., 1944) have reported large yield reductions on severely eroded or desurfaced plots. In Ohio (Uhland, 1949) and southern Piedmont (Langdale et al., 1979), desurfacing experiments showed that reduced amount of topsoil (A horizon material in plow layer) resulted in reduced corn grain yields. In Kentucky, corn grain yields were decreased 12–21% on eroded soils when compared with uneroded soils (Frye et al., 1982). Researchers in Illinois (Olson and Nizeyimana, 1988; Olson and Carmer, 1990) found a positive correlation of grain yields to rooting depth for soils with root-restricting subsoils, but not for soils with nonrestricting subsoils. Yield differences were thought to be due to differences in clay content and lower water-holding capacity. Corn grain yields were related to depth of rooting and available soil water (Olson and Nizeyimana, 1988). Root penetration in the root-restricting subsoils was reduced by high bulk density, low aeration, and lack of structural development.

In the midwestern and western United States, precipitation and available soil water limit crop production when nutrients are in adequate supply. The extent of production loss on eroded soils depends largely on landscape position, rainfall, runoff, internal drainage, and the yield potential of the original uneroded soil. For droughty areas which are highly eroded, residue maintenance on the soil surface can reduce runoff and increase infiltration and hydraulic conductivity in the top 20 cm.

In Nebraska (Jones et al., 1989), responses of crops to landscape position were largely dependent upon soil and topographic features. Best corn production was associated with upper summit and footslope (bottom of hill) landscape positions. Loess soils of Nebraska had greater soil-water recharge over the winter on the footslopes, which provided more plant-available water during the growing season and resulted in higher yields (Hanna et al., 1983). Other researchers (Spomer and Piest, 1982) have compared yields for 16 years for summit, backslope, and footslope positions and found that footslope positions produced higher yields. In North Carolina, landscape position has been shown to be a better parameter than erosion phase of a soil as a predictor of corn grain yield due to associated moisture condition as a reflection of the net effects of surface and subsurface water movement (Stowe et al., 1985).

Loss of organic matter has been shown to affect the use and effectiveness of herbicides (Frye et al., 1985). Herbicides often injure crops on severely eroded soils low in organic matter. Eroded soils (Frye et al., 1982) in the humid southeastern United States usually are more acid and have higher lime requirements than uneroded soils as a consequence of higher buffering capacity associated with the clay mixed into the plow layer.

Effects of Soil Property Changes on Soil Quality

Threshold soil values and productivity parameters can serve as an early warning signal of reduced productive capacity of soils common to the north central United States (Olson, 1992). Specific soil parameters and threshold values were suggested for both topsoil and subsoil layers. Possible changes in the topsoil properties that appear to affect productivity under a high level of management include erosion phase, total porosity, bulk density, aggregation, organic C, infiltration, texture, and coarse fragments. Changes in subsoil properties that affect soil productivity include mechanical strength, structure, aeration porosity, water storage, total porosity, bulk density, hydraulic conductivity, pH, and rooting depth.

Minor and major reductions in inherent soil productivity (Olson, 1992) were considered as a basis for setting threshold values for measurable and observable soil properties or conditions based on current methods of technology and research. Threshold values were determined by correlating soil properties with yields obtained from research fields. These soil parameters and specific threshold values were based on erosion–productivity relationships shown by previous NC-174 research as affected by soil conditions in the north central United States, which can vary with location, crop, soil, management, and climate.

Olson (1992) defined soil quality standards for soil physical properties as related to the topsoil and subsoil layers which affect soil productivity. Values for soil properties were defined as no reduction, minor (less than 15%), and major (more than 15%) reduction in productivity as reflected in corn grain yields. Similar procedures for chemical and biological parameters can be developed. These values could serve as warning signals of reduced productivity.

The objectives of this study were (1) to measure the relationship of corn grain yields to soil physical properties and climate under conditions of high-input management systems and (2) to determine the effects of erosion on the quality and productivity of representative regional soils. The project was designed to minimize the effects of landscape position and agronomic practices on data interpretation.

METHODS

The regional research project (NC-174) was designed to measure the relationship of yield to soil properties and climate under conditions of high-input management systems and to determine the effects of erosion on the soil productivity of representative soils of the North Central Region (Schumacher et al., 1994). Soils were selected from each participating state in the north central United States based on the importance of the soil for a given state and its use as an agricultural soil. The uneroded soils included in the study represent both Mollisols and Alfisols (Table 15.1), the orders that represent the majority of crop production land in this region.

The soils can further be categorized (Table 15.2) by natural vegetation (prairie or forest) and parent material (till, loess, and glacial sediments). The soils can further be grouped according to subsoil conditions for rooting. Soils with and without root-

TABLE 15.1
Classification of the Uneroded Soil Series Included in the North Central United States Study

Soil series	State	Classification
Beadle	South Dakota	Fine, montmorillonitic, mesic Typic Argiustoll
Clarence	Illinois	Fine, illitic, mesic Aquic Argiudoll
Dubuque	Wisconsin	Fine-silty, mixed, mesic Typic Hapludalf
Egan	South Dakota	Fine-silty, mixed, mesic Udic Haplustoll
Grantsburg	Illinois	Fine-silty, mixed, mesic Typic Fragiudalf
Hoyleton	Illinois	Fine, montmorillonitic, mesic Aquollic Hapludalf
Marlette	Michigan	Fine-loamy, mixed, mesic Glossoboric Hapludalf
Miamian	Ohio	Fine, mixed, mesic Typic Hapludalf
Rozetta	Illinois	Fine-silty, mixed, mesic Typic Hapludalf
Sharpsburg	Nebraska	Fine, montmorillonitic, mesic Typic Argiudoll
Tama	Illinois	Fine-silty, mixed, mesic Typic Argiudoll
Ves	Minnesota	Fine-loamy, mixed, mesic Udic Haplustoll

Data from Soil Survey Staff, 1975.

TABLE 15.2
Vegetation and Subsoil Characteristics of North Central United States Soils

Soil series and erosion phase	Natural vegetation	Parent material	Subsoil bulk density (mg m^{-3}) (STD)	Restrictive subsoil	Depth and thickness of restrictive subsoil layer (cm)
Beadle					
Slight	Prairie	Till	1.56 (0.04)	Dense till	85–150[a]
Moderate	Prairie	Till	1.56 (0.04)	Dense till	75–150[a]
Severe	Prairie	Till	1.58 (0.01)	Dense till	59–150[a]
Clarence					
Slight	Prairie	Loess/till	1.68	Dense clay till	97–150[a]
Moderate	Prairie	Till	1.70	Dense clay till	88–150[a]
Severe	Prairie	Till	1.70	Dense clay till	73–150[a]
Dubuque					
Slight	Forest	Loess/residuum	1.33	Limestone	84–150[a]
Moderate	Forest	Loess/residuum	1.36	Limestone	61–150[a]
Severe	Forest	Loess/residuum	1.28	Limestone	38–150[a]
Egan					
Slight	Prairie	Sediment/till	1.60 (0.01)	Dense till	110–150[a]
Moderate	Prairie	Sediment/till	1.61 (0.01	Dense till	98–150[a]
Severe	Prairie	Sediment/till	1.59 (0.01)	Dense till	80–150[a]
Grantsburg					
Slight	Forest	Loess	1.56 (0.03)	Fragipan	100–150[a]
Moderate	Forest	Loess	1.52 (0.06)	Fragipan	85–150[a]
Severe	Forest	Loess	1.58 (0.01)	Fragipan	50–150[a]
Hoyleton					
Slight	Prairie	Loess	1.37 (0.02)	Claypan	48–115
Moderate	Prairie	Loess	1.35 (0.03)	Claypan	42–100
Severe	Prairie	Loess	1.42 (0.03)	Claypan	25–90
Marlette					
Slight	Forest	Till	1.86 (0.08)	Dense till	135–150[a]
Moderate	Forest	Till	1.90 (0.05)	Dense till	92–150[a]
Severe	Forest	Till	1.88 (0.06)	Dense till	56–150[a]
Miamian					
Slight	Forest	Till	1.41	Dense clay till	64–150[a]
Moderate	Forest	Till	1.39	Dense clay till	59–150[a]
Severe	Forest	Till	1.45	Dense clay till	55–150[a]
Rozetta					
Moderate	Forest	Loess	1.31 (0.02)	None	—
Severe	Forest	Loess	1.35 (0.06)	None	—
Sharpsburg					
Slight	Prairie	Loess	—	None	—
Moderate	Prairie	Loess	—	None	—
Severe	Prairie	Loess	—	None	—

TABLE 15.2
Vegetation and Subsoil Characteristics of North Central United States Soils (continued)

Soil series and erosion phase	Natural vegetation	Parent material	Subsoil bulk density (mg m⁻³) (STD)	Restrictive subsoil	Depth and thickness of restrictive subsoil layer (cm)
Tama					
Moderate	Prairie	Loess	1.35 (0.01)	None	—
Severe	Prairie	Loess	1.40 (0.02)	None	—
Ves					
Slight	Prairie	Till	1.57	Dense till	97–150[a]
Moderate	Prairie	Till	1.60	Dense till	69–150[a]
Severe	Prairie	Till	1.58	Dense till	20–150[a]
All root-restricting soils					
Slight					92 ± 26
Moderate					73 ± 18
Severe					46 ± 19

[a] Extending below 150 cm.

restricting subsoils represent the two primary subsoil conditions found on agricultural lands in the north central United States.

Landscape positions used in the study were primarily linear backslope or sideslope (Table 15.3). An attempt was made to place the erosion–productivity plots in similar landscape positions at a given site. Methods for data collection were standardized across the region. Experimental units were larger than 0.01 ha (0.025 acre) and surrounded by at least a 3-ha (7.5-acre) field of corn, which was the crop used at all locations. Three replications were selected for each eroded phase of a soil series. Soil loss by erosion was estimated by depth to the C horizon in relation to slightly eroded soil profiles within the same field and landscape position. Eroded soil phases correspond to the definition from the Soil Survey Staff (Olson et al., 1994b; Soil Survey Staff, 1975, 1993).

Soil measurements were taken outside of the yield rows. Each yield area was greater than 22.8 m² (245 ft²). All plots were fertilized according to soil test recommendations for a given state to achieve a yield goal considered high for the given soil and climatic region (Dahnke, 1988). Yield goals developed for the uneroded treatments (slightly eroded) were used together with soil samples taken annually from each plot to determine fertilizer recommendations specific for each eroded soil phase.

Corn grain yield is reported on a 15.5% moisture basis. Hand harvest methods were used for corn yield determinations. Varieties appropriate for the growing season at the specific experimental sites were used for the study. The tillage systems

TABLE 15.3
Landscape Characteristics of North Central United States Soils

Soil series and erosion phase	Slope gradient (%)	Slope length (m)	Slope shape Down	Slope shape Across	Landscape position
Beadle					
Slight	2	33	Linear	Linear	Backslope
Moderate	7	33	Linear	Linear	Backslope
Severe	9	33	Linear	Linear	Backslope
Clarence					
Slight	3	10	Convex	Linear	Shoulder
Moderate	5	30	Convex	Linear	Upper backslope, including lower shoulder
Severe	6	45	Linear	Linear	Middle backslope
Dubuque					
Slight	14	—	Linear	—	Sideslope
Moderate	12	—	Linear	—	Sideslope
Severe	14	—	Linear	—	Sideslope
Egan					
Slight	1	—	Concave	—	Footslope
Moderate	4	26	Linear	Linear	Backslope
Severe	6	34	Linear	Linear	Backslope
Grantsburg					
Slight	7	30	Linear	Linear	Upper backslope
Moderate	9	60	Linear	Linear	Middle backslope
Severe	10	90	Linear	Linear	Lower backslope
Hoyleton					
Slight	2	20	Linear	Linear	Middle backslope
Moderate	3	50	Linear	Linear	Lower backslope
Severe	4	60	Linear	Linear	Lower backslope
Marlette					
Slight	3	50	Linear	Linear	Backslope
Moderate	4	26	Linear	Linear	Backslope
Severe	4	50	Linear	Linear	Backslope
Miamian					
Slight	1	—	Concave	Linear	Summit
Moderate	2	—	Convex	Concave	Shoulder
Severe	3	—	Convex	Concave	Backslope
Rozetta					
Moderate	11	70	Convex	Linear	Upper backslope including lower shoulder
Severe	12	80	Linear	Linear	Middle backslope
Sharpsburg					
Slight	3	200	Linear	Linear	Sideslope
Moderate	3	200	Linear	Linear	Sideslope
Severe	3	200	Linear	Linear	Sideslope

TABLE 15.3
Landscape Characteristics of North Central United States Soils (continued)

Soil series and erosion phase	Slope gradient (%)	Slope length (m)	Slope shape		Landscape position
			Down	Across	
Tama					
Moderate	9	10	Convex	Linear	Upper backslope including lower shoulder
Severe	11	20	Linear	Linear	Middle backslope
Ves					
Slight	3	—	Linear	—	Sideslope
Moderate	6	—	Linear	—	Sideslope
Severe	6	—	Convex	—	Shoulder

consisted of chisel or moldboard plowing usually in the spring, followed by a secondary disking operation in the spring. No-till systems were used at two sites (Grantsburg and Hoyleton). Irrigation was not used at any of the sites.

Soils at all sites were characterized by horizon from representative pits dug near the experimental plots and included all eroded soil phases. Soil characterization included complete profile descriptions, particle size analysis (Day, 1965), bulk density determinations (Blake, 1965), 0.03- and 1.5-MPa pressure plate determinations for available water (Richards, 1965), pH (Peech, 1965), and organic C measurements (Nelson and Sommers, 1982).

Growing season precipitation (Table 15.4) was used as the climatic variable to separate experimental sites in this study, as this variable was complete for all data sets. Unweighted soil series–erosion phase means for corn grain yield and plant population for each year were analyzed in accordance with a split-plot linear model in which years were considered as replications, soil series were considered as main plots, and erosion phases were considered as subplots, under the assumptions that year effects were random whereas those of soil series and erosion phases were fixed. We performed analyses of variance using the general linear models (GLM) procedure in Statistical Analysis System computer software (SAS Institute, 1985). The least significant difference (LSD) procedure was used ($P = 0.05$) to determine if differences between erosion phases within a soil series or averaged over soil series were statistically significant.

RESULTS

Five years of corn grain yield data for plots on the paired slightly, moderately, and severely eroded phases of nine soils with root-restricting subsoils are summarized in Table 15.5. As expected, significant differences ($P = 0.05$) in the corn yield averages occurred as a result of erosion. Magnitudes of corn grain yield reduction were from 7 to 35% (0.6 to 2.8 mg ha^{-1}). Across all soils with root-restricting subsoils, there

TABLE 15.4
Monthly Rainfall Data by Year for the Growing Season

Soil series	Year	Rainfall (mm)					
		May	June	July	August	Total	Long-term average
Beadle	1984	69	239	50	34	392	340
	1985	82	33	32	81	228	340
	1986	123	176	84	76	459	340
	1987	56	56	144	52	308	340
	1988	85	51	22	131	289	340
Clarence	1984	153	61	70	20	304	391
	1985	71	92	110	111	384	391
	1986	124	175	100	32	431	391
	1987	82	83	80	126	371	391
	1988	62	31	50	64	207	391
Dubuque	1985	124	33	53	84	294	426
	1986	98	137	46	91	372	426
	1987	95	103	168	170	536	426
	1988	22	11	45	73	151	426
Egan	1984	69	239	50	34	392	340
	1985	82	33	32	81	228	340
	1986	123	177	84	76	460	340
	1987	56	56	144	52	308	340
	1988	85	51	22	131	289	340
Grantsburg	1984	142	80	120	41	383	393
	1985	184	182	50	303	719	393
	1986	271	70	102	110	553	393
	1987	82	100	183	40	405	393
	1988	61	20	42	40	163	393
Hoyleton	1984	96	20	32	20	168	390
	1985	81	223	40	74	418	390
	1986	82	61	104	50	297	390
	1987	10	82	133	30	255	390
	1988	41	43	210	62	356	390
Marlette	1985	68	55	51	101	275	302
	1986	86	167	74	97	424	302
	1987	23	62	61	125	271	302
	1988	15	4	60	89	168	302
	1989	125	83	46	172	426	302
Miamian	1992	56	55	181	45	337	417
	1993	20	90	197	22	329	417
Rozetta	1984	112	54	110	30	306	403
	1985	91	113	154	120	478	403
	1986	70	102	101	43	316	403
	1987	71	40	92	90	293	403
	1988	94	40	41	153	328	403

TABLE 15.4
Monthly Rainfall Data by Year for the Growing Season (continued)

Soil series	Year	May	June	July	August	Total	Long-term average
					Rainfall (mm)		
Sharpsburg	1984	123	164	41	24	352	439
	1985	142	57	92	75	366	439
	1986	69	80	110	140	399	439
	1987	156	42	96	217	511	439
	1988	82	26	89	18	215	439
	1989	21	104	104	41	270	439
Tama	1984	132	61	82	20	295	
	1985	11	42	70	73	196	400
	1986	143	115	152	31	441	400
	1987	84	30	42	270	426	400
	1988	40	32	0	95	167	400
Ves	1984	69	198	70	64	401	338
	1985	112	104	59	154	429	338
	1986	62	80	180	93	415	338
	1987	59	51	150	39	299	338
	1989	38	19	21	111	189	338

was an 18% reduction in corn grain yield when comparing the less eroded to the severely eroded phases of a soil series.

Five years of corn grain yield data for plots on the paired slightly, moderately, and severely eroded phases of three soils without root-restricting subsoils are summarized in Table 15.6. As expected, nonsignificant 5-year yield differences were measured between paired erosion phases on the plots of the Rozetta, Sharpsburg, and Tama soils. Magnitudes of corn grain yield reduction were from a 5% decrease to an increase of 7% (decrease of 0.5 mg ha^{-1} to an increase of 0.6 mg ha^{-1}). The increase in corn grain yields for plots on the severely eroded Tama soil could have been the effect of sidehill water seepage (Olson and Carmer, 1990). In dry summers, such as in 1987 and 1988 (Table 15.4), this lateral seepage could have provided a source of water that was available to corn roots and resulted in slightly higher corn grain yields for plots on the severely eroded Tama soil. This water from the summit may have masked a slight corn grain yield reduction due to erosion phase of Tama soil. In a wetter spring (1986), lateral seepage under the plots with severely eroded Tama soil did not provide the previously observed yield advantage (Table 15.6). Across all soils without root-restricting subsoils, there was a 0% reduction in corn grain yield when comparing the less eroded to the severely eroded phases of a soil series.

Declines in plant populations were up to 18% on average for a soil (Table 15.7) and represented a 7% decrease with increasing degree of erosion for the severely eroded phases for a subset of five soils. However, only the fine-textured Clarence

TABLE 15.5
Annual Corn Grain Yield for Erosion Phases and Relative Yield Losses of the Soils with Root-Restricting Subsoils

Year	Erosion phase	Beadle	Clarence[a]	Dubuque	Egan	Grantsburg[a]	Hoyleton[a]	Marlette	Miamian	Ves
						mg/ha^{-1}				
1984	Slight	8.6 (0.3)	—	—	—	—	—	—	—	9.4 (0.4)
	Moderate	8.1 (0.6)	3.3 (0.2)	—	9.0	8.5 (0.7)	5.9 (0.1)	—	—	8.7 (0.3)
	Severe	6.9 (0.2)	2.2 (0.1)	—	5.9	6.2 (0.5)	5.0 (0.5)	—	—	8.0 (0.3)
1985	Slight	8.3 (0.2)	—	10.4 (0.2)	8.7	—	—	6.8 (0.5)	—	10.3 (0.3)
	Moderate	7.9 (0.2)	4.8 (0.8)	9.4 (0.4)	8.8	9.3 (0.1)	8.1 (0.1)	5.6 (0.6)	—	10.5 (0.2)
	Severe	7.5 (0.2)	3.2 (0.2)	9.5 (0.1)	6.7	9.0 (0.1)	6.3 (0.1)	5.1 (0.3)	—	9.9 (0.3)
1986	Slight	8.7 (0.4)	—	11.6 (0.2)	9.2	—	—	8.1 (0.4)	—	10.5 (0.2)
	Moderate	7.3 (0.4)	4.6 (0.0)	10.7 (0.2)	9.6	8.8 (0.3)	5.5 (0.3)	7.3 (0.7)	—	10.1 (0.3)
	Severe	8.2 (0.2)	3.9 (0.3)	11.5 (0.1)	9.3	8.0 (0.1)	4.4 (0.2)	6.0 (0.6)	—	10.3 (0.3)
1987	Slight	8.3 (0.2)	—	12.2 (0.1)	9.1	—	—	8.4 (0.1)	—	8.9 (0.3)
	Moderate	8.3 (0.1)	5.4 (0.4)	11.6 (0.1)	9.9	7.8 (0.5)	5.1 (0.2)	8.5 (0.3)	—	8.7 (0.4)
	Severe	7.6 (0.1)	2.9 (0.5)	11.2 (1.0)	9.6	5.8 (0.5)	4.3 (0.1)	6.9 (0.6)	—	8.2 (0.7)
1988	Slight	2.7 (0.4)	—	2.6 (0.1)	—	—	—	3.5 (0.2)	—	3.1 (0.3)
	Moderate	3.4 (0.2)	0.6 (0.1)	2.5 (0.2)	—	1.9 (0.1)	6.1 (0.1)	3.8 (0.3)	—	2.7 (0.4)
	Severe	2.4 (0.1)	0.0 (0.0)	2.4 (0.1)	—	0.0 (0.0)	5.7 (0.2)	1.8 (0.4)	—	2.4 (0.6)
1989	Slight	—	—	—	—	—	—	9.9 (0.5)	—	8.1 (0.2)
	Moderate	—	—	—	—	—	—	9.9 (0.5)	—	8.1 (0.5)
	Severe	—	—	—	—	—	—	9.3 (0.2)	—	7.3 (0.4)
1992	Slight	—	—	—	—	—	—	—	11.0	—
	Moderate	—	—	—	—	—	—	—	8.5	—
	Severe	—	—	—	—	—	—	—	7.2	—

1993									
Slight	—	—	—	—	—	—	—	6.6	—
Moderate	—	—	—	—	—	—	—	5.6	—
Severe	—	—	—	—	—	—	—	4.9	—
Average Slight	7.3	—	9.2	9.0	—	—	7.3	8.8	8.4
Average Moderate	7.0	3.7	8.6	9.2	7.3	6.1	7.1	7.0	8.1
Average Severe	6.5	2.4	8.6	7.9	5.8	5.1	5.8	6.0	7.7
Relative[b] yield loss (%)	-11[c]	-35[c]	-7[c]	-12[c]	-20[c]	-16[c]	-21[c]	-31[c]	-8[c]

Note: Standard errors of the mean are given in parentheses.

[a] Slightly eroded phase not measured on this soil series.

[b] Site yield failures are excluded from relative yield loss analysis.

[c] Indicates mean relative yield loss from severely eroded phase was significantly different than 0 at the 5% probability level.

TABLE 15.6

Annual Corn Grain Yield for Erosion Phases and Mean Relative Yield Losses of the Soils Without Root-Restricting Subsoils

Year	Erosion phase	Soil series (mg ha⁻¹)		
		Rozetta[a]	Sharpsburg	Tama[a]
1984	Slight	—	6.2	—
	Moderate	11.2 (0.0)	6.0	10.5 (0.3)
	Severe	11.0 (0.6)	6.9	10.2 (0.5)
1985	Slight	—	10.7	—
	Moderate	10.4 (0.2)	9.9	9.5 (0.3)
	Severe	9.8 (0.4)	9.6	10.0 (0.2)
1986	Slight	—	9.0	—
	Moderate	8.5 (0.4)	8.0	10.7 (0.3)
	Severe	7.1 (0.2)	9.4	9.3 (0.3)
1987	Slight	—	3.7 (0.2)	—
	Moderate	9.2 (0.2)	4.0 (0.3)	8.7 (0.4)
	Severe	7.8 (0.2)	3.2 (0.2)	10.0 (0.8)
1988	Slight	—	2.1 (0.6)	—
	Moderate	7.7 (1.0)	2.0 (0.3)	1.9 (0.1)
	Severe	8.2 (0.5)	1.6 (0.2)	4.8 (0.4)
1989	Slight	—	3.5 (0.7)	—
	Moderate	—	3.8 (0.9)	—
	Severe	—	3.8 (0.9)	—
Average	Slight	—	5.9	—
Average	Moderate	9.4	5.6	8.3
Average	Severe	8.9	5.8	8.9
Relative yield loss (%)		−5[b]	−2[b]	+7[b]

Note: Standard errors of the mean are given in parentheses.

[a] Slightly eroded phase not measured on this soil series.

[b] Indicates relative yield loss from severely eroded phase was not significantly different than 0 at the 5% probability level.

and Hoyleton soils had statistically significant plant population decreases ($P = 0.05$). These plant population reductions were a consequence of tillage mixing subsoil materials high in clay into the Ap horizon of severely eroded soils. These subsoil materials lacked durable soil aggregates, resulting in clodding, crusting, reduced soil–seed contact, and reduced seedling emergence. The 5-year plant population average of the five Illinois soils was significantly greater for moderately eroded phases than for severely eroded phases. A significant (up to 22%) decrease in plant populations occurred for Marlette soil and a nonsignificant reduction for all other slightly eroded versus severely eroded phases of soils in the region (Table 15.8).

TABLE 15.7
Plant Population Data for the 5 Years at the Paired Moderately and Severely Eroded Phases of Clarence, Grantsburg, Hoyleton, Rozetta, and Tama Soils

Soil series and erosion phase	Plant population[a] (plants/ha)					5-year average	Change	% change
	1984	1985	1986	1987	1988			
Clarence								
Moderate	35,100	42,200	46,200	57,600	42,500	44,700[b]		
Severe	34,100	28,400	47,400	53,100	38,500	40,300	−4,400	−10
Grantsburg								
Moderate	54,300	51,100	44,200	57,600	56,300	52,600[c]		
Severe	53,100	55,100	42,700	56,600	45,200	50,600	−2,000	−4
Hoyleton								
Moderate	42,700	61,000	57,300	55,600	56,100	54,600[b]		
Severe	35,300	54,300	39,300	45,200	49,900	44,700	−9,900	−18
Rozetta								
Moderate	42,500	48,700	42,500	57,300	56,600	49,400[c]		
Severe	43,500	46,400	41,200	57,300	54,300	48,700	−700	−2
Tama								
Moderate	59,300	56,100	55,100	55,100	60,000	57,100[c]		
Severe	59,000	55,600	57,100	56,300	56,600	56,900	−200	0
LSD (0.05) between erosion phases within series	9,000	9,000	9,000	9,000	9,000	4,000		
Average								
Moderate	46,800[c]	51,800[c]	49,100[c]	56,600[c]	54,300[b]	51,700[b]		
Severe	45,000	48,000	45,500	53,700	48,800	48,200	−3,500	−7
LSD (0.5) between erosion phases averaged over five series	4,000	4,000	4,000	4,000	4,000	1,800		

[a] Mean of two to five plots.
[b] Moderate phase significantly different from severe phase ($P = 0.05$).
[c] Moderate phase not significantly different from severe phase ($P = 0.05$).

There were insufficient years to evaluate the plant population differences for the fine-texture Miamian soils.

In general, clay content increases with increasing degree of erosion (Table 15.9). The organic C decreases with erosion. The magnitude and direction of cation exchange capacity (CEC) change by eroded soil phases vary with the magnitude and change of both clay content and organic C. If organic C is decreasing, the CEC is

TABLE 15.8
Plant Population Data for 2–5 Years at the Paired Slight, Moderate, and Severely Eroded Phases of Other North Central Soils

Soil series and erosion phase	Plant population (plants/ha)								Mean	% change
	1984	1985	1986	1987	1988	1989	1992	1993		
Beadle										
Slight	62,500	60,600	64,200	66,000	41,000	—	—	—	58,900a	
Moderate	62,800	59,300	67,300	64,600	38,000	—	—	—	58,400a	−1
Severe	59,200	61,200	63,500	63,100	39,500	—	—	—	57,300a	−1
Dubuque										
Slight	—	64,000	65,700	72,900	63,300	—	—	—	66,600a	0
Moderate	—	62,900	68,800	71,600	65,100	—	—	—	66,600a	0
Severe	—	66,600	67,300	70,000	65,200	—	—	—	67,300a	+1
Egan										
Slight	59,800	60,400	64,600	56,900	—	—	—	—	60,400a	
Moderate	58,700	60,700	67,300	57,200	—	—	—	—	61,000a	+1
Severe	63,000	59,600	63,300	56,600	—	—	—	—	60,600a	0
Marlette										
Slight	—	43,600a[a]	59,400a	52,000ab	53,600a	57,500a	—	—	53,200a	
Moderate	—	33,900b	48,800b	54,500a	46,800ab	57,700a	—	—	48,400b	−7
Severe	—	33,700b	32,800c	49,000b	34,700b	56,500a	—	—	41,300c	−22
Miamian										
Slight	—	—	—	—	—	—	72,500	64,200	68,400a	
Moderate	—	—	—	—	—	—	67,500	70,400	69,000a	+1
Severe	—	—	—	—	—	—	68,300	67,900	68,100a	0
Average										
Slight	61,200a	57,200a	63,500a	62,000a	52,600a	—	—	—	—	—
Moderate	60,800a	54,200a	63,000a	62,000a	50,000ab	—	—	—	—	—
Severe	61,100a	55,300a	56,700b	59,700a	46,500b	—	—	—	—	—

[a] Value in the same year followed by the same letter is not significantly different at the 0.05 probability level.

TABLE 15.9

Texture, Organic C, and CEC of the Ap Horizon, Depth to C Horizon, and Soil-Available Water Storage Capacity of Selected North Central United States Soils

Soil series and erosion phase	Ap			Depth to C horizon (cm)	Soil-available water capacity (cm)	Change from moderate to severe (%)
	Clay (STD) (%)	Organic C (kg m⁻³)	CEC (STD) (cmol + kg⁻¹)			
Beadle						
Slight	36	25.0	30	85	20	
Moderate	40	26.0	32	75	17	
Severe	31	22.5	23	59	11	−35
Clarence						
Moderate	43 (3)	17.3	24 (3)	88	9	
Severe	48 (2)	14.8	17 (2)	83	5	−44
Dubuque						
Slight	13	20.1	—	113	—	
Moderate	16	23.1	—	101	—	
Severe	17	28.8	—	79	—	
Egan						
Slight	35	20.4	29	110	26	
Moderate	35	17.7	26	98	18	
Severe	40	18.3	25	80	13	−28
Grantsburg						
Moderate	16 (2)	15.6	14 (4)	161	18	
Severe	25 (1)	15.5	18 (3)	122	11	−39
Hoyleton						
Moderate	22 (4)	17.4	19 (5)	90	18	
Severe	30 (2)	17.0	24 (2)	77	14	−22
Marlette						
Slight	8	15.1	7 (0.9)	135	—	
Moderate	14	13.1	9 (2.6)	92	—	
Severe	16	10.5	10 (1.1)	56	—	
Miamian						
Slight	40	47.7	11	101	3.8	
Moderate	55	22.9	14	72	3.4	
Severe	54	19.0	14	35	2.9	−15
Rozetta						
Moderate	24 (1)	17.0	21 (3)	120	23	
Severe	27 (1)	14.9	16 (4)	109	22	−4
Sharpsburg						
Slight	29	28.8	16	120	—	
Moderate	32	24.0	18	114	—	
Severe	37	21.7	22	103	—	

TABLE 15.9
Texture, Organic C, and CEC of the Ap Horizon, Depth to C Horizon,
and Soil-Available Water Storage Capacity of Selected North Central
United States Soils (continued)

Soil series and erosion phase	Ap			Depth to C horizon (cm)	Soil-available water capacity (cm)	Change from moderate to severe (%)
	Clay (STD) (%)	Organic C (kg m⁻³)	CEC (STD) (cmol + kg⁻¹)			
Tama						
Moderate	29 (2)	17.4	22 (4)	120	25	
Severe	28 (1)	13.9	20 (2)	97	23	−8
Ves						
Slight	23	25.6	18	97	—	
Moderate	26	23.4	19	69	—	
Severe	20	18.6	13	20	—	
All soils						
Moderate	29	19.6	19.8	100	16	
Severe	31	18.0	18.4	77	13	
Change (%)	(+7)	(−8)	(−7)	(−23)	(−19)	

reduced, but if clay content is increasing, the CEC increases. For a few soil series these changes are offsetting. For all soil series, the depth to the C horizon decreased with increased erosion. Soil structure in the A and B horizon enhances root penetration and the lack of structure in the C horizon impedes root ramification. As a consequence, the effective rooting depth decreases with increasing soil erosion. The soils without root-restricting subsoils have greater (23–25 cm) water storage capacities, and only minor reductions (1 or 2 cm) in soil water storage occurred with increasing erosion. The available water storage capacity of the soils without root-restricting subsoils (Rozetta and Tama) was shown to decrease from 4 to 8% when eroded from the moderately to severely eroded phase. The soils with root-restricting subsoils (Beadle, Clarence, Egan, Grantsburg, Hoyleton, and Miamian) decreased from 15 to 44% when eroded from the moderately to severely eroded phase. The already low soil water storage of soils with the root-restricting subsoils was significantly reduced with increased erosion. Clay content increased by 7% when soils were eroded from the moderately to the severely eroded condition. The severely eroded phases of all soils had an 8% decrease in organic C, a 7% reduction in CEC, a 23% decrease in rooting depth, and a 19% reduction in soil water storage when compared to the moderately eroded phases of the same soils.

Because the 5-year corn grain yield data for moderately and severely eroded phases of the Illinois soil series (Clarence, Grantsburg, Hoyleton, Rozetta, and Tama) (Tables 15.5 and 15.6) and corresponding plant population data (Table 15.7) showed statistically significant reductions with increasing erosion, we attempted to

TABLE 15.10
Five-Year Analyses of Variance of Corn Grain Yield and
Plant Population Data for Moderately and Severely Eroded Phases

Source of variation	Df	Mean square	
		Yield	Plant population
Total	49		
Year	4	27,402,000[b]	132,837,000[c]
Soil	4	59,060,000[b]	272,344,000[a]
Year × soil = error a	16	5,537,000[b]	67,283,000[b]
Phase	1	6,798,000[b]	153,776,000[b]
Soil × phase	4	1,767,000[a]	37,567,000[a]
Year × phase (soil) = error b	20	52,000	9,292,000

[a] Denotes significance at $P = 0.05$.

[b] Denotes significance at $P = 0.01$.

[c] Denotes nonsignificance at $P = 0.05$.

evaluate the relationship using regression analysis (Table 15.10). The relationship between plant population (X) in plants per hectare and corn grain yield (Y) in milligrams per hectare can be expressed by the following linear equation: $Y = -8.644 + 0.000305X$. The correlation ($r^2 = 0.52$) was relatively poor. Under a high level of management with about 60,000 corn seeds planted per hectare (24,000 seeds per acre) (slightly above the optimum rate), population reductions did not appear to be the primary cause of corn grain yield reductions. Past research (Olson and Nizeyimana, 1988) has suggested that the primary reason for reduced corn grain yield was reduction in water storage capacity of severely eroded soils with root-restricting subsoils. Field observations suggest that when a few plants are missing from one replicate on an eroded phase of a soil series, adjacent corn plants produce either more ears per plant or bigger ears. As a result, corn grain yield from a plot with lower plant population often was equal to and occasionally higher than that for other replicated plots on the same eroded phase of a soil series. Table 15.10 provides the analyses of variance of 5-year corn grain yield and plant population data for the moderately and severely eroded phases of the five Illinois soils. As expected, the effects of years, soil series, and phases were highly significant ($P = 0.01$) for corn grain yield. For plant population, year effects were not significant, but effects of soil series and erosion phases were significant ($P = 0.05$ and $P = 0.01$, respectively).

Minor and major reductions in inherent soil productivity were considered as a basis for setting threshold values for measurable and observable soil properties or conditions based on current methods of technology and research. Threshold values were determined by correlating soil properties with yields obtained from research fields. The threshold soil values and productivity parameters serve as an early warning signal of reduced productive capacity of soils common to the north central United States (Olson, 1992). Specific soil parameters and threshold values were identified for both topsoil and subsoil layers. Possible changes in the topsoil layer properties that appear to affect productivity under a high level of management include erosion phase, organic C, infiltration, and clay content. Changes in subsoil properties that affect soil productivity the most were rooting depth and soil water storage. Values for soil properties were defined as no reduction, minor (less than 15%), and major (more than 15%) reduction in productivity as reflected in corn grain yields (Olson, 1992). These values could serve as warning signals of reduced productivity.

For soils with root-restricting subsoils, the long-term productivity, measured by corn grain yield loss, was greater than the 15% threshold value, which suggests that the soil quality has been degraded. The changes in surface soil properties which appeared to have the greatest impact on soil productivity and quality were the increase in clay content (9 of 12 soils) and decrease in organic C (10 of 12 soils) for the severely eroded soil phases. For fine-texture soils, this resulted in a significant reduction in plant population, which affected corn grain yields. When erosion occurred, the subsoil properties which changed the most dramatically were rooting depth and water storage capacity. This condition resulted in a soil water storage decrease of 15–44% when the soils had root-restricting subsoils.

CONCLUSIONS

The 5-year corn grain yield averages on plots of severely eroded soils with root-restricting subsoils were significantly less (between 7 and 35%) than those on plots of less eroded phases of the same soils. The 5-year plant population averages for plots on severely eroded phase of the fine-textured Clarence and Hoyleton soils were significantly less (10–18%) than those on plots of the moderately eroded phase of these five soils. Five-year corn grain yield averages for paired plots of slightly, moderately, and severely eroded soils with favorable subsoils for rooting showed nonsignificant differences. Plant population differences between eroded phases of soils with favorable soils for rooting were not significant. The primary reasons for the corn grain yield decline were a reduction of the amount of topsoil (A horizon material in plow layer) and the associated effects (reduced plant population) of increased clay content in the Ap horizon, restricted rooting depth, and reduced plant-available water storage. Soils with low initial water storage capacities and root-restricting subsoils showed the largest yield reductions when severely eroded. Relative corn grain yield loss averaged 18% for severely eroded soil phases, when compared to less eroded soil phases, for soils with root-restricting subsoils and 0% on soils without root-restricting subsoils. For soils with root-restricting subsoils, the long-term productivity, measured by corn grain yield loss, was greater than the suggested 15% threshold value, indicating soil quality had been severely degraded. The changes in topsoil properties which appeared to have the greatest impact on soil productivity and quality were the increase in clay content and decrease in organic C. For fine-texture soils, this resulted in a significant reduction in plant population, which affected corn grain yields. When erosion occurred, the subsoil properties which changed the most dramatically were the rooting depth and water storage capacity. This condition resulted in a soil water storage decrease of 15–44% when the soils had root-restricting subsoils.

ACKNOWLEDGMENTS

The following scientists contributed data sets used in the development of this chapter: A.J. Jones, University of Nebraska; B. Lowery, University of Wisconsin; J.B. Swan, Iowa State University; and W. Nelson, University of Minnesota. Supported in part by Cooperative Regional Research Funds NC-174 (Soil Productivity and Erosion).

REFERENCES

American Society of Agricultural Engineers. 1985. Proc. National Symposium on Erosion and Soil Productivity. ASAE Publ. 8-85. American Society of Agricultural Engineers, St. Joseph, MI.

Andraski, B.J. and B. Lowery. 1992. Erosion effects on soil water storage, plant water uptake, and corn growth. *Soil Sci. Soc. Am. J.* 56:1911–1919.

Blake, G.R. 1965. Bulk density. In C.A. Black (ed.). *Methods of Soil Analysis, Part 1.* Agronomy 9. American Society of Agronomy, Madison, WI, pp. 374–390.

Cassel, D.K. and D.W. Fryrear. 1990. Evaluation of productivity changes due to accelerated soil erosion. In W.E. Larson, G.K. Foster, R.R. Allmaras, and C.M. Smith (eds.). *Research Issues in Soil Erosion/Productivity.* University of Minnesota, St. Paul, pp. 41–54.

Chengere, A. and R. Lal. 1995. Soil degradation by erosion of a Typic Hapludalf in central Ohio and its rehabilitation. *Land Degrad. Rehabil.* 6:223–238.

Cihacek, L.J. and J.B. Swan. 1994. Effects of erosion on soil chemical properties in the north central region of the United States. *J. Soil Water Conserv.* 49:259–265.

Copley, T.L., L.A. Forest, A.G. McCall, and F.C. Bell. 1944. Investigations in Erosion, Control and Reclamation of Eroded Land at the Central Piedmont Conservation Experiment Station, Statesville, NC. USDA Tech. Bull. 873.

Dahnke, W.C. 1988. Recommended Chemical Soil Test Procedures for the North Central Region. North Central Region Publication No. 221. North Dakota State University, Fargo.

Day, P.R. 1965. Particle fractionation and particle size analysis. In C.A. Black (ed.). *Methods of Soil Analysis, Part 1.* Agronomy 8. American Society of Agronomy, Madison, WI, pp. 545–567.

Fahnestock, P., R. Lal, and G.F. Hall. 1995. Land use and erosional effects on two Ohio Alfisols. II. Crop yields. *J. Sustain. Agric.* 7:85–100.

Follett, R.F. and B.A. Stewart. 1985. *Soil Erosion and Crop Productivity.* American Society of Agronomy, Madison, WI.

Frye, W.W., S.A. Ebelhar, L.W. Murdock, and R.L. Blevins. 1982. Soil erosion effects on properties and productivity of two Kentucky soils. *Soil Sci. Soc. Am. J.* 46:1051–1055.

Frye, W.W., O.L. Bennett, and G.J. Buntley. 1985. Restoration of crop productivity on eroded or degraded soils. In R.F. Follett and B.A. Stewart (eds.). *Soil Erosion and Crop Productivity.* American Society of Agronomy, Madison, WI, pp. 335–355.

Gollany, H.T., T.E. Schumacher, M.J. Lindstrom, P.D. Evenson, and G.D. Lemme. 1992. Topsoil depth and desurfacing effects on properties and productivity of a Typic Argiustoll. *Soil Sci. Soc. Am. J.* 56:220–225.

Hammel, J.E., M.E. Sumner, and H. Shahandeh. 1985. Effect of physical and chemical profile modification on soybean and corn production. *Soil Sci. Soc. Am. J.* 49:1508–1511.

Hanna, A.Y., P.W. Harlan, and D.T. Lewis. 1983. Effect of slope on water balance under center pivot irrigations. *Soil Sci. Soc. Am. J.* 47:760–764.

Jones, A.J., L.N. Mielke, C.A. Bartles, and C.A. Miller. 1989. Relationship of landscape position and properties to crop production. *J. Soil Water Conserv.* 44:328–332.

Krauss, H.A. and R.R. Allmaras. 1982. Technology masks the effects of soil erosion and wheat yields — a case study in White Man County, Washington. In B.L. Schmidt, R.R. Allamaras, J.V. Mannering, and R.I. Papendick (eds.). *Determinants of Soil Loss Tolerance.* Special Publication No. 45. American Society of Agronomy, Madison, WI, pp. 75–86.

Langdale, G.W. and W.D. Shrader. 1982. Soil erosion effects on soil productivity of cultivated cropland. In B.L. Schmidt, R.R. Allamaras, J.V. Mannering, and R.I. Papendick (eds.). *Determinants of Soil Loss Tolerance.* Special Publication No. 45. American Society of Agronomy, Madison, WI, pp. 41–51.

Langdale, G.W., J.E. Box, R.A. Leonard, A.P. Barnett, and W.G. Fleming. 1979. Corn yield reduction on eroded southern Piedmont soils. *J. Soil Water Conserv.* 34:226–228.

Lindstrom, M.J., T.E. Schumacher, A.J. Jones, and C. Gantzer. 1992. Productivity index model comparison for selected soils in north central United States. *J. Soil Water Conserv.* 47:491–494.

Lowery, B., J. Swan, T. Schumacher, and A. Jones. 1995. Physical properties of selected soils by erosion class. *J. Soil Water Conserv.* 50:306–311.

Mokma, D.L. and M.A. Sietz. 1992. Effects of soil erosion on corn yields on Marlette soils in south-central Michigan. *J. Soil Water Conserv.* 47:325–327.

Mokma, D.L., T.E. Fenton, and K.R. Olson. 1996. Effect of erosion on morphology and classification of soils in the north central United States. *J. Soil Water Conserv.* 51:171–175.

Nelson, D.W. and L.E. Sommers. 1982. Total carbon, organic carbon, and organic matter. In A.L. Page (ed.). *Methods of Soil Analysis, Part 2.* Agronomy 9. American Society of Agronomy, Madison, WI, pp. 539–579.

Nizeyimana, E. and K.R. Olson. 1988. Chemical, mineralogical, and physical property differences between moderately and severely eroded Illinois soils. *Soil Sci. Soc. Am. J.* 52:1740–1748.

Olson, K.R. 1992. Soil physical properties as a measure of cropland productivity. In Proc. Soil Quality Standards Symposium. Wo-WSA-2. USDA Forest Service, pp. 41–51.

Olson, K.R. and S.G. Carmer. 1990. Corn yield and plant population differences between eroded phases of Illinois soils. *J. Soil Water Conserv.* 45:562–566.

Olson, K.R. and E. Nizeyimana. 1988. Effects of erosion on the corn yields of seven Illinois soils. *J. Prod. Agric.* 1:13–19.

Olson, K.R., R. Lal, and L.D. Norton. 1994a. Evaluation of methods to study soil erosion-productivity relationships. *J. Soil Water Conserv.* 49:586–590.

Olson, K.R., L.D. Norton, T.E. Fenton, and R. Lal. 1994b. Quantification of soil loss from eroded soil phases. *J. Soil Water Conserv.* 49:591–596.

Peech, M. 1965. Hydrogen ion activity. In C.A. Black (ed.). *Methods of Soil Analysis, Part 2.* Agronomy 9. American Society of Agronomy, Madison, WI, pp. 914–926.

Pierce, F.J., W.E. Larson, R.H. Dowdy, and W.A.P. Graham. 1983. Productivity of soils: assessing long-term changes due to erosion. *J. Soil Water Conserv.* 38:39–44.

Richards, L.A. 1965. Physical conditions of water in soils. In C.A. Black (ed.). *Methods of Soil Analysis, Part 1.* Agronomy 9. American Society of Agronomy, Madison, WI, pp. 128–152.

SAS Institute. 1985. *SAS User's Guide: Statistics.* SAS Institute, Cary, NC.

Schumacher, T.E., M.J. Lindstrom, D.L. Mokma, and W.W. Nelson. 1994. Corn yield; erosion relationships of representative loess and till soils in the north central United States. *J. Soil Water Conserv.* 49:77–81.

Shaffer, M.J., S.C. Gupta, J.A.E. Molina, D.R. Linden, and W.E. Larson. 1983. Simulating of nitrogen, tillage, and residue management effects on soil fertility. In W.K. Lauenroth, G.V. Skogerbo, and M. Flug (eds.). *Analysis of Ecological Systems: State-of-the-Art in Ecological Modeling.* Elsevier, Amsterdam, pp. 525–544.

Shaffer, M.J., T.E. Schumacher, and C.L. Ego. 1994. Long-term effects of erosion and climate interactions on corn yield. *J. Soil Water Conserv.* 49:272–275.

Soil Survey Staff. 1975. Soil Taxonomy, A Basic System of Soil Classification for Making and Interpreting Soil Surveys. AH-436. U.S. Government Printing Office, Washington, D.C.

Soil Survey Staff. 1993. Soil Survey Manual (Revised). U.S. Government Printing Office, Washington, D.C.

Spomer, R.G. and R.F. Piest. 1982. Soil productivity and erosion in Iowa loess soils. *Trans. ASAE* 25:1295–1299.

Stallings, J.W. 1950. Erosion of Topsoil Reduces Productivity. USDA Report SCS-TP 98. U.S. Government Printing Office, Washington, D.C.

Stone, J.R., J.W. Gilliam, D.K. Cassel, R.B. Daniels, L.A. Nelson, and H.J. Kleiss. 1985. Effect of erosion and landscape position on the productivity of Piedmont soils. *Soil Sci. Soc. Am. J.* 49:987–991.

Uhland, R.E. 1949. Crop Yields Lowered by Erosion. USDA-SCS-TP-75. Washington, D.C.

Williams, J.R., K.G. Renard, and P.T. Dyke. 1983. EPIC — a new method for assessing erosion's effect on soil productivity. *J. Soil Water Conserv.* 38:381–383.

16 Erosion Impact on Soil Quality in the Tropics

R. Lal

INTRODUCTION

Accelerated soil erosion is widely recognized as a severe problem in the tropics (Lal, 1984, 1994a). A casual visitor to the tropics is usually amazed by the extent and severity of visible soil erosion on farmland by water (Figures 16.1 and 16.2), wind (Figures 16.3 and 16.4), and mass movement (Figures 16.5 and 16.6). In fact, several hot spots of erosion-induced soil degradation around the world (Scherr and Yadav, 1996) include tropical regions of Africa, Asia, and South America. However, the link between accelerated erosion and food security is neither established nor widely recognized (Lal, 1987). Some argue that accelerated soil erosion has severe on-site effects on global productivity (Pimentel et al., 1995). Others contend that on-site effects of erosion on productivity are of no major consequence (Crosson and Anderson, 1992). Yet, regions plagued by perpetual food deficit and malnutrition (e.g., sub-Saharan Africa, South Asia) are also prone to severe forms of erosion and attendant soil degradation (Lal, 1988; Dregne, 1990, 1992).

This apparent contradiction may be partly due to our lack of understanding of the mechanisms of erosional impact on productivity, especially in soils of the tropics for which the data are sketchy and incomplete. Impact of soil erosion on productivity varies among soils, land use and farming systems, and ecoregional factors related to climate. Soil erosion may influence productivity through the loss of effective rooting depth, decrease in plant-available water capacity, reduction in soil fertility, or any combination of these factors. If these were not the yield-limiting constraints, even severe soil erosion may have no impact on agronomic productivity (Lal, 1987). The controversy may also be caused by the lack of appropriate methodology in scaling up the data from plot level to a regional, national, or global scale.

1-57444-100-0/99/$0.00+$.50
© 1999 by Soil and Water Conservation Society

(a)

(b)

FIGURE 16.1   Accelerated erosion of farmland by runoff on (a) Vertisols in India and (b) Alfisols in northern Australia.

TROPICAL REGIONS WITH SEVERE EROSION PROBLEMS

Despite the voluminous literature on the risks of soil erosion and its impact on food security and environmental quality, there is a conspicuous lack of credible data on the magnitude (areal extent) and severity (degree in relation to impact on soil

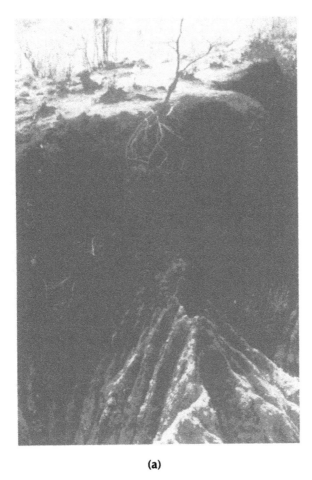

(a)

FIGURE 16.2 Severe gully erosion in Tanzania: (a) an advancing gully head, (b) a virtually ruined landscape, and (c) a gully-infested land in eastern Nigeria.

quality) of soil erosion in the tropics. Sanchez et al. (1982) estimated tropical areas with soil-related constraints to crop production. The data in Table 16.1 show estimates of some erosion-related soil physical constraints. For the tropical region as a whole, 64% of the land area is influenced by drought, 34% by steep gradient (>30%), and 2% by shallow topsoil. All three factors are strongly influenced by soil erosion. Estimates of global soil degradation by Oldeman (1994) are qualitative and based on "structured informed opinion analyses." This methodology has not yet been widely used on an ecoregional basis. Application of this methodology in South Asia, one of the principal global hot spots, has shown that about 82 million ha are affected by water erosion and 59 million ha by wind erosion (Table 16.2). Global estimates show that regions prone to water and wind erosion are largely concentrated throughout the tropics of Asia, Africa, and South/Central America (Table 16.3).

(b)

(c)

FIGURE 16.2
(continued)

FIGURE 16.3   Wind erosion in Niger.

Water erosion is a severe problem in Central America and Asia and wind erosion in Africa and Asia. Asian rivers carry the largest amount of sediments to the ocean (Table 16.3).

FACTORS RESPONSIBLE FOR SEVERE EROSION IN THE TROPICS

Several factors responsible for erosion-induced soil degradation in the tropics may be grouped as exogenous and endogenous factors (Table 16.4). High climatic erosivity, high erodibility of some soils, low-input agricultural systems with little or no

FIGURE 16.4   Wind erosion of farmland in Argentina.

FIGURE 16.5   Landslides of steepland in Indonesia.

investment in conservation-effective measures, and removal of crop residues from farmlands lead to severe soil erosion by both water and wind. Climatic erosivity and soil erodibility are among the exogenous factors that cannot be regulated (Table 16.4).

Exogenous Factors

Important among the exogenous factors affecting soil erosion are climate and soil characteristics.

Rainfall Erosivity

Higher erosivity of tropical than temperate rains (Hudson, 1995; Obi and Ngwu, 1988; Lal, 1984; Aina, 1980; Kowal and Kassam, 1976; Roose, 1977) has been

FIGURE 16.6   Mass movement of hillslopes in Brazil.

attributed to numerous factors. Tropical rains are characterized by high intensity. Hudson (1995) observed that rainfall intensity seldom exceeds 75 mm h^{-1} in temperate climates. In many tropical regions, however, intensities of 150 mm h^{-1} are experienced regularly. Short-term intensities as high as 200 mm h^{-1} have been reported from western Nigeria and elsewhere in the tropics (Lal, 1976, 1984). Morin (1996) observed a high 30-min intensity (I_{30}) at Yavne Station in the southern coastal plains of Israel. For the first half of the rainy season in Israel, from November to January, I_{30} ranged from 35 to 85 mm h^{-1}. For the second half of the rainy season, from February to April, I_{30} was low and ranged from 10 to 35 mm h^{-1}.

   High intensities of tropical rains may partly be due to relatively big drop size. In general, the modal value or median drop size (D_{50}) increases with increase in rain intensity until about 100 mm h^{-1} (Laws and Parsons, 1943; Wischmeier and Smith, 1978). The D_{50} of tropical rains is often observed to range from 2 to 3 mm. In

TABLE 16.1
Erosion-Related Soil Physical Constraints to Production in the Tropics

Constraint	Tropical regions			Total
	Africa	Asia	America	
Drought (%)	67	72	45	60
Steep slopes >30% (%)	22	51	32	34
Shallow topsoil (%)	1	2	3	2
Total land area (10^6 ha)	1555	1205	1879	4638

Modified from Sanchez et al., 1982.

TABLE 16.2
Estimates of Area Affected by Water and Wind Erosion in South Asia

Country	Water erosion		Wind erosion	
	Area (10^6 ha)	% of agricultural land	Area (10^6 ha)	% of agricultural land
Afghanistan	11.2	29	2.1	5
Bangladesh	1.5	15	0	0
Bhutan	0.04	10	0	0
India	32.8	18	10.8	6
Iran	26.4	45	35.4	60
Nepal	1.6	34	0	0
Pakistan	7.2	28	10.7	42
Sri Lanka	1.1	46	0	0
Total	81.8		59.0	

Modified from FAO, 1994.

TABLE 16.3
Magnitude of Soil Erosion by Water and Wind

Region	Sediment transport		Water erosion		Wind erosion	
	Total (10^6 mg yr^{-1})	Yield (mg km^{-2} yr^{-1})	Area (10^6 ha)	% of total area	Area (10^6 ha)	% of total area
Africa	731	47.8	227	46	186	38
Asia	8,025	285.6	441	59	222	30
South America	2,391	133.6	123	51	42	17
Central America	—	—	46	74	5	7
World	17,377	196.1	1,094	56	548	28

Modified from Walling, 1987; Oldeman, 1994.

TABLE 16.4
Major Causal Factors of Soil Erosion in the Tropics

Type of factors	Characteristics
Exogenous	Climate, soil properties, landscape
Endogenous	Deforestation, overexploitation of natural vegetation, overgrazing, excessive tillage, burning, monoculture, lack of crop residue mulch, no-input or subsistence agriculture
Institutional	Lack of institutional support, poor extension services

northern Nigeria, Kowal and Kassam (1976) observed D_{50} in excess of 4 mm, corresponding with rain intensity of about 100 mm h^{-1}. Measurements made at the International Institute of Tropical Agriculture (IITA), Ibadan, Nigeria, showed that most erosive rains exceeding an intensity of 25 mm h^{-1} had a D_{50} of >2.5 mm (Table 16.5). A high D_{50} of >3 mm was generally observed for isolated showers of low total amount. The data in Figure 16.7 show that the D_{50} of 20–30% of rains measured in western Nigeria was >3 mm.

High intensity and big D_{50} are also associated with high energy load. The kinetic energy of these rains can be computed from rainfall intensity and amount (Lal and Elliott, 1994). Lal (1981a) developed the following empirical relationships between kinetic energy of rain and other pertinent parameters:

$$E = 24.5\ R_a + 27.6 \qquad\qquad r^2 = 0.81\ (1\%) \qquad\qquad (16.1)$$

$$E = 18.18\ I_{30} + 18.2 \qquad\qquad r^2 = 0.81\ (5\%) \qquad\qquad (16.2)$$

where E is in J m^{-2}, R_a is rainfall amount (in mm), I_{30} is 30-min intensity (in mm h^{-1}), and r is a correlation coefficient significant at a 5% or 1% level of probability. Similar relations have been proposed by Elwell (1979) for Zimbabwe and Kowal and Kassam (1976) for northern Nigeria.

TABLE 16.5
Drop Size Distribution and Energy Load of Rains Received at IITA, Ibadan, Nigeria

Date	Rainfall amount (mm)	Intensity (mm h^{-1}) Maximum	I_{30}	D_{50} (mm)	Kinetic energy (J m^{-2})
February 15, 1981	6.0	17.0	9.1	2.60	154.0
March 31, 1981	25.4	137.2	38.1	2.60	750.8
May 7, 1981	1.8	10.7	3.6	4.39	67.2
May 21, 1981	38.3	166.1	60.5	2.89	1166.0

Unpublished data of R. Lal.

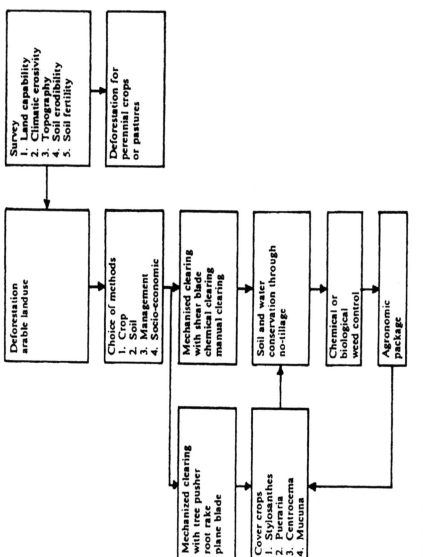

FIGURE 16.7   Sequence of steps needed for proper land development.

TABLE 16.6
Soils of the Tropics and Their Quality-Restrictive Constraints

Soil map unit	Area[a]		Water capacity	Effective rooting depth	Nutrient imbalance	Soil structure	Accelerated erosion
	10^6 ha	% of the tropics					
Oxisols	1100	22.5	2	3	3	1	2
Aridisols	900	18.4	3	2	2	1	3
Alfisols	800	16.2	2	2	1	3	3
Ultisols	520	10.6	2	2	1	2	2
Entisols	490	10.2	1	1	1	2	2
Inceptisols	243	5.0	1	1	1	2	2
Vertisols	200	4.0	2	2	2	3	3
Andisols	57	1.1	1	1	1	1	1
Mollisols	50	1.0	1	1	2	1	1
Histosols	31	0.6	1	1	2	2	2
Spodosols	6	0.1	2	2	2	2	2
Others	503.6	10.3	—	—	—	—	—

Note: 1 = good, 2 = moderate, and 3 = poor.
[a] Data on land area obtained from Van Wambeke (1992).

Soil Quality

Predominant soils of the tropics are Oxisols, Aridisols, Alfisols, Ultisols, Inceptisols, Entisols, and Vertisols, which together account for about 87% of the total land area (Table 16.6). Some of the highly weathered soils (i.e., Oxisols and Ultisols) have low cation exchange capacity, low plant nutrient reserves, and low to moderate plant-available water capacity (Lal, 1990). These soils may also suffer from nutrient imbalance due to toxic concentrations of Al and Mn. Poor soil structure, crusting and compaction, and accelerated soil erosion are severe constraints in Alfisols and Vertisols (Cassel and Lal, 1992). Low available water-holding capacity, frequent drought stress, and root-restrictive layers at shallow depths are also problems in Aridisols and Alfisols (Table 16.6).

Predominant properties affecting soil quality (Lal, 1994b) are (1) soil physical properties including plant-available water capacity, effective rooting depth, soil structure, infiltration capacity, and texture; (2) soil chemical properties including pH, cation exchange capacity, concentration of Al and Mn, and available nutrient capacity; and (3) soil biological properties including soil organic carbon, biomass carbon, and soil biodiversity.

Considering the most important determinants, soil quality may be defined as follows:

$$Sq = f(AWC, SOC, R_d, CEC, Cl),$$ (16.3)

where Sq is soil quality, AWC is available water capacity, SOC is soil organic carbon content, R_d is the effective rooting depth, CEC is cation exchange capacity,

Cl is clay content (%), and t is time. The magnitude of temporal changes in these properties are an important factor in determining soil quality. Accelerated soil erosion can have drastic adverse effects on soil quality through its negative impact on some or all of the parameters listed in Equation 16.3.

Endogenous Factors

There are several endogenous factors that can be managed through improved or scientific technology. The most important among these are deforestation, overexploitation, and excessive grazing (Oldeman, 1994). Some endogenous factors that need to be addressed to improve soil quality and productivity in the tropics are subsistence agriculture, land clearing, and soil management.

Subsistence Agriculture and Soil Quality

A major factor responsible for severe soil erosion is subsistence agriculture based on little or no external input. Subsistence agriculture leads to depletion of soil fertility, decline in soil organic matter content, deterioration of soil structure, low crop stand, low canopy cover, lack of crop residue mulch, and high susceptibility to wind and water erosion.

Shifting cultivation and related bush fallow systems, done properly with necessary built-in natural fallowing for soil restoration and vegetation reestablishment, are highly conservation-effective. Several experiments (Lal, 1974; Okigbo, 1978) have shown that soil erosion risks are minimal whenever shifting cultivation is adopted in its true sense, involving incomplete clearing, mixed cropping, use of crop residue mulch, and a raised-bed (ridges or mounds) system of seedbed preparation. However, the magnitude of soil erosion and runoff losses under traditional farming depends on a range of interacting factors including soil type, terrain, land use, and farming system. Erosion risks are drastically exacerbated with reduction in the fallow period, removal of crop residue, and indiscriminate burning (Lal, 1974). Lal (1981c) observed negligible runoff and erosion from a traditionally cultivated plot during both the cropping and fallow phases (Table 16.7). These data show that with improved farming system and judicious use of science-based inputs, soil erosion risks can be drastically reduced even with intensive cultivation. On a sandy savanna soil, Sabel-Koschella (1987) reported that, in comparison with the no-till and mulch treatments, runoff and soil erosion from traditionally farmed yam were more (Table 16.8). Similar observations were reported by Lal (1986b) from savanna soils in Ilorin, Nigeria. Soil erosion from traditionally farmed plots with ridges made up and down the slope was more than from no-till plots but less than those from disk-plowed plots.

There is a difference between simulating traditional management on research plots versus actual management under on-farm conditions. Traditionally farmers abandon the land when crop yields are too low because of either high incidence of pests or decline in soil quality. The rate of decline in soil quality under traditional management depends on many factors, including soil type, climate, farming system,

TABLE 16.7
Land Use and Farming System Effects on Runoff
and Soil Erosion on Alfisols in Western Nigeria

Farming system	Runoff (mm)	Soil erosion (mg ha^{-1})
Forest control	0	0
Traditional farming	6.6	0.02
Intensive cultivation		
Manual clearing	47.7	5.0
Shear blade clearing	104.8	4.8
Tree pusher/root rake clearing	250.3	20.0

Modified from Lal, 1981c.

and management. Management factors are extremely crucial to maintaining high soil quality.

Methods of Land Clearing and Development

Deforestation is a major factor affecting soil erosion and global land degradation (Oldeman, 1994). The magnitude of adverse effects of deforestation, however, depends on many factors, including the vegetation and methods of land clearing (Hulugalle et al., 1984; Ghuman et al., 1991; Ghuman and Lal, 1991). Lal (1981c) showed that methods of deforestation affect soil quality and its susceptibility to erosion. Therefore, Lal et al. (1986) recommended guidelines for appropriate methods of land clearing and development. If bringing new land under cultivation is inevitable, it should be done with proper planning (Lal, 1986b). Survey and complete inventory of natural resources is the first step (Figure 16.7). This survey should provide information on land use capability, climatic potential and constraints, soil

TABLE 16.8
Farming System Effects on Runoff and Soil Erosion
from a Sandy Savanna Soil in Western Nigeria

Year	Soil erosion (mg ha^{-1})	Runoff (mm)
1983 bare fallow	24.4	131.0
Traditional (yam + maize)	4.9	60.7
No-till (maize)	2.2	68.8
Plow-till (maize)	2.6	85.0
1984 bare fallow	257.9	330.8
Traditional (cassava + maize)	1.5	23.8
No-till (maize)	0.4	6.8
Plow-till (maize)	3.9	101.9

Modified from Sabel-Koschella, 1987.

quality, and soil-related constraints to intensive cropping. The decision as to whether the land should be cleared and by what methods depends on the information generated by these surveys. There is a wide range of factors that are to be considered in the choice of an appropriate method that maintains a favorable soil quality. Important among these factors are vegetation, climatic factors including rainfall erosivity, soil type, terrain characteristics, land use, soil and crop management, farming system, and the area to the cleared (Table 16.9).

Soil Management

Soil and crop management are crucial factors that regulate risks of soil erosion and determine erosion-induced changes in soil quality. Important among soil management techniques are tillage methods, crop residue management system, and soil fertility management including fertilizer use. Regardless of soil erosion, tillage methods are known to have a profound effect on soil quality (Lal, 1986a). Data in Table 16.10 are from a long-term tillage experiment conducted on Alfisols in western Nigeria. These experiments were initiated in 1971, but the fertilizer and mulching variables were introduced in 1979–80. Measurements reported in Table 16.10 were made in 1987 and show drastic effects of tillage methods on soil physical quality. For example, soil sampled from the no-till plot had lower clay, higher silt content, more accumulative infiltration, and more moisture retention at all suctions than plow-till treatments. There were also beneficial effects of crop residue mulch and fertilizer treatment on improvements in soil physical quality, primarily due to high biomass production.

The analyses of data from long-term soil management experiments conducted in India showed that management of soil fertility can influence soil physical properties that regulate erodibility and susceptibility to soil erosion. The data collated by Nambiar (1994) showed a decrease in soil bulk density, an increase in moisture retention at 0.03 MPa suction, and an increase in hydraulic conductivity due to applications of inorganic fertilizers and organic manure (Table 16.11). In comparison with unmanured control, use of NPK fertilizer from 1971 to 1979 slightly decreased soil bulk density (increased total porosity), increased mean weight diameter of aggregates, and increased soil moisture retention at field capacity. There was also an increase in saturated hydraulic conductivity in two out of four sites. In addition to inorganic fertilizers, use of organic manures further decreased soil bulk density in two out of four sites, increased mean weight diameter of aggregates in three out of four sites, increased moisture retention at 0.03 MPa suction in three out of four sites, and increased saturated hydraulic conductivity in three out of four sites. These data strongly support the hypothesis that soil fertility management through the use of inorganic fertilizers and organic manures can improve soil physical properties and decrease risks of soil erosion.

Soil management is probably the most important endogenous factor that determines erosion-induced changes in soil quality. It is important not only because of its impact on productivity and environment quality, but also because it can be regulated through judicious input and appropriate systems of soil and crop management.

TABLE 16.9
Factors to Consider for Deciding the Method and Equipment for Land Clearing

Vegetation	Climatic conditions	Soil	Topography	Intended land use	Soil management	Area to be cleared
1. Tree count or density	1. Rainfall amount and distribution	1. Texture, structure	1. Slope	1. Urban/industrial use	1. Tillage methods	Socioeconomic conditions including pollution problems
2. Size of trees (i.e., diameter and height)	2. Length of the dry season	2. Effective rooting depth	2. Surface depression	2. Agricultural and use	2. Harvesting techniques	
3. Root system	3. Air temperature humidity	3. Moisture retention and transmission properties	3. Nature of slope	a. Plantation crops		
				b. Seasonal crops		
				c. Root crops or grain crops		
				d. Pastures		
4. Undergrowth						
5. Dominant species (hard/soft wood)						

Modified from Lal, 1986b.

TABLE 16.10
Tillage, Mulching, and Fertilizer Effects on Soil Physical Properties
of 0- to 50-cm Depth of an Alfisol in Western Nigeria

Tillage	Mulch and fertilizer treatment[a]	Clay (%)	Silt (%)	Saturation point (%)	0.01 MPa suction	0.05 MPa suction	1.5 MPa suction	2-hr accumulative infiltration (cm)
No-till	M_0F_0	12.1	11.5	35.4	12.5	8.1	5.4	84.6
	M_0F_1	13.5	13.5	36.7	13.7	8.2	7.2	68.6
	M_1F_0	14.1	17.0	43.3	18.1	13.2	4.9	79.2
	M_1F_1	10.8	14.8	39.7	12.3	8.2	5.8	68.3
	Mean	12.6	14.2	38.8	14.2	9.4	5.8	75.2
Plow-till	M_0F_0	13.5	9.7	28.6	8.9	6.7	4.0	30.9
	M_0F_1	15.5	10.7	26.2	8.0	6.5	4.3	20.5
	M_1F_0	14.1	11.2	29.9	9.1	6.9	4.0	24.3
	M_1F_1	18.5	12.1	31.9	11.5	8.8	5.7	34.8
	Mean	15.4	10.9	29.2	9.4	7.2	4.5	27.6
LSD (0.05)								
Tillage (T)		1.2	0.9	1.9	1.5	0.9	0.6	18.7
Mulch (M)		0.8	0.6	1.0	0.7	0.5	0.3	18.2
Fertilizer (F)		0.8	0.6	1.7	1.1	0.8	0.5	18.2
T × M		1.1	0.8	1.5	0.9	0.7	0.5	25.8
T × F		1.1	0.8	2.3	1.5	1.1	0.7	48.2
M × F		1.1	0.8	2.3	1.5	1.1	0.7	48.2

[a] M_0 = residue removed, M_1 = residue returned as mulch, F_0 = no fertilizer, and F_1 = recommended rate of fertilizer.

Unpublished data of R. Lal.

SOIL EROSION AND CROP PRODUCTIVITY IN THE TROPICS

Accelerated erosion has severe adverse effects on crop yield at farm, local, regional, and global scales. The exact magnitude of the effect of erosion on agronomic yield depends on soil profile characteristics, weather conditions during the crop cycle, and management systems. Dregne (1990) estimated that agronomic productivity of some parts of Africa has declined by as much as 20% due to soil erosion and desertification. Productivity losses due to erosion are especially severe in the Maghreb region, Nigeria and Ghana in West Africa, and parts of southern Africa and east African Highlands. Lal (1995) estimated crop yield reduction of 2–5% per millimeter of soil loss and reported that crop yield reduction due to past erosion in Africa ranges from <2% for slight erosion to 40% for severe erosion. For 44 countries in sub-Saharan Africa, Lal (1995) estimated erosion-induced crop yield reduction of 6.2% at the continental scale (Table 16.12). On a plot scale, Lal

TABLE 16.11
Long-Term Manure and Fertilizer Effects (1971–79) on Soil Physical Properties at Several Locations in India

Soil physical properties	Unmanured control				Nitrogen–phosphorus–potassium				Nitrogen–phosphorus–potassium + farmyard manure			
	A	B	C	D	A	B	C	D	A	B	C	D
Soil bulk density (g/m³)	1.46	1.47	1.22	1.18	1.40	1.41	1.21	1.17	1.40	1.42	1.18	1.05
Aggregate stability (mean weight diameter, mm)	0.25	0.31	1.39	—	0.32	0.33	1.52	—	0.57	0.38	1.70	—
Moisture retention at 0.03 MPa (%)	29.9	19.9	34.9	9.1	30.2	20.6	36.0	10.2	30.1	21.8	37.3	12.0
Saturated hydraulic conductivity (cm/hr)	0.088	0.66	0.126	3.09	0.095	0.69	0.124	3.02	0.106	0.70	0.133	2.90

Note: A = Barrackpore, B = Delhi, C = Jabalpur, and D = Bhubneshwar.

Modified from Nambiar, 1994.

TABLE 16.12
Erosion Effects on Grain Production in Sub-Saharan Africa in 1989

Commodity	Total production without erosion (10^6 mg)	Actual production in 1989 (10^6 mg)	Loss in production due to erosion (10^6 mg)
Cereals	59	55	4
Roots and tubers	105	98	7
Pulses	6	5	1

Modified from Lal, 1995.

(1981b) estimated exponential yield decline with increase in soil erosion. Crop yields on eroded soils were highly correlated with soil organic carbon content, total soil N, and moisture retention at 0.01 MPa suction. Dregne (1992) also reported serious productivity losses in Asia, including parts of India, China, Iran, Israel, Jordan, Lebanon, Nepal, and Pakistan. Productivity loss in several of these regions is as much as 20%.

RESEARCH AND DEVELOPMENT PRIORITIES

There is a continued and ever-increasing threat to productivity, food security, and environment quality in the tropics, especially in ecologically sensitive ecoregions characterized by fragile soils (sub-Saharan Africa), high population density (South Asia), and harsh environments (the Himalayan–Tibetan ecosystem, the Andean region). Food security is a major concern in some of these regions, especially in sub-Saharan Africa, where the attainment of food security is closely linked to soil degradation (Lal, 1988).

Soil quality is a major factor to be considered in relation to food security. Soil fertility depletion and loss of plant nutrients from the ecosystem are no doubt among the principal causes of soil degradation in Africa (Stoorvogel et al., 1993). However, the force driving other degradative processes is accelerated soil erosion. Soil erosion adversely impacts crop yield through effects on soil physical, chemical, and biological quality. It affects soil physical quality through decline in soil structure and reduction in effective rooting depth, leading to severe problems of drought stress, crusting, compaction, and poor seedling emergence and crop stand establishment. Soil chemical quality effects of accelerated erosion are due to overall depletion of soil fertility, especially linked to losses of clay and organic colloids, and transport of dissolved and sediment-borne nutrients. Reduction in soil biological quality due to accelerated erosion is also related to rapid decline in soil organic matter content, reduction in microbial biomass carbon, and decrease in activity and species diversity of soil fauna. These degradative trends are set in motion by erosional processes.

Yet, research data on erosion-induced changes in soil quality for principal soils and ecoregions of the tropics are not known. There is no information on the cause–effect relationship between soil erosion and soil quality on the one hand and soil erosion and agronomic/biomass productivity on the other. There is also no information on key soil properties, and their critical limits, as influenced by erosional processes.

It is difficult to identify policy issues when the factual information is sketchy and full of emotional rhetoric rather than scientific fact. The cause–effect relationship between erosion and productivity can only be determined by establishment of some long-term field experiments on principal soils and major ecoregions of the tropics. These experiments should be well planned, monitored over a long time to determine soil quality by standardized procedures, and their continuity ensured so that trends in soil quality can be established with regard to land use and management practices.

REFERENCES

Aina, P.O. 1980. Drop characteristics and erosivity of rainfall in southwestern Nigeria. *Ife J. Agric.* 2:35–44.

Cassel, D.K. and R. Lal. 1992. Soil physical properties of the tropics: common beliefs and management restraints. In R. Lal and P.A. Sanchez (eds.). *Myths and Science of Soils of the Tropics.* SSSA Special Publication No. 29, Soil Science Society of America, Madison, WI, pp. 61–89.

Crosson, P. and J. Anderson. 1992. *Resources and Global Food Prospects.* World Bank Technical Paper No. 184. World Bank, Washington, D.C.

Dregne, H.E. 1990. Erosion and soil productivity in Africa. *J. Soil Water Conserv.* 45: 431–436.

Dregne, H.E. 1992. Erosion and soil productivity in Asia. *J. Soil Water Conserv.* 47:8–13.

Elwell, H.A. 1979. Destructive potential of Zimbabwe rainfall. *Zimbabwe Agric. J.* 76: 227–232.

FAO. 1994. *Land Degradation in South Asia: Its Severity, Causes and Effects Upon the People.* World Soil Resources Reports 78. FAO, Rome, p. 100.

Ghuman, B.S. and R. Lal. 1991. Land clearing and use in the humid Nigerian tropics. II. Soil chemical properties. *Soil Sci. Soc. Am. J.* 55:184–188.

Ghuman, B.S., R. Lal, and W. Shearer. 1991. Land clearing and use in the humid Nigerian tropics. I. Soil physical properties. *Soil Sci. Soc. Am. J.* 55:178–183.

Hudson, N.W. 1995. *Soil Conservation,* 3rd ed. Iowa State University Press, Ames, p. 391.

Hulugalle, N.R., R. Lal, and C.H.H. ter Kuile. 1984. Soil physical changes and crop root growth following different methods of land clearing in western Nigeria. *Soil Sci.* 138:172–179.

Kowal, J.M. and A.H. Kassam. 1976. Energy load and instantaneous intensity of rainstorms at Samaru, northern Nigeria. *Trop. Agric.* 53:185–197.

Lal, R. 1974. *Soil Erosion and Shifting Agriculture.* FAO Soils Bulletin 24. FAO, Rome, pp. 48–70.

Lal, R. 1976. *Soil Erosion Problems on an Alfisol in Western Nigeria and Their Control.* IITA Monograph 1. IITA, Ibadan, Nigeria.

Lal, R. 1981a. *Analysis of Different Processes Governing Soil Erosion in the Tropics.* IAHS Publ. 133, pp. 351–364.

Lal, R. 1981b. Soil erosion problems on Alfisols in western Nigeria. VI. Effects of erosion on experimental plots. *Geoderma* 25:215–227.

Lal, R. 1981c. Deforestation and hydrological problems. In R. Lal and E.W. Russell (eds.). *Tropical Agricultural Hydrology.* John Wiley & Sons, Chichester, U.K., pp. 131–140.

Lal, R. 1984. Soil erosion from tropical arable lands and its control. *Adv. Agron.* 37:183–248.

Lal, R. 1986a. Soil surface management in the tropics for intensive land use and high and sustained production. *Adv. Soil Sci.* 5:1–108.

Lal, R. 1986b. Different methods of land clearing for agricultural purposes in the tropics. In R. Lal, P.A. Sanchez, and R.W. Cummings (eds.). *Land Clearing and Development in the Tropics.* A.A. Balkeme, Rotterdam, Holland, pp. 55–67.

Lal, R. 1987. Effects of soil erosion on crop productivity. *Crit. Rev. Plant Sci.* 5:303–368.

Lal, R. 1988. Soil degradation and the future of agriculture in sub-Saharan Africa. *J. Soil Water Conserv.* 43:441–451.

Lal, R. 1990. Tropical soils: distribution, properties, and management. *Resour. Manage. Optimiz.* 7:39–52.

Lal, R. 1994a. Global overview of soil erosion. In *Soil and Water Science: Key to Understanding Our Global Environment.* SSSA Special Publication No. 41. Soil Science Society of America, Madison, WI, pp. 39–51.

Lal, R. 1994b. Methods and Guidelines for Assessing Sustainable Use of Soil and Water Resources in the Tropics. SMSS Technical Monograph 21, Washington, D.C., p. 78.

Lal, R. 1995. Erosion–crop productivity relationships for soils of Africa. *Soil Sci. Soc. Am. J.* 59:661–667.

Lal, R. and W. Elliott. 1994. Erodibility and erosivity. In R. Lal (ed.). *Soil Erosion Research Methods.* SWCS, St. Lucie Press, Boca Raton, FL, pp. 181–210.

Lal, R., P.A. Sanchez, and R.W. Cummings, Jr. (eds.). 1986. *Land Clearing and Development in the Tropics.* A.A. Balkeme, Rotterdam, Holland, p. 450.

Laws, J.O. and D.A. Parsons. 1943. The relation of raindrop size to intensity. *Trans. Am. Geophys. Union* 24:452–459.

Morin, J. 1996. Rainfall analysis. In M. Agassi (ed.). *Soil Erosion, Conservation, and Rehabilitation.* Marcel Dekker, New York, pp. 23–40.

Nambiar, K.K.M. 1994. *Soil Fertility and Crop Productivity Under Long-Term Fertilizer Use in India.* ICAR, Krishi Bhavan, New Delhi, India, p. 144.

Obi, M.E. and O.E. Ngwu. 1988. Characterization of rainfall regime for predictions of surface runoff and soil loss in southeastern Nigeria. *Beitr. Trop. Landwirtsch. Vet.* 26:39–46.

Okigbo, B.N. 1978. *Cropping Systems and Related Research in Africa.* Occasional Publications Series OT-1. Association for the Advancement of Agricultural Sciences in Africa, Ibadan, Nigeria, p. 81.

Oldeman, L.R. 1994. The global extent of soil degradation. In D.J. Greenland and I. Szabolcs (eds.). *Soil Resilience and Sustainable Land Use.* CAB International, Wallingford, U.K., pp. 99–117.

Pimentel, D., C. Harvey, P. Resosudarmo, K. Sinclair, D. Kurz, M. McNair, S. Crist, L. Shpritz, L. Fitton, R. Saffouri, and R. Blair. 1995. Environmental and economic costs of soil erosion and conservation benefits. *Science* 267:1117–1123.

Roose, E.J. 1977. Use of the Universal Soil Loss Equation to predict erosion in West Africa. In *Soil Erosion: Prediction and Control.* SCSA Special Publication 21, pp. 60–74.

Sabel-Koschella, U. 1987. Field Studies on Soil Erosion in the Southern Guinea Savanna of Western Nigeria. Ph.D dissertation. University of Munich, Germany, p. 180.

Sanchez, P.A., W. Couto, and S.W. Buol. 1982. The fertility capability soil classification system: interpretation, applicability and modification. *Geoderma* 27:283–309.

Scherr, S.J. and S. Yadav. 1996. Land Degradation in the Developing World: Implications to Food, Agriculture and the Environment to 2020. IFPRI Discussion Paper #14, Washington, D.C., p. 36.

Stoorvogel, J.J., E.M.A. Smaling, and B.H. Janssen. 1993. Calculating soil nutrient balances in Africa at different scales. I. Supranational scale. *Fert. Res.* 35:227–235.

Van Wambeke, A. 1992. *Soils of the Tropics: Properties and Appraisal.* McGraw-Hill, New York, p. 343.

Walling, D.E. 1987. Rainfall, runoff, and erosion of the land: a global view. In K.J. Gregory (ed.). *Energetics of Physical Environment.* John Wiley & Sons, Chichester, U.K., pp. 89–117.

Wischmeier, W.H. and D.D. Smith. 1978. Predicting Rainfall Erosion Losses. Handbook 537. USDA, Washington, D.C.

Section V

Conclusions

17 Applying Soil Quality Concepts for Combating Soil Erosion

R. Lal

INTRODUCTION

The concept of soil quality is not necessarily a new idea. The "soil capability" classification system has been used by soil surveyors around the world since the early 1950s if not before (Soil Survey Staff, 1951; Bouma, 1989). In addition, numerous "soil quality" evaluation systems have been used for the purpose of revenue assessment in several ancient cultures around the world (e.g., India, China). Soil quality, according to these systems, was rated in relative terms, with the highest value of 1 for the best quality soil. This type of soil quality assessment system in India, used for centuries at the village level, is based on the relative price concept. High-quality soil (e.g., flat, irrigable, loam texture, dark brown color, no salinity hazard) is rated 1, and the poor quality soil is given a lower rating depending on the specific constraint. The system is still used in rural areas for soil evaluation. There may be similar systems developed and used in other cultures. Therefore, there is a need to study such indigenous systems developed and used on the basis of local knowledge in different cultures around the world.

The evaluation of the current concept of soil quality in the United States began in the late 1980s and is gaining momentum (Arshad and Coen, 1992; Doran and Parkin, 1994; Bezdicek et al., 1996) because of the emerging new paradigm in soil science. Soil scientists are searching for a professional identity and a new role and are in search of applications for soil science to issues other than agriculture and food production. Emerging issues in which soil science can and should play a major role include environmental concerns (e.g., water quality, the greenhouse effect), industrial uses and waste disposal, mineland reclamation, athletic applications, urbanization and disposal of city wastes, reclamation of contaminated and polluted soils, etc.

1-57444-100-0/99/$0.00+$.50
© 1999 by Soil and Water Conservation Society

It is because of these diverse and varied applications of soil science that the concept of soil quality is gaining momentum (Lal, 1993, 1994; Lal and Miller, 1993; Larson and Pierce, 1991; National Research Council, 1993).

Similar to soil quality, the problems of accelerated soil erosion have plagued the earth since the dawn of settled agriculture. The literature on soil erosion and methods of its mitigation is voluminous and diverse, and many ancient civilizations have struggled for millennia with the tough challenge of keeping the soil in place. The challenge posed by accelerated soil erosion is greater now than ever before, because of the high population density, low and rapidly declining per capita arable land area, the need for mechanization that often increases soil erosion risks, and the necessity to grow erosion-promoting row crops for the enormous task of feeding the world population. Despite its importance and severity, scientists are not sure of the on-site impact of erosion on soil quality and productivity and how to quantify erosion-induced loss in productivity.

A possible approach would be to apply the concepts of soil quality to understand the processes of soil erosion, especially with regard to erosion-induced changes in productivity. A successful interaction between these two emerging concepts requires a clear understanding of both, so that erosion-induced changes in soil quality can be quantified to establish the cause–effect relationship between soil erosion and agronomic productivity.

NEED FOR QUANTITATIVE ASSESSMENT OF SOIL QUALITY

Numerous definitions of soil quality being used can be grouped into four categories (Table 17.1):

1. **Inherent soil characteristics**: Evaluation of soil properties has been the basis used by soil surveyors and those interested in soil capability assessment.
2. **Ability to support plant growth**: This criterion is based on the importance of soil quality to agriculture and food security. Biomass productivity and its sustainability are related to soil quality.
3. **Sustaining resource base and improving environment**: This criterion is based on the objective of broadening the scope of soil to natural resources and for application of soil science to issues of production and environmental concerns.
4. **Soil's capacity to perform numerous functions**: These criteria reflect the emerging issues of nonagricultural uses of soil (e.g., industrial waste disposal, athletic and recreational uses of soil, urban waste disposal, restoration of soil degraded by mining, and other industrial uses). This concept broadens the use of soil to perform numerous functions.

One of the major limitations of the soil quality concept lies in its subjectivity, vagueness, and qualitative approach. For the concept to be applied to solve problems of sustainable management of natural resources and environment quality, it is im-

TABLE 17.1
Different Approaches to Defining Soil Quality

Approach	Reference
Inherent soil characteristics	Soil Survey Staff, 1951; Soil Science Society of America, 1987; Bouma, 1989; Arshad and Coen, 1992; Bezdicek et al., 1996
Ability to support crop growth	Power and Myers, 1989; Lal and Miller, 1993; Parr et al., 1992
Ability to sustain resource base and improve the environment	Larson and Pierce, 1991; National Research Council, 1993; Lal, 1993; Doran and Parkin, 1994
Capacity to perform numerous functions	Pierce and Larson, 1993); Bezdicek et al., 1996; Johnson et al., 1997

portant that soil quality be assessed objectively, accurately, and quantitatively. To do so requires identification of key indicators of soil quality and their assessment by standardized methods.

Appropriate indicators of soil quality depend on the specific application or the issues concerned (Table 17.2). Key soil quality indicators that are common to most applications include soil organic carbon content, solum depth, cation exchange capacity (CEC), available water capacity, water infiltration capacity, microbial biomass carbon, and soil biodiversity as indicated by activity and species diversity of soil fauna. It is important to standardize methods of determination of these properties and develop "indices" based on a rating system for specific uses (Lal, 1994).

SOIL EROSION AND PRODUCTIVITY

The debate over whether accelerated soil erosion affects on-site agronomic productivity can only be resolved by quantifying erosion-induced changes in soil quality.

TABLE 17.2
Soil Quality Indicators

Objective/issue	Key indicators[a]
Crop growth and agricultural productivity	AWC, i_c, SOC, CEC and exchangeable bases, EC and total salts, effective rooting depth, pH, soil texture and structure, microbial biomass C, soil biodiversity
Environment quality and natural resource base	SOC, CEC, solum depth, AWC, i_c, microbial biomass C, soil biodiversity, soil texture, pH
Multiple uses	Texture, solum depth, SOC, CEC, i_c, AWC, pH, microbial biomass C, soil biodiversity

[a] AWC = available water capacity, CEC = cation exchange capacity, EC = electrical conductivity, i_c = infiltration capacity, and SOC = soil organic carbon.

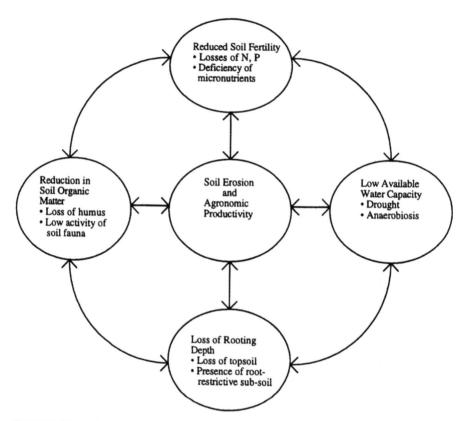

FIGURE 17.1   On-site effects of accelerated erosion on agronomic productivity depend on changes in soil quality.

Indicators of soil quality that determine erosional impact on agronomic productivity shown in Figure 17.1 include the following:

1. **Plant-available water capacity**: Assessment of plant-available water capacity or, better still, the least limiting water range (da Silva et al., 1994; da Silva and Kay, 1997a,b) is extremely important to evaluate the magnitude of the effect of soil erosion on crop yield. Susceptibility to drought is considerably exacerbated on eroded soils due to loss of water as runoff (Figure 17.2) and reduction in water retention pores.
2. **Plant-available nutrient reserve**: In addition to drought, soil erosion affects productivity through reduction in plant-available reserves of N, P, and several micronutrients (e.g., Zn, B, etc.). Plants growing on eroded soils show chlorotic symptoms of nutrient deficiency (Figure 17.3). The CEC is an important indicator of soil fertility and its ability to hold nutrients.

FIGURE 17.2   An important reason for erosion-induced decline in productivity is loss of water runoff.

3. **Soil organic matter content**: Accelerated soil erosion preferentially removes soil organic matter content and deprives soil of one of its most essential constituents. Loss of soil organic matter content adversely affects soil structure, decreases microbial biomass carbon, and reduces activity and species diversity of soil fauna. Eroded soils are prone to crusting (Figure 17.4) and formation of massive soil structure.

4. **Effective rooting depth**: Reduction in rooting depth by the loss of topsoil below the critical level is an important factor affecting plant growth on eroded soil. Loss in productivity can be severe in soils with root-restrictive layers at shallow depth (Figure 17.5).

Therefore, it is important to develop indicators of soil quality that reflect erosion-induced changes in soil quality. Based on the key soil properties listed above, an index of soil quality may be as follows:

$$Sq = f(AWC, CEC, SOC, R_d)_t \qquad (17.1)$$

where Sq is the soil quality index, AWC is the available water capacity, and R_d is the effective rooting depth. It is necessary to establish critical limits of these properties for different uses and intended cropping systems. Based on knowledge of critical limits, a rating system should be developed to quantitatively assess soil quality (Lal, 1994).

(a)

(b)

FIGURE 17.3   Erosion of surface soil leads to loss of plant-available nutrients, (a) as in plowed soil in a runoff plot and (b) on a cornfield on Alfisols in western Nigeria.

FIGURE 17.4   Water into a gully head is fed through runoff from a highly crusted soil with massive structure

FIGURE 17.5   Erosional effects on crop growth are accentuated on soil with root-restrictive subsoil, as in this soil near Morogoro, Tanzania.

CONCLUSIONS

Soil erosion is a severe global problem, especially in ecologically sensitive ecoregions of the tropics and subtropics and sloping lands. However, information on the on-site impact of soil erosion is limited, especially at national and global scales. Quantitative assessment of soil quality is important to an objective evaluation of the on-site effects of accelerated soil erosion. Knowledge of soil quality is important especially with regard to (1) water retention and transmission properties, (2) nutrient retention and availability, (3) soil organic carbon content and its effects on soil biological processes, and (4) soil rooting depth. The latter depends on profile characteristics with regard to horizonation, presence of a root-restrictive layer at shallow depth, edaphological characteristics of the subsoil, and rate of new soil formation (now weathering). On-site effects of erosion on productivity also depend on soil resilience and its ability to restore itself following a major perturbation.

Because global soil resources are finite and easily degraded, it is important to restore soil quality of eroded soils. Choice of appropriate restorative measures depends on the knowledge of soil quality, its temporal and spatial variations, and key properties that govern its dynamic under different management systems.

There are several important researchable issues for improved understanding of soil quality: (1) fine-tuning basic concepts and standardizing definitions and terminology, (2) developing and standardizing methods of quantifying soil quality, (3) identifying key soil properties for different types of soil quality and establishing their critical limits and threshold values, and (4) developing simple and routine methods of soil quality evaluation under field conditions.

Soil erosion will remain a major issue for generations to come. Successful mitigation of soil erosion requires an objective and reliable assessment of on-site effects in terms of loss in productivity. Loss of agronomic productivity should be evaluated at local, regional, and national scales, and scaling procedures should be developed to aggregate the data from plot scale to regional and global scales.

A quantitative relationship should be developed between soil erosion and soil quality so that the cause–effect relationship can be established. These relationships differ among soils, land use and management systems, and ecoregions. Long-term field experiments should be established to develop the necessary database to establish such relationships. These long-term experiments are urgently needed for tropical and subtropical regions where the population pressure is high, soils are fragile, climate is harsh, and the effects of erosion on productivity and environment are drastic and often irreversible.

REFERENCES

Arshad, M.A. and G.M. Coen. 1992. Characterization of soil quality: physical and chemical criteria. *Am. J. Altern. Agric.* 7:5–12.

Bezdicek, D.F., R.I. Papendick, and R. Lal. 1996. Introduction: importance of soil quality to health and sustainable land management. In J.W. Doran and A.J. Jones (eds.). *Methods for Assessing Soil Quality.* SSSA Special Publication No. 49. Soil Science Society of America, Madison, WI, pp. 1–8.

Bouma, J. 1989. Using soil survey data for quantitative land evaluations. *Adv. Soil Sci.* 9: 177–213.

da Silva, A. and B.D. Kay. 1997a. Estimating the least limiting water range of soils from properties and management. *Soil Sci. Soc. Am. J.* 61:877–883.

da Silva, A. and B.D. Kay. 1997b. Effect of soil water content variation on the least limiting water range. *Soil Sci. Soc. Am. J.* 61:884–888.

da Silva, A., B.D. Kay, and E. Perfect. 1994. Characterization of the least limiting water range of soils. *Soil Sci. Soc. Am. J.* 58:1775–1781.

Doran, J.W. and T.B. Parkin. 1994. Defining and assessing soil quality. In J.W. Doran, D.C. Coleman, D.F. Bezdicek, and B.A. Stewart (eds.). *Defining Soil Quality for a Sustainable Development.* SSSA Special Publication No. 35. Soil Science Society of America, Madison, WI, pp. 3–21.

Johnson, D.L., S.H. Ambrose, T.J. Bassett, M.L. Bowen, D.E. Crummey, J.S. Isaacson, D.N. Johnson, P. Lamb, M. Saul, and A.E. Winter-Nelson. 1997. Meaning of environmental terms. *J. Environ. Qual.* 26:581–589.

Lal, R. 1993. Tillage effects on soil degradation, soil resilience, soil quality and sustainability. *Soil Tillage Res.* 27:1–8.

Lal, R. 1994. Methods and Guidelines for Assessing Sustainable Use of Soil and Water Resources in the Tropics. SMSS Technical Monograph 21. Soil Management Support Services, USDA-NRCS, Washington, D.C., p. 78.

Lal, R. and F.P. Miller. 1993. Soil quality and its management in humid subtropical and tropical environments. In Proc. XVII Int. Grassland Congr., Palmerston North, New Zealand, pp. 1541–1550.

Larson, W.E. and F.J. Pierce. 1991. Conservation and enhancement of soil quality. In Evaluation for Sustainable Land Management in the Developing World. IBSRAM Proc. 12, Vol. 2. Int. Board on Soil Res. and Management, Bangkok, Thailand.

National Research Council. 1993. *Soil and Water Quality: An Agenda for Agriculture.* National Academy Press, Washington, D.C.

Papendick, R.I. and J.F. Parr. 1992. Soil quality: the key to sustainable agriculture. *Am. J. Altern. Agric.* 7:2–3.

Parr, J.F., R.I. Papendick, S.B. Hornick, and R.E. Meyer. 1992. Soil quality: attributes and relationship to alternative and sustainable agriculture. *Am. J. Altern. Agric.* 7:5–11.

Pierce, F.J. and W.E. Larson. 1993. Developing criteria to evaluate sustainable land management. In J.M. Kimble (ed.). Proc. 8th Int. Soil Management Workshop; Utilization of Soil Survey Information for Sustainable Land Use, May 1993. USDA-SCS, National Soil Surv. Center, Lincoln, NE, pp. 7–14.

Power, J.F. and R.J.K. Myers. 1989. The maintenance or improvement of farming systems in North America and Australia. In J.W.B. Stewart (ed.). *Soil Quality in Semi-Arid Agriculture.* Can. Int. Dev. Agency, Ottawa, Canada.

Soil Science Society of America. 1987. *Glossary of Soil Science Terms.* SSSA, Madison, WI.

Soil Survey Staff. 1951. Soil Survey Manual. USDA, U.S. Government Printing Office, Washington, D.C.

Index